D1308619

TOXIC SUBSTANCES
IN THE ENVIRONMENT

ENVIRONMENTAL SCIENCE AND TECHNOLOGY
A Wiley-Interscience Series of Texts and Monographs
Editors: JERALD L. SCHNOOR AND ALEXANDER ZEHNDER
A complete list of titles in this series appears at the end of this volume.

TOXIC SUBSTANCES IN THE ENVIRONMENT

B. MAGNUS FRANCIS

Institute for Environmental Studies
University of Illinois at Urbana-Champaign
Urbana, Illinois

A WILEY-INTERSCIENCE PUBLICATION

JOHN WILEY & SONS, INC.

New York · Chichester · Brisbane · Toronto · Singapore

Library of Congress Cataloging in Publication Data:
Francis, B. Magnus, 1943–
    Toxic substances in the environment / Bettina Magnus Francis.
      p. cm. — (Environmental science and technology)
    Includes bibliographical references and index.
    ISBN 0-471-50781-4
    1. Pollution—Environmental aspects. 2. Pollution—Health
aspects. 3. Toxicology. I. Title. II. Series.
QH545.A1F72   1993
363.73—dc20   92-26896

Printed in the United States of America

10 9 8 7 6 5 4 3

For George

CONTENTS

FOREWORD

During the twentieth century, applied science has given us a great variety of chemical products with specific target actions that are immensely useful and sometimes indispensible to present day society. Nevertheless, the unintended results of their overenthusiastic and injudicious use have often been counterproductive in producing a wide spectrum of adverse consequences to man and the biota of the planet. Exploration of the benefit/risk interfaces of these environmental micropollutants is the province of environmental toxicology and forms the subject matter of this book.

Instructive examples, presented as "case histories" in the book, include the insecticide DDT, the "antiknock" substance tetraethyl lead, the synthetic estrogen diethylstilbestrol, and the chlorofluorocarbon refrigerants. None of these substances is intrinsically pernicious and indeed DDT has unquestionably prevented more illnesses and saved more lives than any other man-made chemical. Tetraethyl lead was instrumental in the efficient development of the high-compression internal combustion engine, diethylstilbestrol was the first successful cancer chemotherapeutant, and the chlorofluorocarbon refrigerants have improved the habitability of vast areas of the world. Yet in turn, each of these substances has been characterized by environmental toxicologists as a deleterious micropollutant with unintended and undreamed of consequences that followed the relatively massive use patterns of DDT as a broad spectrum insecticide, diethylstilbestrol as animal growth supplement, tetraethyl lead as an antiknock additive in gasoline, and the chlorofluorocarbons as refrigerants/propellents.

The question is posed as to whether regulation permitting appropriate restricted use is preferable to outright ban. Clearly, careful weighing of benefit/risk has now become an essential responsibility of the environmental toxicologist. Rigorous evaluation of newly developed chemicals, based on environmental chemistry, human and animal toxicology, and ecotoxicity must be made before massive use is encouraged. The key word is "balance" and it is to be applied to use patterns, to health effects evaluations, and to socioeconomic judgments.

This book provides the *modus operandi* for student entree into this perplexing arena. It is exceedingly well written, replete with histories, and displays a refreshing candor that stimulates the application of logic rather than hysteria to judgments about the fruits of technology as they impact on an increasingly crowded earth whose inhabitants are bewildered about environmental concerns. Study of this book should provide students of all persuasions with stimulus toward career opportunities in environmental toxicology. The book will also be an important resource for professionals in areas impinging upon environmental toxicology, such as regulatory officials, journalists, lawyers, and the concerned laity.

ROBERT L. METCALF

University of Illinois, Urbana-Champaign

SERIES PREFACE
Environmental Science and Technology

The Environmental Science and Technology Series of Monographs, Text-books, and Advances is devoted to the study of the quality of the environment and to the technology of its conservation. Environmental science therefore relates to the chemical, physical, and biological changes in the environment through contamination or modification, to the physical nature and biological behavior of air, water, soil, food, and waste as they are affected by man's agricultural, industrial, and social activities, and to the application of science and technology to the control and improvement of environmental quality.

The deterioration of environmental quality, which began when man first collected into villages and utilized fire, has existed as a serious problem under the ever-increasing impacts of exponentially increasing population and of industrializing society. Environmental contamination of air, water, soil, and food has become a threat to the continued existence of many plant and animal communities of the ecosystem and may ultimately threaten the very survival of the human race.

It seems clear that if we are to preserve for future generations some semblance of the biological order of the world of the past and hope to improve on the deteriorating standards of urban, suburban, and rural public health, environmental science and technology must quickly come to play a dominant role in designing our social and industrial structure for tomorrow. Scientifically rigorous criteria of environmental quality must be developed. Based in part on these criteria, realistic standards must be established and our technological progress must be tailored to meet them. It is obvious that civilization will continue to require increasing amounts of fuel, transportation, industrial chemicals, fertilizers, pesticides, and countless other products; and that it will continue to produce waste products of all descriptions. What is urgently needed is a total systems approach to modern civilization through which the pooled talents of scientists and engineers, in cooperation with social scientists and the medical profession, can be focused on the development of order and equilibrium in the presently disparate segments of the human environment. Most of the skills and tools that are needed are already in existence. We surely have a right to hope a technology that has created such manifold environmental problems is also capable of solving them. It is our hope that this Series in Environmental Sciences and Technology will not only serve to make this challenge more explicit to the established professionals, but that it also will help to stimulate the student toward the career opportunities in this vital area.

JERALD L. SCHNOOR
ALEXANDER ZEHNDER

PREFACE

There once was a legendary prince, from the beautiful island of Serendip, whose adventures gave us the word *serendipity:* fortunate chance. Serendipity led me to accept a temporary job doing a literature search on the fate of pesticides in soil, under the aegis of Dr. James Sanborn. That job—my introduction to toxicology—led to four years of working for Professor Robert L. Metcalf, and that led to my teaching the course "Toxic Substances in the Environment," which has resulted at long last in my writing this book. But human activities are rarely linear, and it would take the "cobbly worlds" of Clifford Simak—the alternative realities about which theoretical physicists speculate—to explore the possible routes by which I might have come to write this book.

Even in the halcyon days of the 1950s, when entomologists feared that they would become obsolete along with the pests that the miracle insecticides controlled so well, I cannot remember insecticides being used in our house. Weeds were invariably hand-pulled. My father read Rachel Carson's *Silent Spring* in its original form in *The New Yorker*. Only the smallest wobble of reality would have led me to *Silent Spring* then, introducing me to environmental toxicology directly. Instead, my student job in Professor Milan Kopac's laboratory at New York University included cleaning mouse cages. One of his students was studying thalidomide. I added a fascination with chemically induced birth defects to a long-standing interest in genetics and became a developmental toxicologist.

In those same years, the increasing resistance of insects to insecticides, and the entomologist's recognition of interdependence among species, led Metcalf to search for biodegradable alternatives to DDT, and then to develop his "terrestrial-aquatic model ecosystem"—the first microcosm—as a way to identify ecological toxicity *before* releasing chemicals into the environment by the kiloton. When he returned to Illinois from Riverside in 1968, he brought the microcosm with him.

In 1969, Professor Herbert Carter, the vice chancellor for academic affairs, discussed the best way for the University of Illinois to "expand and transfer knowledge of the environment to the public" with Bob Metcalf and with Professor Ben Ewing of the civil engineering department. The result was the Institute for Environmental Studies, with an official mission of undertaking interdisciplinary graduate research. Formally established in 1972, the institute embodied the idea that environmental research is necessarily inter- disciplinary and multidisciplinary: that it requires collaboration among biolo- gists, economists, sociologists, chemists, engineers, and psychologists. When the institute began hiring faculty in 1975, the first new faculty members were a chemist and a sociologist. Although the institute did not (and still does not) award degrees, its founders designed a series of courses that would provide a broad overview of environmental problems. One of these, ''Toxic Substances in the Environment,'' the namesake of this book, was taught initially by Metcalf.

I knew none of this when I began working for Jim Sanborn. I had still not read *Silent Spring*. I assumed vaguely that DDT was ''bad'' but knew little more. The next 14 months were a crash course in environmental toxicology. I reviewed the literature on 42 pesticides, ranging from household staples such as malathion and captan to little-known fungicides such as dodine. I learned that some scientists never tested an innocuous chemical, whereas others never tested one with harmful effects. I read *Silent Spring*. And I met Profes- sor Metcalf.

In the history of science there are giants whom we remember for a single achievement, a momentous discovery. But there are others who, over a long career, steadily produce one superb piece of work after another. They come early to an idea, carry out seminal studies that give rise to an entire field, and then move on. Few of us, building on their work, know the totality of their achievement, or appreciate the scope of their genius. Such a man is Robert Lee Metcalf, whose fascination with the relationship between chemical struc- ture and biological activity has made him a pioneer in pesticide chemistry, ecotoxicology, and integrated pest management.

Jim Sanborn left Illinois almost as soon as the literature review was completed. A year later I joined Metcalf's lab, looking at structure–activity relationships among organophosphorus esters that induce delayed paralysis. By then ''Toxic Substances'' was being taught by Professor James Johnston, a new faculty member at the institute. Johnston was a biochemist by training, and his research focused on the carcinogenicity of contaminants in drinking water. To the case histories of pesticide bioaccumulation and metal pollution that Metcalf had taught, he added lectures on environmental carcinogens and mechanisms of carcinogenesis. I joined the institute faculty in 1981, and when Jim Johnston left in 1983 I became the third person to teach ''Toxic Sub- stances in the Environment.'' My job was made infinitely easier because Jim left me not only a syllabus but also his notes: 30 pages of concentrated information on toxicology. Metcalf had compiled fact sheets on metals and pesticides: these were also passed along. This book has grown from that nucleus.

In the 1980s, environmental toxicology was preoccupied with the human health effects of environmental pollutants. Most of the notoriously bioaccumulative pesticides had been banned or severely restricted. Falcons and pelicans were returning to their habitats as DDT residues declined, and it seemed that the major ecological problems had been solved. In contrast, environmental carcinogens made headlines with alarming regularity, as did the dumps that often leaked them into air or water. Equally important, my own interest in developmental toxicology and neurotoxicity was consistent with emphasizing the health effects of toxic substances.

The students disagreed. By ones and twos they asked, politely, when we would talk about the environment, given the title of the course. My husband returned from a visit to Austria and reported that the pines above the Brenner Pass were dying and that air pollution was suspected to be the primary cause. Birds feeding on Great Lakes fish were producing malformed offspring or none at all. The Everglades were drying up, and the last California condors were captured in a desperate effort to keep the species from becoming extinct. Once again "Toxic Substances in the Environment" changed. By decade's end, when global warming, the ozone layer, and loss of biodiversity had emerged as the most frightening consequences of human activity, roughly half my lectures dealt with the ecological consequences of releasing chemicals—toxic or not—into the environment. The remainder of the course reflects the unending human fascination with our own affairs, and still looks at the health effects of chemicals. Since the same mechanisms that act on people frequently act in other species, even in plants and microbes, even the study of human health effects can shed light on ecological consequences of chemicals. It is this perspective on environmental toxicology that this book presents.

In addition to the mentors who shaped my career, I must thank the many people who helped me during the writing of this book. Charlotte Fink and Eileen McCulley moved large sections of class notes from manuscript to computer files in the old-fashioned way: by typing. Tori Corkery converted graphs, charts, and tables to computer files, making it possible to use many without redrawing them. My colleagues Gary Johnson and Henry Van Egteren taught me a modicum of economics, and something of how economists think, which I hope not to have distorted too badly.

Last but far from least, I wish to thank the students. Their varied interests have shaped the course and forced me to rethink my own ideas. Chemistry students want to see chemical structures; ecology majors are impatient of the emphasis on cancer; toxicologists write term papers on global warming. Graduate students write highly technical papers on incineration or chemical destruction of hazardous wastes and undergraduates write highly emotional arguments for (and against) nuclear power. Student activists who work with people living at the edge of leaking dumps report their observations; would-be activists find that there can be another side to a story. Students employed by federal and state agencies turn case histories into term papers (with permission from their superiors). Foreign students have used the term paper to gather information which they could apply on returning to their own countries. One student proposed research in her term paper that eventually became two publications. Through their questions and their term papers, the students shape the course and thus have shaped this book.

B. Magnus Francis

TOXIC SUBSTANCES
IN THE ENVIRONMENT

1

INTRODUCTION

Generations have trod, have trod, have trod;
And all is smeared with trade; bleared, smeared with toil;
And wears man's smudge and shares man's smell: the soil
Is bare now, nor can foot feel, being shod.

—Gerard Manley Hopkins[1]

That the environment poses risks to humans is hardly a new idea. Only the shape of the threat has varied. Humans have always had to cope with heat and cold, storms and floods, wild animals and poisonous plants. Medieval villages were clustered together as much for protection against wild animals as against human enemies. Throughout history, famine has followed environmental cues: unusual weather, outbreaks of pests, or plant diseases. Great periodic plagues of yellow fever and bubonic plague decimated populations. Lesser plagues like typhoid and infant diarrhea undramatically but consistently controlled populations in the intervals.

Throughout most of human history, therefore, environmental concerns emphasized dangers to human life and health, and the first efforts of public health agencies were directed against contagious disease. Nonetheless, people have been aware for millennia of human-made environmental dangers. From the time of Pliny the Elder in the first century of the Roman Empire, citizens have complained about the degradation of their environment, the damage wrought by industrial practices, and the consequences of toxic substances in their food. Not surprisingly, the first laws regulating toxic substances dealt with adulterated food, and fears of adulterated food continue to produce the largest headlines.

With the partial conquest of water- and foodborne epidemics of contagious disease, our species has expanded to every corner of the globe. With increasing human populations, the pressure on other species and on the earth itself are becoming apparent. Today, roads carve the deepest tropical rain forest, dams harness rivers from the Nile to the Colorado. The human population is at 5 billion and still rising. At long last, emphasis is shifting from the risks that the environment poses for our species to the risks that our species poses

---

[1] From "God's Grandeur," by Gerard Manley Hopkins, in *Poems of Gerard Manley Hopkins*, 3rd ed., Oxford University Press, New York, 1948, p. 70.

to our environment. Accelerating extinction of species is a truism: ecologists *assume* that half the species now alive will be extinct within 30 years. Many will be gone before they have been named. The existence of the *greenhouse effect,* in which the climate is altered by generation of excess carbon dioxide, is not in doubt, although the severity of its consequences is disputed. Forests in the Third World are ravaged by indiscriminate cutting, with accompanying degradation of soils, silting of rivers, and desertification. Half of the wetlands in the United States have been drained, and ninety percent of our old growth forests have been cut for lumber. Forests in Europe and North America are dying under the onslaught of the many pollutants lumped under the heading of "acid rain." Wilderness has become rare, and must be cherished, where once it was to be conquered.

In the past 10 years, it has also become clear that the human species has less control over its environment than many thought. The idea that humans could destroy the global ecosystem first arose in the late 1940s, during the development of the hydrogen bomb. Some physicists were concerned about the theoretical possibility that such an explosion could trigger the fusion of hydrogen in water. A second false alarm, in the late 1960s, was that DDT might kill enough oceanic algae to decrease global oxygen levels.[2] Today the occurrence of chemically induced changes in global climate is underway. The crises include not only massive pollution-induced damage to forests across Europe and North America, but also the rise in temperature due to increasing atmospheric levels of carbon dioxide and the thinning of the stratospheric ozone layer due to the release of chlorofluorocarbons. All of these global changes were well underway when first recognized, and each results from the release of chemicals into the environment as a result of human activity.[3]

At the same time, the elimination or control of most of the common infectious diseases has given prominence to the toll exacted by cancer, heart disease, nerve damage, and birth defects. Causes of these diseases are undoubtedly complex, but chemicals have been implicated, often in well-publicized stories of industrial malfeasance or governmental carelessness.[4] Although the proven examples involve highly specific toxic effects of a very small number of chemicals, a nagging uncertainty exists in many people's

[2] N. S. Fisher, "Chlorinated hydrocarbon pollutants and photosynthesis of marine phytoplankton; a reassessment," *Science* 189:463–465, 1975.

[3] As discussed in Chapter 3, there are scientists who dispute the anthropogenic origins of each of these global changes. There remains a small possibility that much of the forest damage is due primarily to cyclic natural causes we do not fully understand, rather than to pollution; that global warming is apparent rather than real, and that observed phenomena represent part of the normal chaos of weather patterns; that the decreased levels of ozone in the stratosphere can be effectively controlled by present regulations.

[4] Examples of these highly publicized cases are presented in the case histories in Chapter 5 (PBBs), Chapter 8 (leptophos), and Chapter 12 (DES).

minds about the role played by synthetic chemicals in disease processes. Psychologists say that fear increases if a risk is potentially fatal, if it is not assumed voluntarily, if the consequences are unknown, or if the risk is not easily reduced.[5] For most people, all of these elements combine to make the hazards of toxic chemicals in the environment very frightening indeed.

SCOPE OF THE SUBJECT

The study of toxic substances in the environment necessarily includes several disciplines. *Environmental chemistry* focuses on the presence and fate of chemicals in the environment and on their transport between air, soil, and water. *Environmental toxicology* focuses on the effects of chemicals in the environment on organisms. *Ecology* focuses on interactions between organisms, and between organisms and their environments, whether or not human influences are present. Each of these disciplines can be subdivided into narrower areas of specialization, depending on the classes of chemicals, the definition of environment, the organisms, and the type of effects on organisms that are of interest. An example is a dichotomy that occurs within toxicology. Toxicologists who study the health effects of chemicals on humans or on domestic animals usually investigate effects on individual organisms. These toxicologists studying health effects typically have backgrounds in paramedical fields such as pharmacology. In contrast, ecological toxicologists investigate effects of chemicals on populations or species and on interactions between organisms in relatively natural habitats. Ecotoxicologists often have backgrounds in entomology or ecology. Obviously, differences in training and interests are even greater between disciplines.

Despite differences in the backgrounds and interests of ecologists, environmental chemists, and environmental toxicologists, the subjects are inextricably intertwined. Awareness of toxicity is necessary before concern arises about a pollutant: one must at least suspect that a hazard exists before acting to minimize it. If a chemical is considered harmless, its discharge into rivers or into the air will be condoned. Therefore, identification of the effects of chemicals on living organisms is critical to controlling chemical pollution. However, if no one realizes that a specific chemical is present in the environment, no effort will be made to identify its potential hazard or to control its presence. Therefore, an understanding of the transport, transformation, and persistence of chemicals in the air, the soil, and the water is an essential element of understanding the effects of toxic substances in the environment. When chemical production increased exponentially in the years immediately following World War II, it was implicitly assumed that chemical wastes

[5] See Chapter 13

would stay buried, that pesticides would disappear from the fields when they lost their pesticidal activity, and that effluents from chimneys could be diluted into harmlessness by making the chimneys high enough. The events of the past four decades have taught us otherwise. We now know that our discards can return in the air we breathe, the food we eat, and the water we drink.

Given the presence of a chemical in the environment, it is to a considerable extent irrelevant whether the initial identification of its hazard be in humans, birds, rodents, or fish. Just as miners carried canaries into mines to warn them of carbon monoxide, wild plants and animals can warn us of the presence of poisons. Whereas the canary only warned of acute danger, ecological toxicology can warn of subtle problems involving interactions between organisms and between classes of organisms. This may be an expensive warning, however, since it is not always possible to remove a chemical from the environment once it is released. The alternative—studying toxicity in laboratory animals—can often prevent ecological disasters. Perhaps the most significant advance of the past 40 years is that the environmental fate of many chemicals must now be examined even before they are marketed.

Once a chemical has been shown to have an untoward effect on a species or an ecosystem, the mechanism by which the effect occurs should be identified. It is a beginning to see that a bird population is declining, and a large step forward to conclude that there is a correlation between spraying an insecticide and that decline. It is a quite different matter to explain how and why the birds died and whether the same process might occur in humans or in fish. All three types of information are necessary to predict and prevent similar problems in other species, including our own. But identification of mechanisms is best carried out in the laboratory, where careful controls can minimize confounding variables. In summary, the chemist, the ecologist, and the toxicologist must all cooperate in the study of toxic substances in the environment.

In discussing toxic substances in the environment, it is also necessary to remember that most chemicals have a use as well as a hazard.[6] Banning a chemical eliminates the hazard, but also the benefits of its use, and therefore carries a cost of its own. *Risk assessment* and *risk management* are the effort to balance these benefits and risks within the framework of a society's goals.

Unfortunately, it is often much easier (although rarely easy) to estimate the environmental fate of a chemical and its effects on organisms than to

---

[6] There are exceptions. Dioxins are contaminants produced during the synthesis of chlorinated phenols. They are not used commercially, but the extreme toxicity of several members of the class, notably 2,3,7,8-tetrachlorodibenzodioxin or TCDD, has made their control a regulatory priority. Several related classes of compounds are also byproducts of chemical production: chlorinated dibenzofurans, which result from heating polychlorinated biphenyls, and chlorinated diphenyl ethers, which are thought to be byproducts of chlorophenol production.

determine the full costs of its use and the corresponding costs of not using it. Breathing asbestos damages lungs, but asbestos makes the best fireproof clothes, theater curtains, and insulation. How many deaths from fire do we balance against how many deaths from lung disease? Does it matter that the people who will die from asbestos-induced diseases are not necessarily the same people who are at risk from fires? Another example is the insecticide DDT, which is one of the safest insecticides to spray, but which is ecologically harmful. When DDT was banned, the highly toxic parathion replaced it in many uses. Deaths and illnesses among farm workers, pesticide applicators, and children increased precipitously. But as DDT residues in the environment decreased, sensitive avian species recovered. How does one compare extinction of species with human deaths? Were there alternatives to banning DDT? Were there safer insecticides that should have been introduced instead of parathion? Should growers have been forced to buy safer alternatives even though parathion was significantly cheaper? Should the manufacturer have been forced to switch to a less toxic analog? Or should the government have subsidized safer insecticides? These questions are not issues of toxicity, but of equity and of costs to society. As such, they fall into the realm of social sciences: economics, public policy, law.

In a participatory democracy, the laws governing risk will result (to a considerable extent) from the perceptions of public officials about the risks "people" accept or reject. If "people" think that their food contains filth, legislators will pass laws requiring food inspection.[7] If "people" decide that smoking in public places is obnoxious, smoking will be banned in malls, restaurants, and government buildings. But who influences what "people" think? How does a population come to a consensus about the risks it will or will not accept? Psychologists and sociologists examine the ways in which individuals and groups, respectively, form opinions. Sociologists and psychologists also examine ways in which opinions can be shaped or changed. Thus, understanding how a society deals with toxic substances requires some understanding of sociology and psychology. These subjects are far removed from the biologist's and chemist's laboratory, yet they are central to any understanding of public policy and public response.

Moreover, the major controversies over the management of environmental contaminants arise from determining which pollutants to manage and how to manage them. Such disputes can rarely be settled in the laboratory. When biologists disagree over the danger posed by a chemical, they only occasionally dispute the laboratory data (e.g., the number of tumors among rats

---

[7] For example, the Food and Drug Administration's authority to inspect food originated in the outcry following publication of Upton Sinclair's *The Jungle* in 1906. Sinclair, whose polemic was aimed at miserable working conditions, said "(I) aimed at the public's heart and by accident I hit it in the stomach."

exposed to a suspect carcinogen, the number of bald eagles in the Midwest). Disputes typically arise in deciding the *significance* of the data. Some of the disagreements are somewhat related to the data (e.g., is the biological model underlying the statistical analysis correct? Are there other populations of the endangered species?). Others are concerned with technical aspects of risk assessment (e.g., Is a valid model used to extrapolate the rat data to humans? Can a persistent chemical be managed so sensitive species are not exposed?). But the most intractable disagreements are rooted in value judgments. How much money should be spent to protect people from very small probabilities of harm? How much should we spend to clean up an ecosystem so eagles breed successfully? (So sports fishers can eat their catch? So commercial fishing can continue?) Is there a value to wilderness apart from the products it may generate? Should we damage our economies by cutting carbon dioxide emissions now, or wait until any consequences of global warming become obvious (and irreversible)?

Because such decisions involve very large amounts of money, it is easy to impute venal motives to proponents of either side. It is also regrettably easy to take the argument to "the public" by playing on people's emotions. An outcry then results from those who are (or feel they are) threatened by the perceived pollution; a countercry comes from those who depend (or think they depend) on maintenance of the status quo for their jobs. Examples of such differences are legion. Equally contested is the relative importance of risks associated with life-style decisions compared to risks imposed by society. Many Americans worry about pesticide residues in food, a risk that most experts say is negligible, but eat diets high in fat, despite the incessant publicity connecting high fat diets with heart disease and colon cancer. People continue to smoke, even though one in ten people who smoke a pack a day will die of lung cancer, and even more will die prematurely of emphysema or heart disease. Are the experts then right in considering the public to be ignorant? Are the experts "blaming the victim" while ignoring industrial malfeasance? Are ordinary people at the mercy of shrill extremists?

Or is the disparity due to different perceptions of risks and benefits? Is there a difference between risking eventual colon cancer by eating steak (which I taste and enjoy) or by eating a food additive (which I neither taste nor see) that is profitable for the manufacturer? Environmental psychology and environmental sociology analyze the factors that determine the acceptability of risk and how this acceptability can be influenced. Environmental economics considers issues such as discounting a future cost for a present gain, whether the cost is monetary or health-related. These issues are as important for public decisions as the epidemiologic data that identify the risk. Finally, it is important to consider not only the absolute costs and benefits of a policy, but also to look at the *equity* with which costs and benefits are distributed; that is, to consider not only the magnitude of costs and benefits, but also who pays the cost and who reaps the benefit.

RATIONALE FOR THIS BOOK

It is obvious that neither a single book nor a single author can do justice to such a broad range of subjects, and this book is not intended as a systematic exposition of any single disciplinary science. Rather, it is an introduction to a subject that spans multiple disciplines: toxic substances in the environment.

Survey after survey reaffirms that people everywhere are concerned about the environment and concerned about toxic substances in the environment. Meanwhile, an assortment of environmental disasters makes headlines almost weekly. There is alar in apples and it may cause cancer, if the second-hand smoke doesn't. The oil-well fires in Kuwait may stifle the monsoons, causing famine in India. The greenhouse effect will cause oceans to rise and inundate Florida. Global warming is a myth. Decaffeinated coffee may cause cancer, if it isn't the dioxin in the filters. Dioxin isn't a carcinogen. Songbirds are declining in the eastern forests and amphibians are declining all over the world. Acid rain only affects 10% of the lakes in the Adirondacks.

The cacophony of threats and rebuttals is self-defeating. The episodic reporting of environmental crises provides no way of discriminating between acceptable and unacceptable risks or between immediate and remote hazards. Without understanding the scientific underpinning of the headlines, one cannot judge the magnitude of the risk and cannot determine a prudent course of action for oneself or for society. Once informed, individuals can make their own choices about many risks. Even for risks requiring international action, such as global warming, individual choices can have significant impact, while organized action can move governments. Long before townships realized landfill space was running out, people began recycling. Those efforts led to community recycling centers and demonstrated to politicians that people *will* recycle, if given the chance. When the flame retardant Tris became known as a probable carcinogen, sales of Tris-treated baby clothes plummeted well before Tris was banned.

Unfortunately, the manipulators of public opinion also recognize the power of popular anger and popular fear. Environmentalists realize that it is easier to raise money to fight cancer-causing chemicals than to explain the relationships among ecosimplification, the beef industry and modern agribusiness, and pesticide production. Industries understand that it is far easier to dismiss environmentalists as wild-eyed radicals than to reorganize a manufacturing process to waste less or pollute less. It is always easier to call an opponent names than to attack complex structural problems in our society, and in the short run it is very effective. But in the long run it is self-defeating and leads only to the same cynicism among targets of fear-mongering as among the lobbyists who generate it. It is the purpose of this book to provide information about toxic substances in the environment and about the hazards they pose or are unlikely to pose so you can judge the risks for yourself and decide which risks are acceptable and which are unacceptable.

2

ECOSYSTEMS AND ECOTOXICOLOGY

The "control of nature" is a phrase conceived in arrogance, born of the Nean-
derthal age of biology, when it was supposed that nature exists for the conve-
nience of man. . . . It is our alarming misfortune that so primitive a science
has armed itself with the most modern and terrible weapons, and that in turning
them against the insects it has also turned them against the earth. . . .

—Rachel Carson[1]

INTRODUCTION

The effects of a chemical begin with its effects on individual organisms,
but the effects of pollutants on ecosystems cannot always be predicted by
describing toxicity to individual organisms. Toxic effects on populations
often differ greatly between species and depend as much on the nature of
the most sensitive portion of the population as on the precise degree of
toxicity to individuals of the species. Numerous examples can be drawn
from the history of pesticides. For example, adult birds are, individually,
not more sensitive to DDT than are mammals.[2] But relatively low levels of
DDT disrupt the breeding of birds and not of mammals, so the former are far
more severely affected. Even among birds, however, considerable variability
occurs. Early efforts to demonstrate the reproductive effects of DDT in the
laboratory failed, because the hatchability of chicken eggs is hardly affected
by levels of DDT that cause disastrous eggshell thinning in falcons. Standard
laboratory tests of chemical toxicity, restricted to the few "model" species
that are easily kept in captivity, can only begin to identify the effects of
chemicals on distantly related wild species. Moreover, only healthy animals
are used in laboratory studies, and they are fed optimum diets. This is very
different from the situation for wild animals, especially when species are
already under stress from climate, predators, or other factors. *Ecology* is

---

[1] Rachel L. Carson, *Silent Spring*, 1962. A paperback version of *Silent Spring*, which first
appeared in the *New Yorker*, is available, published by Fawcett Crest Books, New York.
[2] Median lethal doses of DDT (in mg/kg) for representative species of birds and mammals are:
mouse, 135; dog, 300; monkey, 200; chickens, 300; pigeons, 800; mallards, 3300.

9

the study of relationships between species, a subject of study in its own right. Nonetheless, some of the simpler aspects of ecology must be understood in order to discuss environmental toxicology.

SOME CONCEPTS IN ECOLOGY

An *individual* is composed of subunits (so to speak) consisting of cells, which are in turn organized into tissues and organs.[3] All of these individual components have, of course, evolved to coordinate and cooperate; the functioning and survival of the individual depends on the simultaneous cooperation of all its parts, and individual organisms have evolved by the joint selection of the parts of the whole.

Individual organisms belong to a *species*. A species is a group of organisms capable of perpetuating their kind by interbreeding with each other but not with members of other groups. A species is typically separated from closely related species by some physical barrier that has prevented interbreeding for long enough that the two groups (races or subspecies) have diverged genetically to the point that they cannot mate and/or cannot produce fertile offspring if they do mate. In some cases this physical separation may later be removed. Once divergence has proceeded far enough, however, even physical proximity does not allow genetic material to be interchanged, and the groups are truly distinct species.[4] In general, the overall needs of a species in a given habitat differ, at least to some extent, from the requirements of every other species in that habitat. The difference may consist of different nesting preferences or differences in the size of prey that can be eaten. Each species is said to occupy a *niche,* complementing other species, and each niche has a *carrying capacity,* or maximum number of organisms of a particular species that can be supported.

Most species with extensive ranges are divided into *populations*, consisting of individuals of the same species in geographical proximity. Deer on one side of a large river are one population; those on the other side are a separate population. Within a population, individuals are in competition with each other because they have identical requirements for food and shelter. In a very real sense, the survival or success of one individual in a population is at the expense of others. Populations are defined by their age structure, mortality rate, and replacement rate. Within a species, individuals of different ages, including at the least individuals that are already reproduc-

---

[3] Some of the subcellular organelles actually appear to have been free-living organisms that became symbionts: mitochondria are the best characterized candidates.

[4] Species may be subdivided into demes, which are subunits that actually interbreed. Chihuahuas and Great Danes represent two canine demes that rarely, if ever, interbreed directly. However, gene exchange among these wildly different breeds of dogs is possible via intermediate breeds as well as through artificial insemination. Horses and donkeys, despite similar size and despite their ability to mate and produce offspring (mules), are distinct species because the progeny of their matings are almost always sterile.

ing and their juvenile replacements, serve to maintain the population's stability. It is possible to destroy a species without killing a single individual if one merely prevents the species from breeding. In fact, the total destruction of all members of the species over a small part of its range is usually less harmful than steady pressure on a single segment of the population (by age or reproductive capacity) over the whole of its range. For example, DDT is not very toxic to adult birds but upsets reproduction of entire species because it causes eggshells to be thinner than normal. This effect, which is more pronounced in some species (raptors, pelicans) than others (chickens, pigeons), is more harmful to hawks than shooting all the hawks in one county would be. After a local slaughter, hawks can recolonize from adjacent areas; upsetting reproduction across large areas prevents replacement of lost birds.

The carrying capacity of a niche very often depends on the food supply, but suitable nesting sites may be a limiting factor for some species.[5] A given supply of sunflower seeds supports a certain number of sunflower seed eaters; a given insect population supports a certain number of insectivores. Species are often in competition with each other for the same food, but the overlap is usually only partial. Both squirrels and cardinals love sunflower seeds. Cardinals also eat smaller seeds, but not nuts. Squirrels eat nuts and larger seeds, such as maple seeds. Thus the two species compete only for a portion of their food supply. Animals often do not recognize the competition posed by other species that share their foods. This can be seen at backyard bird feeders: the cardinal that will not tolerate another cardinal in his territory (even though there is more than enough bird seed for two) is undisturbed by the squirrel that eats most of the birdseed. In dry environments, the availability of water may be the most critical determinant of carrying capacity. This is seen in semiarid regions, where each plant has a bare area around it. Quite simply, the availability of water determines the spacing of the plants.

A *community*, in its turn, consists of species of diverse structure and function and depends on this diversity for maintenance. For stability, a community must include food species, or prey, as well as predators. A given species may be both predator and prey, eater and eaten, and neither is limited to the warm-blooded animals. Grasshoppers prey on plants and are preyed on by insectivorous birds, which in turn are eaten by hawks. Cows and grasshoppers therefore occupy a similar niche in the field: both are consumers, herbivores, competing for a single type of food. All animals are consumers, since they obtain energy by eating other organisms. Green plants are primary producers, since they convert the sun's energy into organic material that is food for other species.

---

[5] The Kirtland's or Jackpine Warbler (*Dendroica kirtlandii*) is a species endangered by its insistence on building its nest only in jackpine trees (*Pinus banksiana*) between 6 and 18 feet tall. Clear-cutting of large tracts of timber, with subsequent replanting of a single species of trees, means that Kirtland's Warbler does not find suitable nesting sites when trees are too young or too old. It was never found outside a restricted area of the lower peninsula of Michigan, but was once rather common there. It is now endangered.

An *ecosystem* is the pool of communities in a given area and depends on the continued functioning of diverse elements to maintain the balance of the system. The term *ecosystem* is also used for organisms that interrelate (e.g., predators and their prey). One often talks of *food chains,* which are the sequence of organisms that prey on one another, such as

grass → cow → humans

grass → insect → fish → bird

The bird and the human are "at the top" of their respective food chains. It is actually more informative to think of webs instead of chains, however, because the cow's excrement also fertilizes the grass, and there is more than one species that preys on the insect, including other insects (Figure 2.1). The study of the relationships among individuals, populations, and communities is ecology and is a major area of research within biology. We have only touched on a few simplified aspects.

Stability within an ecosystem depends on diversity. Therefore, as a first approximation, the existence of many species with unlike life spans, and a redundancy of predators and prey, leads to a stable ecosystem. Stability is threatened if there are few species, if they have similar life spans, or if many of the species have unique habitat requirements. In most cases that we know of, permanent ecosystem alterations have occurred either because a new species without natural controls was introduced (especially into a "fragile" ecosystem that contains relatively few species, or is limited by climatic conditions), or because habitat destruction eliminated a "key" species.[6] Ordinarily, however, the loss of a single species will not cause increasing cascades of disruption in an ecosystem. Alternative predators and prey will replace the extinct species.

Soil bacteria provide an example of a stable ecosystem: most soils contain many species of nitrifying and ammonifying microorganisms, all with very short generation times and each with different sensitivities to fungicides. Therefore, even large doses of fungicides often do not interfere with the total nitrifying (or ammonifying) capacity of the soil for long, because the resistant species will reproduce more rapidly as the sensitive species are eliminated and their food supply becomes more abundant. On the other hand, the functional stability of populations of soil bacteria may mask major changes in species composition. Such changes may be as great as the shift in bird populations caused by DDT, remaining unnoticed only because we do not have favorites among microorganisms.

---

[6] Habitat destruction alone endangers many individual species. In the United States, the California condor (*Gymnogyps californianus*) and the Jackpine Warbler are nearly extinct, while the snail darter (*Percina tanasi*) was initially preserved by preventing completion of the Tellico dam in Tennessee, even though construction was nearly completed, because closing the dam would have destroyed the only known habitat of this obscure species. Eventually, a compromise was reached whereby snail darters were moved into creeks not affected by the dam. Additional populations of snail darters were also found, so the dam was constructed at the expense of some, but not all, of this small fish's expanded habitat.

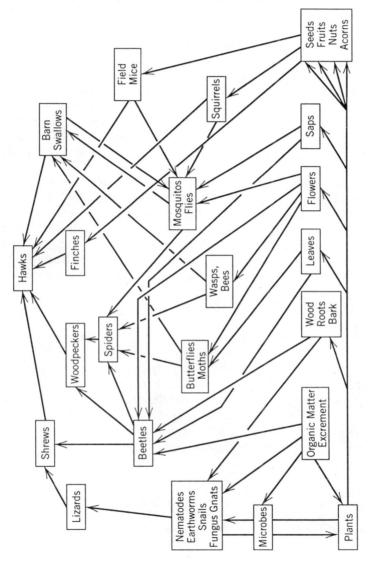

Figure 2.1 Food web for fields and woods. This diagram is greatly simplified, since many species are lumped into single categories (e.g., "beetles") or simply omitted (e.g., fleas). Many connections are also greatly simplified (mosquitos are eaten by bats as well as by many species of bird) or left out entirely (microbial webs).

At the other end of the continuum, the giant panda (*Ailuropoda melano-leuca*) now has a very restricted range, a mere fragment of a once extensive domain covering most of southeast Asia. The decrease in its range and the increased human cultivation of the area have eliminated large tracts of the panda's food supply, bamboo. The panda also reproduces slowly. As a result, pandas may well become extinct despite all efforts at protection, feeding, and captive breeding programs. The extinction of the panda will be due to a combination of causes. Human population pressure on the panda's habitat, in the form of cultivating land formerly left wild, has resulted in a scattered habitat consisting of isolated pockets of mountainous wilderness, often separated by cultivated fields. Such a habitat reduces the available food supply and therefore the number of individuals that can survive. It makes it less likely that pandas will find mates at suitable seasons. The same habitat destruction has led the panda to depend on a very small number of species of bamboo.

It is characteristic of bamboo that each species blooms and then dies in virtual synchrony across its entire range, regardless of the chronological age or size of individual plants. In 1985–1987 the two major species on which the panda depends entered the bloom/dying part of their cycle. Even recently germinated bamboo specimens, raised in Europe or the United States, bloomed in synchrony with their species in China. Such blooms and die-offs must have occurred many times since the panda evolved as a species, and the synchrony would not matter if there were many available species of bamboo. But because the panda's range has been drastically decreased by human expansion, and because this reduced habitat does not contain alternative species of bamboo, and because pandas in the wild apparently cannot change their diet, the species may not survive, where more opportunistic or more fecund species might have little difficulty. In the absence of human pressure on its habitat, on the other hand, a larger panda population on a larger range would probably forage enough different species of bamboo that enough animals would survive the period of scarcity while still reproductively active.[7] It is the combination of habitat destruction and restricted diet that may well combine to destroy the giant panda.

---

[7] The question of how many animals are "enough" to maintain a species is problematic and undoubtedly varies between species. It is now being investigated with DNA "fingerprinting" techniques, which can be used to determine the degree of genetic variability between individuals within a species. Some endangered species of deer demonstrate considerable genetic variability among relatively few individuals. The evidence is that this genetic variability will allow zoos to overcome the inbreeding depression resulting from repeated breedings among the few available specimens. The cheetah (*Acinonyx jubatus*), in contrast, has little genetic diversity, suggesting that it has gone through population crashes before. The lack of genetic variability is correlated with a high incidence of infertility, poor pregnancy performance, and high mortality among those kittens that are born. Outbreeding to seemingly unrelated cheetahs does not improve the reproductive performance, since the same deleterious genes are present. Conservation biologists are dubious about the survival of cheetahs. The effects of inbreeding on gene frequencies are described in Chapter 10.

In contrast to the panda, other species can change their "life-styles" to survive habitat destruction. In the early 1950s, the eastern peregrine falcon (*Falco peregrinus*) disappeared from most of its range because DDT essentially prevented falcon eggs from hatching. Once DDT residues had declined to a point that falcon eggs could hatch, reintroduction of the species to many parts of its earlier range was successful. In addition to the forests of New England and Wisconsin, New York and several other cities now boast breeding populations of peregrine falcons. Less particular about its habitat than many other species, the peregrine falcon discovered that skyscrapers are a reasonable facsimile of cliffs and canyons, and that pigeons are as good a dietary staple as field mice.

Sometimes the elimination of a single species will drastically alter an ecosystem. Such a species is called a *key species*. An example is the wild fig (*Ficus* sp) in tropical forests. Its fruit is critical to small mammals and to monkeys. The mammals serve, in their turn, as food for carnivores. Without the fig, leopards starve. Note that the key species are rarely those dramatic (or cuddly) animals that are the focal point of massive efforts at conservation. The fig, not the leopard, is the key species; the leopard, not the fig, makes the better symbol. On the other hand, the highly specialized animals at the top of the food chain do require well-integrated ecosystems to survive. Thus although the fig may be more important to the jungle than is the leopard, any conservation scheme that keeps leopards happy will almost certainly protect the rest of the ecosystem.

Key species may also maintain plant communities. In the prairies of North America, the buffalo and fire combined to preserve vast sweeps of grass from encroaching forests. In African savannahs, grazing serves the same purpose of controlling trees that fire served in the Great Plains. It was long thought that the key grazing species was the elephant, since it eats trees that are too old for most herbivores, and it is true that the elephant may be the key to savannah maintenance in some areas. In many areas, however, it is the lowly porcupine that is the key species. By eating bark, it kills young trees. Because the young trees die, the grasslands remain open; the grass-eaters find food. In the Chihuahuan semidesert areas of the North American southwest, kangaroo rats (*Dipodomys* species) appear to be the species that maintain the desert shrubland and prevent grasslands from dominating. When the kangaroo rats were excluded from fenced plots, tall perennials and annual grasses colonized the spaces between shrubs, and other changes in plant species also occurred.[8]

Because of the multiplicity of partially independent communities that make up an ecosystem, analysis of the functional parts is extremely difficult. All too often, components are identified only after they become extinct. *Maculinea arion,* the Large Blue butterfly of Europe, was known to feed on thyme plants as a caterpillar, and was known to depend on the presence of certain

[8] J. H. Brown and E. J. Heske, *Science* 250:1705–1707, 1990.

ants, but the latter relationship remained unclear when the last Large Blue was seen in England in 1979. Since then, biologists studying Swedish colonies of Large Blues demonstrated that after the last moult, the caterpillars fall to the ground at dusk and secrete a milky fluid. This fluid is attractive to *Myrmica sabuleti*, a red ant,[9] which carries the caterpillars to its nest, where it remains until the following spring. In the ant nest, the caterpillar feeds on ant eggs for 10 months, then pupates. Although several species of red ants will collect Large Blue caterpillars, good survival rates are seen only in nests of *M. sabuleti*. This particular species, however, flourishes only on sunny slopes covered with very short turf. Such short turf results from grazing. When myxomatosis killed off the rabbits of southern England, and sheep were put to graze on better land, the ant's turf was invaded by scrub. The ant died out; the Large Blue followed. This story, unlike many others, may have a happy ending, since both species exist elsewhere in Europe. The Nature Conservancy and the World Wildlife Fund are cooperating to reintroduce the Large Blue to England.[10]

Equally bizarre is the relationship between the dodo (*Raphus cuculatus*) and the dodo tree (*Calvaria major*). It took nearly 300 years after the last dodo was killed before botanists realized why the dodo tree was following its namesake into extinction. In the 1970s, with only 13 aging trees left, it was shown that the seeds of the dodo tree cannot germinate unless the shell is first weakened by passing through the digestive system of a bird. Since dodos became extinct, few birds large enough to swallow *Calvaria* seeds exist, and none are native to the habitat of the dodo tree. Scientists initially force-fed turkeys with *Calvaria* seeds but are now abrading the seeds artificially.

Even when individual relationships between a single species and a single extraneous factor, or between two species related in a predator–prey relationship, are well studied, the system as a whole may be completely misinterpreted. We are still unable to predict how complex systems will respond to the loss (or addition) of one species of plant or animal. Nonetheless, our species has altered ecosystems from the time we first ceased being hunter-gatherers and cultivated land to grow crops. The habit of changing ecosystems is so old, in fact, that we rarely realize the extent to which humans have changed the face of the globe. Italy just north of Rome was wooded until about 2000 years ago. A warming climate, agriculture, and the resulting erosion led to the semiarid landscape we call "mediterranean." Iceland was forested until the Viking voyages required timber for shipbuilding. Today, the cold winters and savage winds make it extremely difficult for young trees to survive.

Even the definition of habitat destruction becomes interesting as the perspective lengthens. For example, almost all of New England was forested

[9] Other sources give the nurse species of red ant as *Formica rufus*.
[10] From "Thyme to come home," *The Economist*, April 22, 1989, p. 80.

before Europeans came. With the abandoning of farms in New England after World War II, the woods began to creep back. Meanwhile, deer populations had adjusted to open meadows. Undoubtedly, certain plant communities similarly adjusted to the open fields produced by grazing. Is reforestation then merely the restoration of the original habitat? Or is it disruption of the new habitats? What is the significance of imported tree species, which change the makeup of the new forests? What of imported species such as sparrows and gypsy moths? An even more controversial question of natural habitat arose when trees were planted on the open, grassy hills of northern England. It is known that these slopes were densely wooded in medieval times, but modern inhabitants vehemently opposed changing the "natural" open vistas, which were initially produced by logging for shipbuilding during England's expansionist period, and which are still maintained by grazing sheep.[11]

ALTERING ECOSYSTEMS

Humans alter ecosystems by selectively destroying species, by importing new species, by destroying habitats, and by introducing xenobiotics.

Selectively Destroying Species

Hunting can exterminate a food species: neither dodos nor the passenger pigeon (*Ectopistes migratorius*) exist any longer. The bison or American buffalo (*Bison bison*) was nearly exterminated by wanton slaughter, with only choicest portions (sometimes only the tongue) being eaten. Conservation efforts managed to prevent the buffalo from becoming extinct, and today there are not only buffalo herds in many national parks, but some commercial herds as well. The North American parakeets were driven to the edge of extinction because their feathers were fashionable. Neither the Carolina parakeet (*Conuropsis carolinensis carolinensis*) nor the Louisiana parakeet (*Conuropsis carolinensis loudoviciana*) recovered.

Ironically, attempts at deliberate destruction rarely lead to extinction of the targeted pest species, which are pests because of their large numbers and prolific reproductive capacity.[12] Secondary effects on nontarget species

---

[11] In fairness to the inhabitants, their opposition is also based on opposition to tree "plantations" that replace diverse habitats with a single monoculture, providing neither food nor appropriate shelter to most species now living in the area. At least in Scotland, the incentives for reforestation have recently been altered to take into account the preservation of existing habitats.

[12] The only good example of total eradication of a pest species from a given area is the screwworm program. The screwworm fly (*Cochliomyia hominivorax*) lays its eggs in open wounds. Damage to cattle is considerable. Massive releases of sterile males, coupled with treatment of infected cattle and surveillance at the Mexican border to prevent reentry, actually exterminated screwworms from the southern United States. Larger programs aimed at the imported fire ant (*Solenopsis invicta*) and gypsy moth (*Porthetria dispar*) failed abysmally.

are quite common, however. Destroying, even temporarily, the insects in a given region may well destroy the food supply of other species. For example, a river basin that is sprayed with insecticides may well lack sufficient food for salmon to survive. Even when insect populations recover within weeks or months, the slower-breeding predators may not recover for years.[13] Similar results are seen when insecticides are applied intensively to crops. As predatory insects, mites, and spiders are killed off, species that were previously controlled by predators become pests. With a few exceptions, such *secondary pests* account for most of the insecticides applied to commercial apples. Such predator–prey imbalances also occur when mammalian predators are destroyed. The elimination of wolves as a significant predator in the continental United States led to explosions of deer populations, followed by starvation of deer herds in severe winters. In the absence of severe weather, overgrazing by deer herds can damage forests and grasslands, especially in semiarid environments; the deer also wreak havoc on vegetable and flower gardens.

Importing New Species

The importation of new species has occurred throughout history and prehistory as humans have roamed over the globe. Not all importations are deliberate, and not all are deleterious. Deliberate importations usually considered beneficial include the introduction of the potato and tomato to Europe. Similar advantages accrued from the introduction of most food crops and ornamental species to the Americas: rice from China, wheat from Europe, roses from China via Europe. We take these importations for granted, since we have "always" lived with them.

Many, but not all of the most destructive deleterious introductions have been accidental, such as the introduction of the Colorado potato beetle and the highly allergenic ragweed to Europe. Accidental importations also include Dutch elm disease, which was brought into the United States from Europe on shipments of elm products. The disease followed truck routes from ports of entry through the Northeast into the Middle West. Because the American elm (*Ulmus americanus*) was the dominant shade tree of midwestern towns, the beetle that transmits the disease was initially fought with massive applications of insecticides, leading to secondary effects on suburban ecosystems. Another accidental import, the Russian thistle (*Salsola kali tenuifolia*), probably arrived and moved across the United States with threshing equipment. At times, however, affected farmers attributed the thistles' spread to malicious actions by immigrant groups. Finally, it should be noted that not all imports are either obviously deleterious or obviously beneficial. The purple thyme (*Thymus vulgaris*) that covers much of the northern Catskills may have

---

[13] For detailed case histories of this phenomenon, I recommend Rachel Carson's *Silent Spring*, published in paperback by Fawcett Crest Books, New York, 1964.

hitchhiked on the wool of sheep. It is a pretty ground cover, eaten by cattle and yielding a strong honey, but of minimal economic importance.

Deliberate importations that are now considered harmful have had varied rationales. Many occurred because a species was considered aesthetically pleasing. The English sparrow was imported by a romantic who wished to bring to the New World all the birds mentioned by Shakespeare. A single reference to starlings[14] led him to import this species, which proved success-ful enough to be a major competitor to native species.

Deliberate introductions are also made for seemingly sound economic reasons and become deleterious because they prove to have unforeseen environmental consequences. Even casual perusal of wildlife or conservation magazines will turn up dozens of examples. One dramatic example is the gypsy moth (*Porthetria dispar*). Its importer wished to produce cheaper silk by creating a hybrid with the silk moth (*Bombyx mori*) that would not require a diet of mulberry leaves. In the latter goal he succeeded: when the gypsy moth escaped from his laboratory, it proved to eat an unrestricted diet of most of the hardwoods of the Northeast. Itself distasteful to almost all predators, and relatively resistant both to insecticidal control and to efforts at natural control, it ravaged the forests for 100 years. Only in 1988 did a fungal disease, introduced decades ago in a seemingly forlorn effort to control the moth, take hold. This disease provides the first hope in decades that the gypsy moth might be controlled. Similarly praiseworthy aims accompanied the kudzu vine (*Pueraria lobata*) to the United States: it was imported as an alternative source of fodder for cattle on marginal land. Although cattle will eat kudzu if they are starving, they prefer almost any other fodder. The vine cannot be controlled by limited foraging and is not readily controlled at all. In fact, it does very well in warm climates and is choking many fields and forests in the southeastern United States.

In some cases, the intended purpose of a deliberate importation is achieved successfully but is no longer desired. The *Melaleuca* tree (*Melaleuca quin-quenervia*) was originally imported to Florida by botanists who sought a species that would aid in draining the Everglades, the swampy grassland that once covered much of the Florida peninsula. The *Melaleuca* is native to Australia, transpires large quantities of water, and adapted exuberantly to the Everglades. Since its importation, however, the Everglades have been shrinking due to extensive canalization of northern and central Florida, to increased agriculture, and to draining of land for residential purposes. Moreover, wetlands have been recognized as valuable ecological resources. The earlier determination to drain swamps has been succeeded by efforts to maintain them, especially the unique ecosystem of the "endless grass" called

---

[14] E. L. Dachschlager, in a letter to *Smithsonian*, p. 18, November 1990, quotes from *Henry IV, Part One*: "Nay, I'll have a starling shall be taught to speak/Nothing but 'Mortimer,' and give it him . . ." and concludes that "Because of this one reference, *Sturnus vulgaris* is now found all across America."

the Everglades. The *Melaleuca* is particularly destructive to the Everglades because its seeds sprout after being exposed to fire, and the Everglades depend on periodic fires for control of (native) trees. Once sprouted, *Melaleuca* produces nearly impenetrable thickets. Additional invaders of the Everglades are the Australian Pine (*Casuarina* spp.), which grows rapidly in almost any soil and easily crowds out native trees, and the Brazilian Pepper (*Schinus terebinthifolius*), which produces such a profusion of seedlings that native pine seedlings are shaded out and cannot compete. Both are escaped ornamentals. Many ecologists fear that the Everglades are doomed and that the best one can expect is that the National Park Service can salvage small islands to remind us of their original glory.[15]

Somewhere between planned introductions and truly accidental "hitchhikers" are the careless introductions that thrive beyond expectation because they have no natural enemies in their new home. Remarkable havoc has been wreaked in Australia by rabbits, which overran large tracts of semi-arid pasture, eating most of the forage of both sheep and native animals. In the United States, descendants of the prospectors' burros are thriving in the Grand Canyon, at the expense of native sheep and other grazers that compete for the same scarce forage. Among plants, prickly pear cactus (*Opuntia* spp.) in Australia and Klamath weed (*Hypericum perforatum*) in the northwestern United States grew at the expense of pasture grasses; in both cases, a semblance of control was achieved by finding a predator in the native terrain and importing it. Long-isolated ecosystems are particularly vulnerable to imported species, because their species are in stable equilibrium. Thus such common species as dogs, goats, and snakes are driving numerous native species in the Galapagos, Hawaii, and other island ecosystems to extinction.

Habitat Destruction

Many species are extremely selective in what they eat, where they will live, and above all, in their reproductive requirements. Human alteration of the environment can drastically alter the ability of a species to survive by destroying any one of the elements needed for its survival. An alternative but equivalent statement is that humans alter or destroy the carrying capacity of the area for some—often native—species in order to increase its carrying capacity for imported species, often humans or their crops. Obvious examples are:

· The cutting down of forests, which deprives arboreal species of shelter and which may alter the seasonal water balance of the watershed as far

---

[15] Of course, the same gloomy prognosis was made for Lake Erie, the smallest and most polluted of the Great Lakes, 25 years ago. Remedial measures were taken despite sage pronouncements that they were too little and too late, and Lake Erie is now reasonably healthy. Although still polluted, it is fishable and swimmable in most places. A ban on phosphate-containing detergents has decreased the rate of eutrophication, not only in Erie but also in other lakes.

as the mouth of the stream. In the event that the forest is on steep slopes, erosion may proceed very rapidly, leading to silting of reservoirs and harbors downstream, or altering water quality so that native species cannot survive.

· The draining of swamps, which not only eliminates species living in the swamp but may also alter the flooding pattern of adjacent streams or the salinity of adjacent bays, as well as the forage for animals in adjacent forests.

· The elimination of a natural control, as when the first settlers controlled prairie fires, with the result that the grasslands of Wisconsin were forested within two generations.

Much has been said recently about the cutting and burning of the Amazon rain forest and the loss of diversity associated with it. Less is heard of the fact that the United States has lost 90% of its own temperate rain forest, and that only the existence of a photogenic endangered species, the Northern Spotted Owl (*Strix occidentalis*), may enable environmental groups to prevent clear-cutting of the remnants. In the last 150 years over 90% of the prairie in the United States has been destroyed by plowing it under. Besides causing the massive wind erosion of the 1930s dust bowl, this wholesale removal of deep-rooted plant cover has driven numerous species to the edge of extinction. Finally, we have drained 70% of our wetlands: not only the tidal salt marshes that serve as breeding grounds for many species of fish and aquatic invertebrates, but also inland wetlands such as the oxbows of prairie rivers. It is estimated that 87% of wetlands have been lost to farming or tree planting, although this percentage can be expected to change as urban sprawl continues and as more and more people seek out vacation homes along the seashore.[16] To put this loss into perspective: 500,000 acres of wetlands are plowed or otherwise destroyed each year. Along the lower Mississippi alone, 100 acres of marsh are lost each day. A hundred years ago the Illinois River was the third richest source of river fish in the United States. Today, commercial fishing on the Illinois is nonexistent. Moreover, even where wetlands remain, they become less suitable for wildlife as human activities disrupt their ecology. In 1978, only 4% of South Dakota's sunflowers were sprayed with pesticides; by 1984, 65% of sunflower acreage was sprayed. Spray drift and runoff from sprayed fields means that aquatic invertebrates were killed off and birds found food less easily. Less available food means less time on nests, less care for nestlings, and greater mortality.

[16] Examples of wetland destruction are all too easy to find. In the *New York Times* of May 9, 1992, it is reported that EPA head Reilly approved of building a golf course that required draining of a marsh on the Crystal River at the northern tip of the lower peninsula of Michigan, even though Valdas Adamkus, EPA administrator for the region, considered the marsh a critical habitat. The developer was a strong supporter of the Republican governor of Michigan, John Engler, who was George Bush's campaign director in Michigan in 1988.

Some direct contamination from aerial spraying can also be assumed and is documented whenever money is available to look.

In short, we are not more virtuous than the poor nations that are called the Third World: we were just quicker.[17] In referring to the collapse of biodiversity that most ecologists consider inevitable, Michael Robinson of the National Zoological Park in Washington, D.C. says: "We are facing the 'enlightenment fallacy.' The fallacy is that if you educate people in the Third World, the problem will disappear. It won't. The problems are not due to ignorance and stupidity. The problems of the Third World derive from the poverty of the poor and the greed of the rich. The problems are those of economics and politics. Inescapably, therefore, the solution is to be found in the same arenas."[18]

Addition of Xenobiotics to the Environment

The thoroughly modern practice of adding xenobiotics to the environment has a long history. The Romans salted the soil of Carthage, their ancient enemy, to prevent the growth of crops. "Scorched earth" policies, where armies burn crops, uproot trees, or otherwise deliberately destroy the land's fruitfulness, have been commonplace in western warfare. Sherman's march to the sea in the U.S. Civil War was one example. More recently, American airplanes destroyed millions of acres of forest in Viet Nam with herbicides, ostensibly to destroy the vegetation that provided cover for the enemy. The long-term ecological effects on river ecosystems and on the mangrove swamps have not yet been catalogued. The oil well fires of Kuwait, although they did not alter the global climate, coated large areas of land and water with petroleum distillates, many of which are acutely or chronically toxic to animals, and all of which can be expected to affect photosynthesis and respiration in plants. Meanwhile, the peacetime use of insecticides, fungicides, and herbicides has provided some of the best documented case histories of the consequences of disturbing single elements of an ecosystem.

Even in the absence of deliberate pollution, the ordinary activity of human beings moves, concentrates, and deposits the products and wastes of commerce unless rigorous efforts are made to control waste disposal. Significant pollutants include mine tailings, the contents of waste dumps, and the discharge of effluents into rivers. Environmental pollution by these materials is as old as the first human settlements. In fact, the primary difference between modern and ancient pollution is that there are more chemicals to

---

[17] J. D. Williams and R. M. Novak, in "Vanishing species in our own back yard: extinct fish and wildlife of the United States and Canada," list 15 piscine, 1 anuran, 3 reptilian, 38 avian, and 39 mammalian species or subspecies that have become extinct in the 50 states of the United States *since 1492*: that is, since Europeans came to the Americas (pp. 133–137 in *The Last Extinction,* edited by L. Kaufman and K. Mallory, MIT Press, Cambridge, Mass., 1986).
[18] Roger Lewin, "Damage to tropical forests, or why were there so many kinds of animals?" *Science,* October 1986, p. 149.

pollute with, both in quantity and in variety; and that we may segregate them in dumping areas when we know they are hazardous. The increasing number of people has increased the overall quantity of wastes, and rising affluence means that more things are used, and therefore discarded, per person. The airplane has greatly increased the number of acres that can be treated with pesticides, and huge modern farm machines have made it practical to plant 50 acres as a single field, which makes monocultures (and their dependence on heavy pesticide applications) economically advantageous.

Effects of Chemicals in Local Ecosystems

Until Rachel Carson wrote *Silent Spring* in 1962, few people thought much about the ecological effects of chemicals. And for another 10 years, the emphasis in ecotoxicology was on the effects of pesticides, not of chemicals in general. In studies of pesticides, interactions between xenobiotics and ecosystems were traditionally described in terms of effects on the pest, or target, species. Harmful effects were generally described in terms of undesirable effects on a single nontarget species. Thus the critical concern in the early pesticide literature was the efficiency with which a given compound kills a given pest and its toxicity to laboratory mammals as a model for humans. Toxicity to food animals and to beneficial wild species, such as fish or honeybees, was also investigated. But the norm was to investigate effects on directly exposed species or individuals. Only after it became apparent that persistent pesticides such as DDT moved through food chains was attention given to the concept of *ecosystem toxicity,* in which the relationships between species, rather than mortality within a single species, is the focus of interest.

The release of a toxic substance, and the resulting effects on the ecosystem into which it is released, can be modeled in several different ways. The most general model is shown in Figure 2.2. The various factors (transport, uptake, organism response, ecosystem response) will vary with different chemicals and different ecosystems, as shown in Figure 2.3.

Ecosystem response can be generalized for specific categories of chemicals. For example, a single application of an insecticide could cause the patterns shown in Figure 2.3. Levels of the chemical in the environment rise sharply after application and decrease gradually. The rate of decrease depends on persistence, transport, and degradation. The overall biomass of the ecosystem would first fall, as insects die. Even if not directly affected, insectivores would also decrease in number because of the scarcity of prey. Then, because so many insects are herbivores (plant-eaters), the overall biomass of the system would increase as plants thrive. Finally, as levels of the chemical decrease in the system, insect populations begin to recover (increase) and plant biomass to decrease. Eventually, the system probably returns to its original levels.

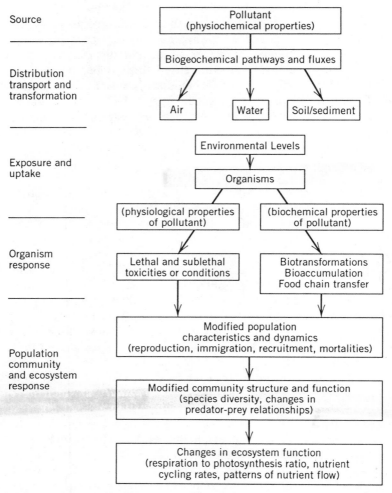

Source

Distribution
transport and
transformation

Exposure and
uptake

Organism
response

Population
community
and ecosystem
response

Figure 2.2 Parameters that must be determined in estimating the effects of a specific chemical on a specific ecosystem. [From D. W. Connell, *Ambio* 16(1):47, 1987.]

Such general models are useful for understanding the concepts of ecosystem effects but add little to our ability to predict the effects of releasing a particular agent. In actuality, each aspect of ecological effects must be determined separately for each chemical—and probably for each ecosystem. The insecticide DDT (dichlorodiphenyltrichloroethane; Figure 2.4) is the archetypal pesticide, both in its effect on insects and in its side effects. DDT provides several excellent case histories of the chain of events that can result from the release of toxicants into the environment.

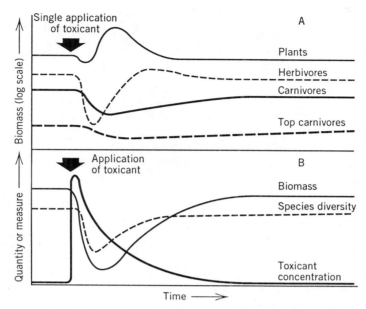

Figure 2.3 Modeling the environmental fate of chemicals: effects of the single application of an insecticide on plants, herbivores, and carnivores. [After D. W. Connell, *Ambio* 16(1):48, 1987.]

CASE HISTORY: DDT AND THATCH IN MALAYSIA

Initially synthesized by Othmar Zeidler in 1874, the insecticidal activity of DDT was identified by Mueller in 1939. It was the first of the new insecticides: the first of the effective, selective agents that became household commonplaces in the next decade. There is no doubt that DDT saved millions of lives. Metcalf[19] writes:

> DDT's insecticidal properties were discovered in a world just plunged into the catastrophe of World War II. Military sanitarians had grim memories of the

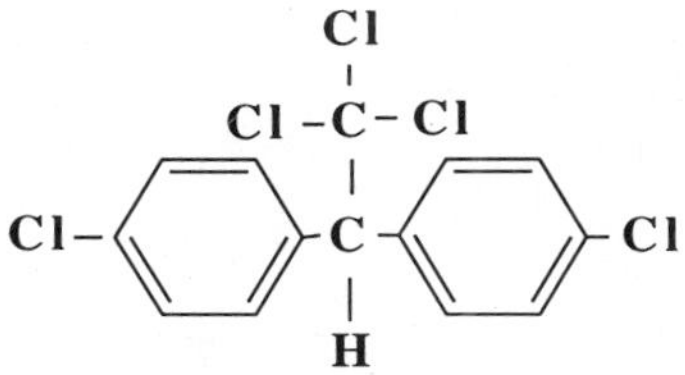

Figure 2.4 Structure of DDT.

[19] R. L. Metcalf, "A century of DDT," *Journal of Agricultural and Food Chemistry* 21:512, 1973.

ravages of louse-borne typhus which, in World War I, had been a critical factor in the fortunes of the Central powers in the Balkans and on the Eastern Front. The rapid enlargement of the conflict of World War II into Greece, North Africa, and later to the South Pacific focused attention on mosquito-borne diseases such as malaria, dengue fever, and filariasis. Thus there was immediate interest in a cheap synthetic chemical which it was safe to apply directly to human bodies and was so persistent that it killed mosquitos for months after application. DDT's use resulted in the conquest of typhus in Naples in 1943 and in the eradication of malaria from the Latina Province of Italy and in Sardinia beginning in 1945. For the first time in history, man had a safe, cheap, and durable weapon which could not only control but could often eradicate enclaves of vector-borne diseases. . . . The use of DDT in public health introduced "death control" on a global scale, and it is estimated that DDT has saved approximately 50 million lives and averted more than 1 billion human illnesses.

The onslaught against flies, mosquitos, lice, fleas, and other insects was so successful that entomologists seriously contemplated their own obsolescence. Instead, pesticides have provided the world with ample evidence of John Donne's aphorism that "no man is an island, entire unto himself." The side effects of massive pesticide use have altered relationships between species in numerous ecosystems. Because it is relatively nontoxic in acute exposure[20] and persistent enough to be transmitted through food chains, DDT provides many examples of secondary effects as it moves through the ecosystem.

One elegant illustration of the multiple effects that can result from a simple attempt to eliminate a pest species occurred in Malaysia, where spraying DDT against malaria mosquitos not only caused the roofs to collapse, but required an air-drop of cats because the village was threatened with plague.[21] In this case the DDT was being used in the World Health Organization's crusade against malaria, and effectively killed most of the mosquitos, achieving its goal of preventing the transmission of malaria. The spray incidentally also killed many other insects living in the village. Among these were the wasps that preyed on straw-eating moth larvae that inhabited the thatched roofs of the village. Without the wasps, the moths multiplied, and the roofs of the village collapsed. On the other hand, the roaches that inhabited the huts were relatively resistant to DDT and accumulated high levels in their bodies without dying. Geckos—small tropical lizards that live in the thatch and eat assorted insects—are not resistant to DDT and died from eating the DDT-laden roaches. Moreover, the dying geckos were easy prey for the village cats, which ingested toxic levels of DDT with the lizards. The cats, like the geckos, died. Rats began to multiply in the villages. Rats carry the fleas

---

[20] Although the median lethal dose of DDT is about the same as that of the organophosphorus ester insecticide diazinon, DDT is far less hazardous because it does not penetrate unbroken skin. DDT must essentially be ingested to be toxic. That is why it was possible to use DDT against louse-borne typhus: people in Naples were "dusted" with the insecticide to kill lice. This could not be done with the organophosphates because they are absorbed through skin.

[21] Most of the information in this case history comes from an article by G. Harrison, "Operation cat drop," in *Natural History,* December 1968.

that transmit bubonic plague, so the World Health Organization had to fly in cats to prevent an epidemic of plague.

This particular form of ecosystem disturbance was the result of the deliberate release of a toxic substance to kill a specific pest, or target species. Because chemicals selected for toxicity to one species are often unselective, ecological disturbances are typical of widespread use of pesticides. Obviously, however, such ecological upsets can also be managed or prevented once the possibility of their occurrence is recognized. Modern pest control stresses the need to know the life cycle of the pest and of beneficial organisms that interact with it.

In this connection it should be stressed that single instances of misuse do not necessarily mean that a chemical should never be used. The eventual banning of DDT was based on its persistence and consequent global bioaccumulation, which resulted in contamination of every species and inevitable risk to sensitive populations. The Malaysian incident provides a dramatic illustration of the unforeseen relationships between beneficial and destructive species. If its warning had been heeded, the profligate use of persistent insecticides might have been stopped and the bioaccumulation of DDT minimized. Development of newer insecticides might have included consideration of the hazards of undue persistence, and the persistent insecticides like DDT could have been reserved for restricted public health use. For over 30 years, DDT was indeed used to control malaria-carrying mosquitos in more than 100 countries. But of the 4 billion pounds of DDT used worldwide since 1940, over 80% was used in agriculture. Moreover, DDT became a household commonplace for mothproofing carpets and for controlling roaches and flies in the house and pests of ornamentals and vegetables in the garden. A few entomologists did voice concern over widespread DDT use, but with little effect.[22] Not until Rachel Carson's *Silent Spring* appeared in 1962 were public questions raised about the environmental consequences of widespread pesticide use. Then, in 1971, almost all uses of DDT in the United States were banned because of essentially world-wide contamination and of bioaccumulation through food chains. Uses of most other bioaccumulative insecticides were also severely restricted during the 1970s: aldrin and dieldrin in 1974, heptachlor in 1976, chlordecone in 1978, chlordane in 1976 (and even more severely in 1987). But as the first, the most used, and the least expensive of the modern insecticides, DDT remains a target for all the emotions associated with the ongoing debate over the role of insecticides in crop protection: the symbol of a time when it seemed that total control of insects was within our grasp and of the bitter failure of that dream.

Today we know that there is no easy solution to the problems posed by insects and other pest species. In 1950, at the beginning of the "Age of Pesticides," the United States lost an estimated 30% of its harvests to pests. Forty-two years later, we still lose 30% of our crops to pests. The pests have changed, the pesticides have changed, our soils and our tissues carry residues of a dozen persistent chemicals: the toll remains the same. Unfortunately, we are slow to apply the lessons learned from pesticides to other chemicals. Any substance that does not degrade rapidly can cause a cascade of effects in an ecosystem. The consequences

---

[22] The history of DDT use is described by R. L. Metcalf in "A century of DDT," *Journal of Agricultural and Food Chemistry* 21:511–551, 1973, and (somewhat more briefly) in "Changing role of insecticides in crop protection," *Annual Review of Entomology* 25:219–256, 1980.

to humans are not always as indirect as in Malaysia: more often, we too are part of the food chain, as were the people of Minamata, Japan.

CASE HISTORY: MINAMATA DISEASE

In Japan, a factory making acetaldehyde released mercury-containing effluent into Minamata Bay. The inorganic mercury used in the factory was methylated (Figure 2.5), probably during the manufacturing process itself, although inorganic mercury can be methylated by microorganisms. It is estimated that up to 600 tons of mercury was released into Minamata Bay between 1932 and 1970. In the bay the methylmercury *bioaccumulated* (a phenomenon discussed more fully in Chapter 5) in shellfish and in fish. These were, in turn, eaten by the fishermen, their families, and the cats of the fishing villages. Changes in the waters of Minamata Bay were seen as early as 1950, when birds began to drop into the sea, shellfish and seaweed died, and children could catch octopus with their bare hands. The cats began dying in 1953, first exhibiting bizarre neurological symptoms. By 1958, few cats remained in the area.

By 1956, enough villagers had become sick that doctors realized that they were confronted by an outbreak of "an unclarified disease of the central nervous system." The human symptoms resembled those seen in cats: salivation, convulsions, staggering. Deaths occurred. Pregnant women were seemingly spared, but their children were born with cerebral palsy and severe mental retardation. The epidemic continued for 10 years before the cause was publicly identified as being due to the mercury effluents. At least 798 people died or suffered permanent neurological damage from postnatal methylmercury exposure. At least 40 cases of congenital methylmercury poisoning were documented, giving the name *Minamata disease* to congenital organic mercury poisoning. The seemingly inverse relationship between maternal and congenital symptoms was eventually recognized to be due to the greater sensitivity of the fetus: maternal exposure that caused adult illness led to infertility, miscarriages, or stillbirths. To some extent the fetus also serves as a depot for maternally accumulated mercury, and the relatively minor symptoms in the mothers of children with Minamata disease were due to this sequestering of maternally ingested mercury in the fetus.[23]

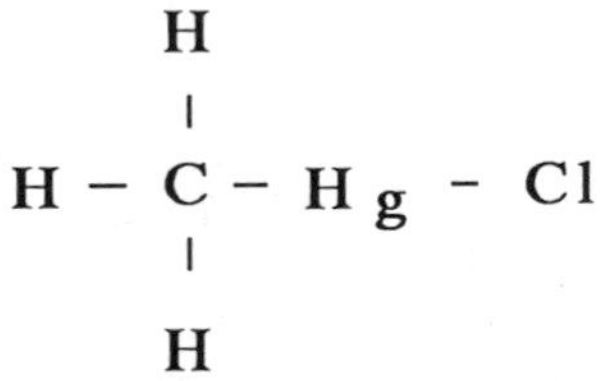

$$H - \underset{\displaystyle H}{\overset{\displaystyle H}{\underset{|}{\overset{|}{C}}}} - H_g - Cl$$

Figure 2.5 Structure of methylmercury.

---

[23] The story of Minamata disease is eloquently told in *Minamata,* by the photographer W. E. Smith and A. Smith, Holt, Rinehart and Winston, New York, 1975. The accompanying pictures were taken during the many months that the Smiths lived in Minamata.

SUMMARY

The nature of the chemical release differed between Malaysia and Minamata. There was carefully targeted use of a pesticide in Malaysia, careless dumping of wastes in Minamata. Nonetheless, the same underlying pattern was seen. A chemical moved through the food chain and was accumulated in some of the organisms in the food chain, eventually injuring or killing more sensitive species. In both cases, the effects were limited to a small geographical region, however, and did not violate the comfortable concept that pollution is really a quite local phenomenon. It remained reasonable to assume that the solution to pollution is dilution. Once DDT residues were found in Antarctic penguins, the full extent of DDT's presence in the environment became known. It was then acknowledged that the pollution was global in nature. At that time some suggestions that chemicals could cause global damage were made, and when DDT was shown to decrease photosynthesis by oceanic algae, it led to fears that net global oxygen synthesis might be affected. Fortunately, this was not the case. Unfortunately, once the hypothesis was disproved, its existence was used by skeptics to argue that environmentalists were hysterical or even unscrupulous, and that most other environmental concerns were also without foundation.

The global consequences of pollution are difficult to identify and even more difficult to prevent, especially when the sources of the pollution are diffuse. In a very real sense, every organism, every species, and every location on earth is part of a single ecological system, or ecosystem. The existence of ecosystem-wide effects has been recognized in principle for years; in the early days of the current environmental movement, the term "spaceship Earth" was coined to emphasize the limits to our environment and the need to preserve resources, as well as the finite nature of the resources and their endless recycling. In the last decade, however, the immediacy of the interactions between seemingly isolated parts of the globe have become apparent as we begin to understand the natural rhythms of weather and climate.

WEATHER

The existence of planet-wide interactions—the global ecosystem—can occasionally be seen. The most notable examples are caused by perturbations in the weather. In 1982–1983, for instance, the phenomenon known as *El Niño* occurred in South America. El Niño is a weather change that occurs around Christmas (hence *El Niño,* the little boy, or the child Jesus). The trigger is the warm southward current that flows off the coast of Peru each winter. In an El Niño year, this current penetrates farther south, causing heavy rains in coastal Peru. In the nineteenth century, the occurrence of an El Niño signaled an abundant harvest, because the rains that came to the arid semidesert were beneficial to farmers. Cotton could be grown where it was too dry in other years; flocks flourished on the abundant pastures. Sea life decreased in these "abundant years," however, and the seabirds moved elsewhere.

In this century, the occurrence of El Niño has come to be considered a disaster rather than a blessing, because Peruvian income depends more on fishing, an industry that generates exports, than on subsistence farming. El Niño is associated with a failure of the anchovy catch off the coast. The immediate cause of the failed anchovy catch is the presence of unusually warm water off the coast of western South America, which causes the anchovy to move to colder water. The failed catch affects the lives of the fishermen and the economy of their countries. El Niño also affects the economy of soybean-producing countries such as the United States and Brazil, because soybeans are the replacement protein for anchovy meal in many animal feeds. When the anchovy catch fails, midwestern farmers thrive because the price of soybeans rises.

The Peruvians who named El Niño thought it occurred approximately every 10 years. Only recently have meteorologists identified the trigger: an alteration in the temperature of the southwestern Pacific Ocean. The first clue to the global nature of El Niño was the recognition of the role of the ocean current in global weather patterns. El Niño occurs when the warm current of water is stronger than usual and so penetrates farther south. In the 1960s, oceanographers realized that the current extends thousands of miles into the ocean and reflects the circulation of global ocean waters. Subsequently, oceanographers realized that the occurrence of El Niño corresponded to years in which monsoons failed in India, causing disastrous droughts, and that the changes in the ocean circulation were actually responses to altered global winds.

Linking El Niño to the monsoons firmly identified the global nature of the phenomenon, because monsoons have been recognized as part of global wind and water patterns since the work of Gilbert Walker, director of India's observatories between 1904 and 1930. Walker saw the relationship between monsoons and the winds of the southern Pacific. He named the cyclic alteration of these winds, and of the rain that altered with them, the *southern oscillation*. When trade winds are weak and the water temperature of the southern Pacific is high, surface pressure does not vary much across the tropical ocean. In such years there is drought over the western Pacific, floods over the eastern tropical Pacific, and El Niño occurs in Peru. Thus it is now thought that the alteration in ocean temperature that occurred in 1982–1983 was the cause not only of El Niño, but also of the very cool spring in the midwestern United States and of the severe drought in the south Pacific and Australia. The opposite pattern, consisting of abnormally low temperatures in the southern Pacific, is accompanied by strong trade winds and by droughts in North America. This has sometimes been called *La Niña,* in contrast to the El Niño phenomenon.

In fact, meteorologists now consider the Pacific Ocean to be perhaps the most important short-term influence on worldwide climate. Recognition of all the patterns is very new and much is still speculative. Critical to the pattern, however, is the temperature of surface water in the Pacific Ocean along the equator, west of South America. If the water is unusually warm, about 7°F above normal, the dramatic pattern known as El Niño is seen, with flooding in western South America and droughts in Australia (Figure 2.6). Water temperatures of 4°F above average, which occurred in 1986–1987, are associated with relatively normal weather and would not have been considered an El Niño year by the anchovy fishermen.

Whatever the maximum temperature of the warm episode, it is typically succeeded by a cool cycle. Trade winds blowing from the east drive warm surface water toward Asia, where the increased warm moisture triggers heavy monsoons (Figure 2.7). Near South America, an increased upwelling of cold water occurs, ending El Niño. This cold water is then driven westward by the trade winds, while skies in the eastern Pacific remain relatively clear and water is warmed by sun. Alternations between cold and warm phases of the cycle occur in spring when trade winds are the weakest, and each cycle lasts 2 to 3 years. The apparent 10-year cycle of El Niño was an artifact, due to identification of only severe episodes. In milder episodes, the anchovies swarm as expected. Similarly, only severely cold episodes are recognized by disastrous flooding or droughts.

The 1988 drought in much of the central United States was probably a result of the cold water in the eastern Pacific, which (in unknown ways) caused the convergence zone of the trade winds to move north.[24] The convergence zone is a major source of thunderstorms and rain and tends to overlie the warmest water in the region. The northward movement of the convergence zone may have been the reason that the jet stream over the United States moved northward in the late spring of 1988, producing a large high-pressure system without rain over the central plains. As moisture evaporated and was not replaced, the lower humidity itself decreased the chance of rain, since not enough evaporation occurred to trigger rain-bearing clouds.

Ironically, for all our long dreams to control the weather, the changes we have already wrought go unnoticed by most people. There is ample evidence that cities, because of their tremendous volumes of concrete, have so much capacity to retain heat that they change the climate within their boundaries and for some distance around them. Similarly, the large numbers of airplanes in the skies over the United States are thought to release enough water vapor to decrease transmission of sunlight significantly. The resulting midday haze has crept over us so gradually, however, that claims of formerly clearer skies will be attributed to nostalgia, if not senility. But weather has always been unpredictable and changeable from year to year. Only by tracking

---

[24] The convergence zone is the area where the trade winds from the northern hemisphere collide with trade winds from the southern hemisphere.

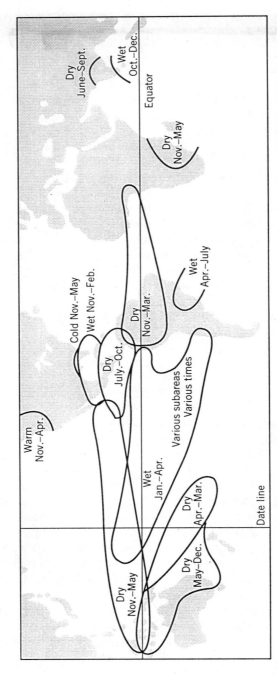

Figure 2.6 Global climatic anomalies in an El Niño. (From E. M. Rasmusson, *American Scientist* 73:169, 1985.)

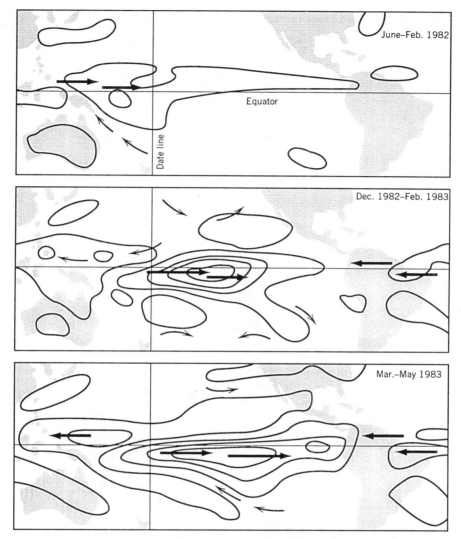

Figure 2.7 Wind patterns in an El Niño, 1982–1983, showing gradual eastward expansion of the El Niño phenomenon. Arrows indicate winds near the surface, while the contours represent anomalous atmospheric convection. (From G. Philander, *American Scientist* 77:451, 1989.)

weather over decades can we be sure that last year's blizzard was part of a long-term pattern, or, conversely, that the scorching summer was atypical. Data gathered over the past two decades are now showing a pattern of change: a change in climate due not to normal cycles of the earth but to chemicals released into the atmosphere. Meanwhile, two powerful influences on weather are known: one natural, one anthropogenic. Volcanos have repeatedly been shown to affect weather around the world; nuclear war has been predicted to do so even more drastically.

CASE HISTORY: VOLCANOS AND NUCLEAR WINTER

While increased levels of CO_2 and methane are expected to raise the global temperature (see Chapter 3), a nuclear war would probably cause severe cooling of the climate, possibly for several years. The evidence for this hypothesis is fortunately indirect. It has been known for a century that volcanic explosions are associated not only with spectacular sunsets, but also with abnormally cool weather. The explosion of Tambora, an Indonesian volcano, in 1815 was followed by a year in which New England had frost every month of the year. This was called the "year without a summer." The cold weather and unseasonal frosts destroyed crops in much of the northern United States. In centuries past, when food moved only a few miles, there would have been starvation.

The effects of volcanos on climate are due to the massive quantities of vapors and of particles they spew into the atmosphere, and the tremendous height to which these particles rise. A handful of dust thrown into the air comes down very quickly and very close to its source. The particles kicked out by a snow blower go a little farther. Those spewed out by volcanos can go into the stratosphere and circumnavigate the globe repeatedly. This stratospheric dust blocks some of the sun's light, causing a cooling of the earth. The refraction of light rays also improves the sunsets.[25]

There is rarely more than one major volcanic eruption in a year. It is estimated, however, that even a small nuclear war, using perhaps 10% of the stockpiled arsenal of the United States and the Soviet Union, would equal many volcanos. The resulting layer of dust in the stratosphere could block more sunlight than did Tambora in 1815 or Krakatoa in 1883 and might well cause crop failure from cold in all temperate regions. Crops in tropical regions would suffer from reduced light levels. Following the disruption of communications and transport that a nuclear war would entail, the crop failures would be far more devastating than in a normal year. The nuclear winter scenario need not postulate an initial breakdown of societies or nations; rather, it might be the cause of such breakdowns even in areas that survived the massive deaths and subsequent radiation poisoning that would occur in areas where bombs fell.[26]

---

[25] Some volcanic eruptions are dirtier than others: the Mexican volcano El Chichon produced glorious sunsets for several years. The dust spewed into the stratosphere by El Chichon is credited with preventing record high global temperatures in the El Niño year of 1983. In contrast, Mt. St. Helen's produced relatively little soot and did not affect sunsets. In 1991 the explosion of Mt. Pinatuvo in the Philippines again threw dust into the stratosphere.

[26] All of the predictions about the effects of nuclear war on climate are based on computer models. Assumptions about the number of bombs released, the distribution of explosions across the globe, and even the altitude at which the bombs explode would alter the amount and distribution of dust and debris in the atmosphere. The season at which explosions occurred might alter the effects. Three articles that describe the kinds of effects that might occur are

    P. J. Cruzen, "The global environment after nuclear war," *Environment* 27:6–12, 34–37, October 1985.

    R. P. Turco, O. B. Doon, T. P. Ackerman, J. B. Pollack, C. Sagan, "Climate and smoke: an appraisal of nuclear winter," *Science* 247:166–176, 1990.

    N. Myers, "Nuclear winter: potential biospheric impacts in Britain," *Ambio* 18:449–453, 1989.

3

AIR POLLUTION AND GLOBAL ECOSYSTEMS

Senescente mundo, *when the hot globe*
Shrivels and cracks
And uninhibited atoms resolve
Earth and water, fruit and flower, body and animal soul,
All the blue stars come tumbling down.
Beauty and ugliness and love and hate
Wisdom and politics are all alike undone.

—Thomas Merton[1]

GLOBAL AIR POLLUTION

Air pollution has long been recognized as a potentially lethal form of pollution. In this century alone, there have been killer smogs in London and in Donora, Pennsylvania. Moreover, air pollution has often provided the visible evidence of environmental disasters: blowing soil carried as far as Washington, D.C. epitomized the Dust Bowl of the 1930s. In the past decade, air pollution has been identified as the source of gases causing global warming and depletion of the ozone layer.

Table 3.1 illustrates the differences between air, soil, and water pollution. Pollution of soil is, in the improbable absence of transport by, through, or into water, a very local phenomenon. Water pollution, although including widescale transport, is nevertheless quite predictable, compared to air pollution. Air is poor in organisms, but its enormous capacity for mixing makes the atmosphere far more homogeneous than even the world's oceans. Air pollution is still unpredictable and can readily be global in nature.

Sources of air pollution include natural sources, such as volcanos, dust storms, pollen, and volatiles from trees. Although these substances can be critical to the local (spatial or temporal) ecology, they are self-limiting, at

[1] From "Senescente Mundo," in *The Tears of the Blind Lions*, Thomas Merton, New Directions, New York, 1949, p. 31.

Table 3.1 Comparison of Water, Air, and Soil as Media for Pollutants

	Biotic Content	Transport by Medium		Risk to Population	Sink for Pollution	Uniformity of Global Dilution
		Distance	Mass			
Air	+	+++	+	+	+	+++
Water	++	++	+++	+++	+++	++
Soil	+++	+	++	+++	+	+

Source: The table was constructed from a talk by D. Calamari at the NATO Advanced Study Institute at Lago di Garda, Italy, October 1986. Proceedings of this meeting, including Calamari's talk (but not the table) appear in *Toxicology of Pesticides: Clinical and Regulatory Perspectives*, L. G. Costa, C. L. Galli, and S. D. Murphy, eds., Springer-Verlag, Berlin, 1987.

least in the sense that they are part of the terrestrial pattern. It is nonetheless true that the eruption of major volcanos causes climatic changes that may persist for years, and the alterations of earthquakes and winds are obviously quite permanent. Even the very local health effects of pollens and volatiles can be drastic for those with allergies. Nonetheless, we will restrict our discussion to the anthropogenic sources of pollution—if only because those sources can be controlled.

Anthropogenic pollution consists primarily of solids and gases, although the gases include vapors of substances more familiar as liquids. Agricultural pollutants consist primarily of particles arising from soil erosion; the major gaseous components are volatilized pesticides. Industrial pollutants include both particulates, often in the form of soot, and numerous gases. Hydrogen fluoride, for example, was identified as an environmental pollutant around 1900. Specific toxic gases, despite their local importance, are not the most serious aspect of air pollution, however. Ozone and other oxidants are thought to account for 90% of gaseous pollutants, when all sources are considered. Other major pollutants are the several oxides of nitrogen (NO_x), including NO, N_2O, and NO_2; and sulfur oxides (SO_x), especially sulfur dioxide, SO_2. Carbon dioxide (CO_2) is a normal constituent of the atmosphere, present at a level of approximately 4% of atmospheric molecules at sea level. It is essential to life, being the carbon source for plant photosynthesis.[2] In the context of concern over global warming, however, carbon dioxide can also be considered a pollutant (pp. 53–60). Finally, organic volatiles interact with the inorganic constituents of air pollution. These volatiles include methane (CH_4), with sources as varied as swamps, oil wells, and ruminant digestion; alkanes from automobiles and airplanes; terpenes from forests; and a host of industry-specific gases such as trichloroethylene from dry cleaning.

---

[2] It is also necessary for mammalian respiration: if the CO_2 concentration in the blood is too low, the brain does not trigger the inhalation reflex. Climbers in the Himalayas must carry CO_2.

Entry of pollutants into the atmosphere occurs in the form of gases, of particles, or of aerosols, by evaporation of liquids or by co-evaporation of dissolved solvents from water, and by wind erosion from soil or with soil. Removal of pollutants from the atmosphere occurs by precipitation of pollutants dissolved in water droplets (rain, fog) or adsorbed on particles (soot, soil, snow, hail). Gases and particulates are the major forms of air pollution, but the role of water cannot be overemphasized. In many cases gases are dissolved in water droplets, and many of the reactions critical to the formation of smog, for example, may be accelerated if the pertinent gases are dissolved. Similarly, adsorption of particles on water droplets, or of gases on particles, facilitates certain reactions between the particles and gases and/or water.

GASES AS POLLUTANTS

Gases are less obvious forms of air pollution than are particles. Most gases are colorless, the only common exception being NO_2; many are also odorless. Unlike particulates, gases do not tend to settle out of the air. They are often very reactive, leading to multiple interactions during their residence time in the atmosphere. Residence times in the atmosphere are measured in *weeks* for liquids and for solids less than 10 μm in diameter, but in *years* for gases. The precise residence time depends on water solubility and on interaction with other gases in the biosphere.

Atmospheric transfer within the troposphere is latitudinal. Circumnavigation occurs in 2 to 4 weeks at midlatitudes and in 3 to 6 weeks at the equator. Other than specific gases associated with particular industrial processes,[3] most gaseous pollutants are the result of combustion. The major elements found in air pollution are, therefore, carbon, nitrogen, hydrogen, oxygen, and sulfur, present primarily as carbon monoxide (CO), carbon dioxide (CO_2), sulfur dioxide (SO_2) and to some extent its trioxide (SO_3), ozone (O_3), and the several oxides of nitrogen, usually characterized as NO_x. Some of these gases are relatively inert (CO_2); others, such as those that interact to form photochemical smog, are reactive. Air pollutants also include aerosols and gases interacting, as in the formation of acid rain, and inert aerosols. Ozone (O_3) is overwhelmingly important in air pollution in the troposphere, not only because it is highly toxic at levels greater than approximately 120 μg/m^3, but because it is highly reactive and interacts with other pollutants. In the stratosphere, ozone is also critical but plays a benign role (see below).

---

[3] Hydrogen fluoride (HF) was probably the first toxic gaseous air pollutant identified as a health hazard. Early in the twentieth century it was shown that cattle grazing downwind from a chimney that was emitting HF developed fluorosis. A recent disaster due to gaseous industrial effluent was the release of methyl isocyanate, an intermediate in the synthesis of the carbamate insecticide carbaryl, in Bhopal, India on December 3, 1984. The accident resulted from errors made by poorly trained and essentially unsupervised workers on the night shift. It killed over 2000 people and injured over 200,000.

At the minimum, combustion of carbonaceous material generates CO and CO_2. Impurities in the initial material generate NO_X and SO_X. Aldehydes and other complex carbon-containing volatiles will also be emitted unless extreme care is taken that combustion goes to completion. Major reactions in the troposphere include those outlined below.

1. *Generation of Atomic Oxygen*

$$NO_2 + light_{(h\nu\ <\ 440\ nm)} \rightarrow NO + O\cdot$$

2. *Ozone Formation*

$$O\cdot + O_2 + M \rightarrow O_3 + M$$

where M is a third molecule, typically N_2 or O_2.

3. *Ozone Destruction*

$$O + O_3 \rightarrow NO_2 + O_2$$

4. *Peroxy Radical Formation (where R is a hydrocarbon)*

$$RH + \cdot OH \rightarrow R\cdot + H_2O$$
$$R\cdot + O_2 \rightarrow RO_2$$

5. *Source of Hydroxy Radicals (at wavelengths <320 nm)*

$$O_3 + \cdot OH \rightarrow 2(\cdot OH) + O_2$$
$$O_3 + H_2O \rightarrow 2(\cdot OH) + O_2$$

Because reaction (5) requires ultraviolet (UV) light, it occurs during the daylight hours of pollution episodes.

6. *Generation of Organic Radicals (frequently from aldehydes)*

$$R\cdot + O \rightarrow \cdot RO_2 \qquad \text{(a peroxy radical)}$$

7. *Peroxyacetyl Nitrate Generation*

The generation of organic radicals (reaction 6) is rate limiting. The remainder of the reactions occur very rapidly. Equilibrium between ozone generation and ozone destruction occurs at approximately 25 part per trillion volumes (pptv) of NO_2.

If sulfur is present in the reaction mixture, SO_2 is formed, with subsequent conversion to H_2SO_3 and H_2SO_4. Approximately half the SO_2 and one-third of the NO_X in the atmosphere come from power plants. In the event that conditions are right, NO will be converted to NO_2:

$$NO + RO_2 \rightarrow NO_2 + RO$$

followed by reactions 1 and 2:

$$NO_2 + h\nu_{(>440\,nm)} \rightarrow NO + O\bullet$$
$$O\bullet + O_2 + M \rightarrow O_3 + M$$

This sequence is extremely important in the stratosphere, where maintenance of the ozone layer essentially requires the presence of NO_2.

Sources of Gases in Air pollution

Automobiles generate some NO_X, but the more recent models of catalytic converters are designed to convert these back to N_2, and to destroy the small quantities of HCN that are formed. Because only 0.03% of gasoline is sulfur, very little SO_2 or H_2SO_4 is formed by automobiles.[4] Most SO_2 comes from the burning of coal (sulfur content ranges from <1% to >4%) and fuel oils. Catalytic converters also prevent the generation of aldehydes by combusting hydrocarbons completely, to CO_2. Without catalytic converters, cars release about 100 to 200 mg/mile of aldehydes; this is reduced to 10 to 20 mg/mile by the catalytic converter.

Additional elements of automobile-caused air pollution are the evaporation of fuel from the gas tank and engine when the engine is not running and the escape of gas from the crankcase ("blowby") during operation. Both are controlled by current emission control equipment. Thus, *per automobile,* the contribution of American cars to air pollution has been decreasing greatly, and at least through the 1980s, was significantly less than that of European cars in Europe. (Most European countries do not yet require catalytic converters, but imports should conform to American standards.) The results can be seen in areas with fairly stable levels of population and industrial output (Figure 3.1).

Despite the potentially decreasing contribution of automobiles to air pollution, the increase in the number of automobiles has prevented any effective cleanup of air in many parts of the United States. Despite the strictest laws

---

[4] Sulfur concentrations may be considerably higher if sulfur-containing additives are used to improve octane rating, as has occurred since the phasing out of leaded gasoline. Considerable H_2S may be generated by these additives in the presence of the catalytic converter, especially while the engine is warming up, leading to auto exhaust that smells like rotten eggs.

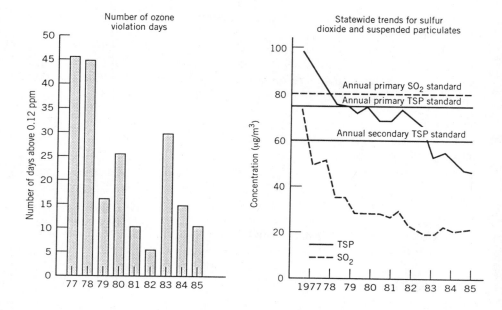

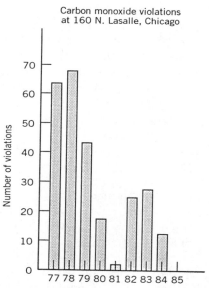

Figure 3.1 Decreasing levels of air pollutants in Illinois: effects of catalytic converters and the Clean Air Act? [From Sweitzer, *Environmental Progress* 11(6):6, 1986, published by the Illinois Environmental Protection Agency.]

in the country, Los Angeles has not materially *decreased* smog levels in decades. Also militating against further improvement of air quality is a serious weakness of the catalytic converters: when poisoned by lead, they allow greater emission levels than would a well-engineered car that was not designed for a catalytic converter.[5] Therefore, the large number of people who buy leaded gasoline, which until recently was still cheaper than unleaded gasoline, ruin their catalytic converters, effectively sabotaging pollution control.[6] For the same reason, pollution occurs when the pollution control equipment on older cars is not repaired or when cars are brought into the United States with inadequate or merely with less durable pollution control equipment. EPA regulations require that pollution control equipment be installed on cars imported to the United States and that pollution control equipment last for 50,000 miles. But many cars are driven more than 100,000 miles, and evasion of the import regulations is thought to be substantial. Moreover, although some states have mandatory yearly inspections that include pollution control equipment, many parts of the country do not require cars to be inspected. The rationale is that air in those localities is clean enough that automobile exhausts do not endanger health—that is, there isn't enough smog around to worry about. This attitude clearly does not take into account the global effects of air pollution, but still draws on the old fallacy that dilution is adequate to control pollution.

Industrial air pollution is most commonly blamed for obviously poor air quality and is often a major source of problems. In several cases, industrially generated air pollution has resulted in killer "smogs." Such episodes require temperature inversions and/or natural barriers that trap the polluted air. These conditions are often achieved in mining and steel milling towns, which are typically located on rivers and are often surrounded by mountains.

Industrial pollution other than that caused by the combustion of fossil fuels can lead to massive local air pollution. The variety of pollutants that can be generated is limited only by the ingenuity of the organic chemist. Besides the poisoning of Bhopal, India by methyl isocyanate in 1984,[7] problems have been caused by elemental pollutants such as arsenic, uranium, and lead in the vicinity of smelters because of the permanence of the materials. For both organic and inorganic pollutants, the initial solution has often been dilution. When higher smokestacks decrease the local levels of a pollut-

---

[5] However, because the platinum catalyst is poisoned by lead, catalytic converters have required the phasing out of leaded gasoline, resulting in the decrease of a major source of lead pollution and of childhood lead poisoning.

[6] Some European countries avoided this problem by pricing unleaded gas below that of leaded gasoline. Since Europeans use very high taxes on gasoline to encourage energy conservation, this social use of pricing structures is possible. In the United States, where powerful oil and consumer lobbies have prevented any increase in the gasoline tax for many years, social pricing is not used. Unleaded gas was more expensive than leaded gas for years, resulting in the use of leaded gasoline in catalytic converter–equipped cars in an estimated 20% of automobiles on the road in 1983.

[7] "Disaster in India: agonies and questions," *New York Times*, December 10, 1984.

ant, few of the former victims ask where the mess is going. How successful dilution is depends on the pollutant, but recent events demonstrate that it is not a permanent solution, even for organic chemicals.

The air pollution engendered by combustion results in the phenomenon called *smog,* a contraction of the presumed precursors, *smoke* and *fog.* Fog is an innocent bystander in the process, but the same atmospheric conditions that cause fog to linger over a valley cause smog to develop, so the nomenclature is not amiss. Smogs occur when pollutants are trapped in the lower atmosphere by topographic and meteorologic conditions: for example, when a temperature inversion, in which cool air is layered over warmer air, prevents the normal rising of the warmer air. Air pollutants then accumulate and under the influence of sunlight (see below) interact to form the visible air pollution called smog. Smogs are particularly likely and especially severe when:

1. The topography ensures that air above the city is easily trapped: for example, by mountains to the leeward side (e.g., Los Angeles); by a ring of mountains around the city (e.g., Denver, Albuquerque, and Donora in the United States; Mexico City; Innsbruck, Austria).
2. Local industrial processes and/or heavy automobile traffic generate many of the active molecules needed to interact with UV light to form photoactive smog (e.g., smelting, oil refining, automobiles).

Home heating has been credited with the famous "London fogs" of the early twentieth century: a result of the burning of coal in numerous inefficient home stoves. More recently, increased use of wood stoves in mountain areas (e.g., Denver, Flagstaff) has led to noticeable smoke odors, and even visible palls, over residential areas. Although many of us consider a slight odor of wood smoke to be pleasant, it is as harmful as less nostalgic pollutants.

In sum, a key ingredient of air pollution is incomplete combustion, of whatever material and for whatever purpose.

AIR POLLUTION AND OZONE IN THE STRATOSPHERE

Despite its many deleterious effects on living organisms, ozone has an essential function in the stratosphere. A layer of ozone serves to protect the earth's surface from a large fraction of solar ultraviolet (UV) radiation. Ozone is both formed and destroyed in the stratosphere—as well as in the troposphere (lower region of the atmosphere)—by interaction with NO_X. The critical reactions in the absence of chlorofluorocarbons (discussed below) are:

$$NO + RO_2 \rightarrow NO_2 + RO$$

followed by reactions 1 and 2 from page 38:

$$NO_2 + h\nu_{(>440\,nm)} \rightarrow NO + O\bullet$$
$$O\bullet + O_2 + M \rightarrow O_3 + M$$

The stratosphere lies at an altitude of 12 to 50 km above the earth's surface. The much-discussed stratospheric "ozone layer" is actually a minor component (a few parts per million) of the molecules of the upper atmosphere. It serves, however, to absorb a significant fraction of the UV radiation that would otherwise impinge on organisms and that would otherwise cause—at the least—greatly increased rates of skin cancer in humans. Additional dele-terious effects of increased UV radiation include effects on some marine organisms, damage to some crops, and alterations in the climate of the earth. Absorption of solar radiation by ozone leads to significant heating of the atmosphere; in addition, ozone both absorbs and emits terrestrial infrared radiation, thus influencing the global climate to some extent.

Therefore, although one cannot predict with certainty the severity of the consequences of stratospheric ozone depletion, there is considerable evidence that consequences exist and that they might not be easily reversible. Like the greenhouse effect that is in progress from increasing levels of carbon dioxide in the atmosphere, like the partially irreversible consequences of deforestation, the effects of ozone depletion may permanently alter our environment before we know that it will.

The measurement of ozone in the atmosphere depends on ozone's absorp-tion of solar UV radiation, first demonstrated in the 1880s, and used ever since. Dobson recognized the connection between ozone concentrations and weather, and also observed the seasonal variability of ozone in the atmo-sphere, before 1930. In the last decade, however, concern over the depletion of the ozone layer has risen sharply (Figure 3.2).

In 1985, the British Antarctic Survey provided data showing that ozone had disappeared from the Antarctic stratosphere for the past five springs, leaving an "ozone hole" that replenished itself in the summer but recurred at a larger size the next spring. The seasonal appearance of the ozone hole is linked to the development of polar stratospheric clouds in the winter. These clouds are of two types. Type I consists of diaphanous sheets of frozen particles of nitric acid trihydrate, the particles having a diameter of approximately 1 μm.[8] Type II clouds consist of particles of 5 to 100 μm diameter, which are thought to consist either of frozen water or of type I particles coated with water. Type II particles are heavy enough to fall several kilometers downward within weeks, whereas type I particles are light and fall slowly. Both types of stratospheric clouds adsorb nitrogen, but type I clouds, when they melt in summer, release the trapped nitrogen at the same level of the stratosphere, while type II clouds may have "dropped" the

---

[8] A micrometer (μm) is 0.001 millimeter.

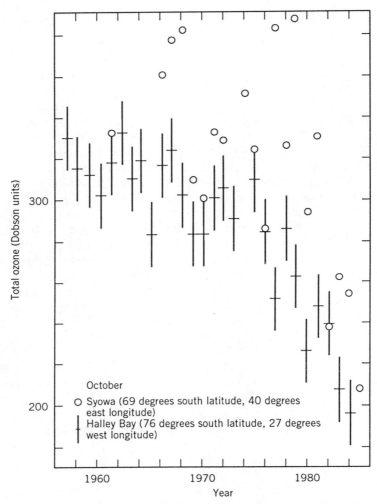

Figure 3.2 Monthly averaged column abundance of ozone, measured each October since the 1960s over the Antarctic stations of Halley Bay and Syowa. (From S. Chubachi and R. Kajiwara, *Geophysical Research Letters* 13:1197, 1986.)

nitrogen by several kilometers. If the lower altitude is below the altitude at which chlorine destroys ozone, the trapping effect of nitrogen on Cl• will not occur.

The major "culprits" in the loss of ozone are thought to be the chlorofluorocarbons (CFCs; Figure 3.3). Major agents include methylchloroform, CH_3CCl_3; carbon tetrachloride, CCl_4; $CFCl_3$; CF_2Cl_2; and (to a lesser extent) $CF_2ClCFCl_2$, and $CHClF_2$. Brominated analogs are also of concern (e.g., CF_2ClBr, $CBrF_3$) but are thought to be only 20% as harmful as chlorinated analogs. Pure fluorocarbons, without any hydrogen substituents, are not thought to affect the ozone layer at all. The CFCs have been used heavily

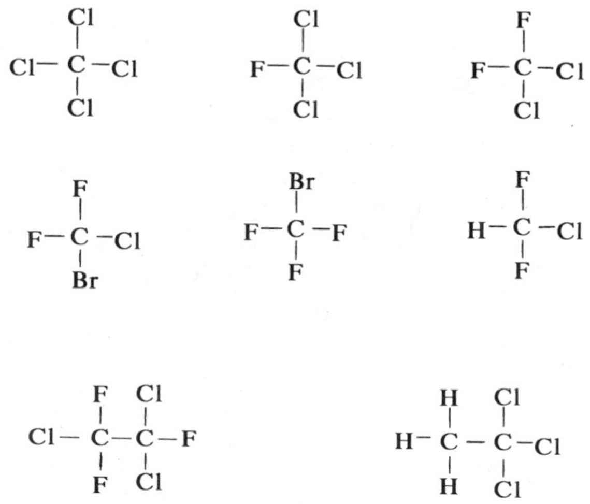

Figure 3.3 Structures of commercial chlorofluorocarbons.

as aerosol propellants, as refrigerants, as foam blowing agents, and as solvents. Most of them are inert and therefore not degraded under normal atmospheric conditions near the earth's surface. Most heavily used CFCs are not very toxic: those that are toxic are no longer widely used in consumer products. For example, carbon tetrachloride is a strong hepatotoxin. It was once the major solvent used in dry cleaning, but it is now only used industrially, and emission is holding steady at 100 kilotons (200,000,000 lb) per year. For the "nontoxic" CFCs, little incentive existed to prevent their release into the atmosphere, and until worry about the ozone hole led to their regulation in the late 1980s, the emissions of many CFCs were increasing between 6% per year (methylchloroform) and 16% per year ($CF_2ClCFCl_2$). It is estimated by the World Meteorological Organization in Geneva that approximately 650 kilotons of CFCs is released into the atmosphere each year.

The critical reactions between chlorofluorocarbons and ozone in the stratosphere are shown for one common CFC, CFC-12:

$$CCl_2F_2 + UV \rightarrow Cl\bullet + CClF_2$$

The atomic chlorine combines extremely rapidly with O_3:

$$Cl + O_3 \rightarrow ClO + O_2$$

ClO is also chemically reactive and will combine with atomic oxygen, $O\bullet$, to form atomic chlorine and molecular oxygen:

$$ClO + O\bullet \rightarrow Cl\bullet + O_2$$

Since atomic oxygen is plentiful in the stratosphere, these reactions result in almost all of the chlorine in CFCs becoming involved in ozone destruction. An atomic chlorine is again generated and can interact with another molecule of O_3. It is estimated that each atom of chlorine participates in these reactions an average of 100 times before being removed from the cycle as, for example, hydrochloric acid (HCl). HCl can diffuse downward into the troposphere, bringing the cycle to an end for one chlorine molecule. On the other hand, some chlorine reservoir molecules, such as HOCl and $ClONO_2$, are highly reactive and tend to regenerate atomic Cl. Thus a single Cl atom has the potential to destroy an average of 100,000 ozone molecules.

Among the uses of CFCs is as "foamers" in the making of Styrofoam: a lightweight plastic beloved for coffee cups, for meat trays in the supermarket, for packing fragile items, and for fast-food hamburger boxes. Some natural sources of CFCs also exist: methyl chloride (CH_3Cl) is not only generated by biomass burning and synthesized for use as a paint remover and solvent, but is also generated by fungi and by oceans.

Another molecule that interacts with ozone is *carbon dioxide* (CO_2), which does not react chemically with ozone, but is radiatively active, playing a major role in the heat balance of the stratosphere and thus influencing the highly temperature-dependent reactions of ozone. CO_2 is present at approximately 350 ppm in the atmosphere but has been increasing at the rate of 0.3% per year since (at least) 1958. Major sources of increase are biomass combustion (coal, oil, wood) and possibly the decreased utilization of CO_2 by trees as a result of deforestation.

Methane, the most abundant hydrocarbon in the atmosphere, also interacts with ozone. Current atmospheric levels are approximately 1.7 ppm. Estimates based on air trapped in polar ice cores suggest that atmospheric CH_4 concentrations have doubled since the seventeenth century. Atmospheric methane is increasing on the order of 1 to 2% per year. Major sources include bacterial action: for example, in swamps[9] or in the rumen of cattle.

Nitrous oxide (N_2O), present in the atmosphere at 307 ppb, up from approximately 285 ppb in the seventeenth century, is another molecule that interacts with ozone. The atmospheric level of N_2O is increasing at 0.2 to 0.3% per year. It is the main source of nitrogen oxides in the atmosphere, and its increase is thought to be anthropogenic, although natural sources include bacterial nitrification.

---

[9] The spontaneous ignition of "swamp gas," methane, is thought to be the source of will-o'-the-wisps. Never having seen a will-o'-the-wisp, I do not know if this is plausible. Sightings of unidentified flying objects (UFOs) near Ann Arbor, Michigan, in the late 1960s were attributed to "swamp gas" by Air Force UFO "experts." The explanation did not remotely match what people had seen, and so was counterproductive. Believers had proof that the military experts were covering up real phenomena; nonbelievers knew of a very bright display of northern lights and a pilot who buzzed a development with his landing lights on. The moral is: a stupid lie about weird phenomena only encourages superstition.

The effects of these various molecules on the ozone layer are as follows (Figure 3.4):

1. Chlorofluorocarbons (CFCs) will reduce ozone, particularly in the stratosphere, and are seen as the major determinant of ozone balance in the near future.

$$X + O_3 \rightarrow XO + O_2$$
$$XO + O \rightarrow X + O_2$$

2. Carbon dioxide, because it causes heating of the atmosphere, will increase the generation of ozone, since ozone production increases with increasing temperature in the middle and upper stratosphere. The net effect is a slight increase in stratospheric ozone levels.
3. Methane increase will also increase ozone concentrations, but will do so primarily in the troposphere, where nitrogen oxide levels are high. However, methane also accelerates the formation of inactive Cl (e.g., HCl) from active Cl (e.g., ClO), thus protecting the ozone layer.
4. Nitrous oxide will reduce the ozone density in the stratosphere.

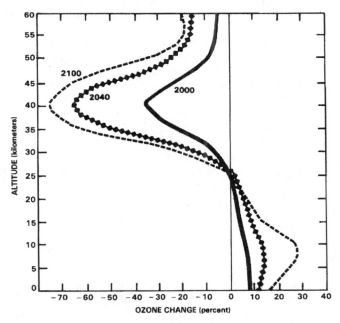

Figure 3.4 Projected alterations in ozone density at different altitudes of the atmosphere over the next century. [From G. Brasseur, *Environment* 29(1):11, 1987.]

CASE HISTORY: FLUOROCARBONS

Most environmental damage is local. Even the worldwide contamination with DDT and PCBs varies from place to place, with "hot spots" near spills or sites of production. Although contamination with these persistent pollutants has endangered many isolates, broadly distributed species are rarely driven to extinction by pollution in the absence of habitat destruction. Terrestrial pollutants are most localized and ecologically least difficult to control, since soil is a relatively immobile medium. Even the world's oceans, although interconnected and eventually communicating, vary in composition and pollution even within a few miles. Truly global effects, in which the localized discharge of a pollutant can have consequences for the entire planet, are almost always airborne. As described in Chapter 2, air circulation, and consequently the distribution of airborne pollutants, is truly global. The chlorofluorocarbons are an example of a nontoxic pollutant that may nonetheless have disastrous effects.

Chlorofluorocarbons (CFCs) were invented in 1930. They are simple to manufacture (Figure 3.5), and the single-step reaction produces very pure products, so they can be made cheaply. They are stable, nontoxic, and make excellent refrigerants because of their low boiling points, heats of vaporization, and specific heats. They do not transmit heat well, so they can be used as the gas in insulating foams, and their low permeation rate means that the foams are stable. As cleaning fluids, their low surface tension and low viscosity mean that they can wet very small spaces (such as the cracks and crevices in computer chips), and their high vapor density limits their evaporation. They are particularly appropriate for consumer uses, because they are remarkably nontoxic and also are not inflammable. Unfortunately, their stability—highly advantageous in many uses—is environmentally disastrous.

The first evidence that CFCs accumulate environmentally came in 1972, but their fate was slow to be determined, since they migrate to the stratosphere before decomposing. By the late 1970s, evidence for damage to the ozone layer was sufficiently convincing that most CFC-containing aerosols were banned in the United States. Other uses—notably automobile air conditioning—continued to grow. Through the early 1980s, evidence for decreasing levels of stratospheric ozone, and for the implication of CFCs in the ozone loss, accumulated. In 1985, identification of an actual hole in the stratospheric ozone layer over the Antarctic suggested that action on CFCs was urgent.

It is now generally conceded that CFCs are so inert in the lower atmosphere that they can float unhindered into the stratosphere, where the chlorine in the CFCs combines with ozone to generate molecular oxygen and ClO. The latter is then broken down by an oxygen radical (O•), regenerating a Cl• that can break down another molecule of O_3. The average residence time of CFCs in the atmosphere is between 60 and 75 years, depending on latitude. This means both that decreases in generation will not produce immediate benefits, and that it is critical to cut releases as fast as possible. At the present rate of emission, a 10-year delay in starting reductions means that ozone recovery would be delayed by 50 years.

At the least, decreases in the ozone layer will lead to increases in skin cancer in humans. The possibility of deleterious effects on other species has not been determined, although there are indications that krill in the Antarctic could be affected. A decrease in krill would have devastating effects on food chains in a fragile ecosystem such as the Antarctic. Terrestrial plants in marginal areas (on

MANUFACTURE OF CFC-12:

$$\boxed{CCl_4} + 2\ HF \longrightarrow \boxed{CF_2Cl_2} + 2\ HCl$$

POTENTIAL ROUTES TO FC-134a:

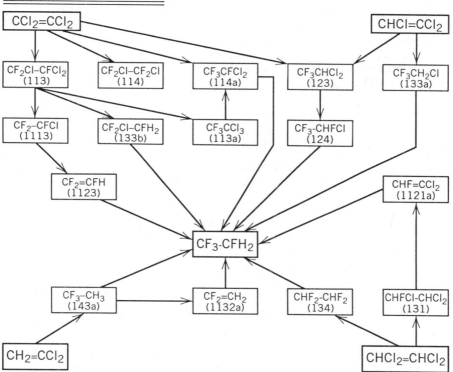

Figure 3.5 Production of chlorofluorocarbons: CFC-12 is easily produced in a single-step synthesis, whereas its less harmful alternatives require more complicated syntheses. (Adapted from R. Pool, *Science* 242:666–668, 1988.)

mountains, at the edge of their range) could also be affected. Finally, fragmentary research suggests that crop plants may be more sensitive to UV than previously thought, and that injuries may increase in the presence of other stresses, such as those resulting from climate change (see below).

Because they are inert and essentially harmless at ground level, CFCs have found many uses. As Freon, they are the cooling agents in refrigerators (9% of production) and in automobile air conditioners. They are the foaming agents in the rigid foams used to insulate refrigerators and to protect fast foods, and in the flexible foams used for mattresses, sofa cushions, and car seats (40% of production). They are essential and so far irreplaceable in cleaning computer chips.[10]

---

[10] However, Apple Corporation announced in the summer of 1991 that it has a non-CFC cleaner for computer chips and will make it available to other companies.

They serve as dry cleaning solvents and were very popular as propellants in spray cans.

Many of these uses can be eliminated by fiat. The fast-food industry, under pressure from publicity, is already phasing out CFC-based foams, and most nonmedicinal aerosols in the United States advertise the absence of CFCs. In furniture construction, most manufacturers are willing to make less puffy cushions, provided that legislation will prevent their less scrupulous competitors from taking advantage of such altruism.[11] Refrigeration is somewhat more difficult. Alternatives to CFCs are less efficient, meaning that more energy would be needed to run the units, or are poisonous (e.g., ammonia). No substitutes for uses in the electronics industry are yet known, although simple thrift could decrease consumption by up to 40%. HCFs, fluorocarbons that contain less chlorine, do not damage the ozone layer nearly as much, and FCs, fluorocarbons containing no chlorine, are apparently also inert in the upper atmosphere. The latter have not been tested sufficiently to be sure that they are environmentally benign, however, and manufacturers are understandably reluctant to gamble that the former will not be banned together with the more aggressive CFCs.

It is ironic, of course, that concern about decreases in stratospheric ozone is matched by concern over increasing levels of ozone in the lower atmosphere, or troposphere. Stratospheric ozone, comprising about 10% of the total atmospheric ozone, absorbs UV radiation and so protects humans from skin cancer and animals and crops from UV damage. Near the ground, ozone is a definite health hazard, and even moderate increases in ozone levels may significantly reduce crop yields. Because of its reactivity, increasing levels of ozone also result in increasing photochemical activity (smog and acid rain) and cause or aggravate breathing difficulties in sensitive individuals (see Chapter 4).

More than any of the other pollutants discussed in this book, CFCs require international regulation. Their effects are truly global. Their extreme persistence in the lower atmosphere means that they will eventually disperse evenly in both the troposphere and stratosphere, even if production were restricted to a single factory. Thus all of the economic benefits of production can accrue to a producer, while health and environmental costs are spread among all nations of the globe. A company or a nation (especially in the tropics, where the least ozone depletion is seen) could benefit enormously from such *externalization* of costs (see Chapter 13). On the other hand, altruistic nations, willing to forego the advantages of CFCs for the global good, would suffer consequences based on their location rather than on their behavior. Control of CFCs, therefore, depends heavily on equalizing the rewards (and consequences) of compliance with international agreements.

The first international agreement on protecting the ozone layer was the Vienna Convention of 1987, which provided for information exchange, research, and monitoring. Thirty months later, the Montreal Protocol set up a *dynamic* mechanism for protecting the ozone layer. Fifty-one nations and the European Community signed the Montreal Protocol, regulating five major CFCs (CFC-11, CFC-12, CFC-113, CFC-114, CFC-115) and two halons (halon-1211 and halon-1301). Carbon tetrachloride and methyl chloroform were not regulated.

---

[11] This is an example of an externality: a cost associated with a product that is not worked into the cost of the product. This topic is dealt with more fully in Chapter 13.

Signers of the Montreal Protocol agreed to freeze production of the regulated CFCs at 1986 levels within one year and of halons within three years after the Protocol came into force. Stepped decreases in production, import, and export follow in 1994 and 1999. The reductions are based on the ozone-damaging potential of the chemicals produced (e.g., a molecule of halon-1211 has an ozone-depleting potential three times as great as CFC-11 or CFC-12) rather than on overall CFC and halon tonnage. Developing countries with current consumption levels of less than 0.3 kg per capita were granted a 10-year grace period before they initiate reductions. In this way the Montreal Protocol obtained the cooperation of developing countries, who have only recently begun to benefit from the uses of cheap, nontoxic CFCs and who can least afford the changeover to more expensive and perhaps more toxic replacements.

Perhaps the most important feature of the Montreal Protocol is that control measures are designed to change as new data are obtained. If there is evidence for an increasing rate of ozone loss, the rate at which CFCs are phased out can be increased, without requiring new negotiations. This feature has already been called on, because the Arctic ozone layer has been decreasing more rapidly than expected and ozone "holes" have penetrated farther into the temperate zones than expected at the time of the Montreal Protocol.

In sum, the Montreal Protocol provides a working model for control of international environmental problems. Like the treaties protecting Antarctica from exploitation, it offers hope that the nations of the world can cooperate on global problems. There is, of course, a price. Many environmentalists consider that far too slow a pace of CFC phaseout was set at Montreal. They cite not only the dangers of increased ultraviolet radiation, but also the role CFCs play in global warming. But most people would probably agree with Dr. Mostafa Tolba, the Executive Director of the United Nations Environment Program, who quoted Benjamin Franklin's endorsement of the U.S. Constitution: "I consent, Sir, to this Constitution because I expect no better." Faint praise, perhaps, but probably the only reasonable approach to international environmental accords.[12]

CLIMATE

Weather is the cycle of precipitation, temperature, and sunlight that repeats yearly. Variations in weather occur between years, but keeping records for 100 years (coupled with computer models that assume that the pattern repeats each year) leads to quite good weather forecasting. The models are now good enough that if a cold air mass is stationed just so over the Pacific and a warm air mass just so over the Gulf of Mexico, the meteorologist (whose computer knows where the jet stream is, what season it is, and what happened

[12] A summary of the Montreal Protocol is found in J. Koehler and S. A. Hajost's article, "The Montreal Protocol: A dynamic agreement for protecting the ozone layer," *Ambio* 19:82–86, 1990. A summary of then-available scientific information about CFCs and the ozone layer is presented in M. McFarland and J. Kaye's article, "Chlorofluorocarbons and ozone," *Photochemistry and Photobiology* 55:911–929, 1992.

the last time these factors combined) can usually predict whether we will have rain, snow, or an ice storm. Because weather is infinitely variable, predictions cannot be absolute. There is always a border zone where a very small change in temperature may have disproportionate consequences. For example, in a winter storm that produces heavy snow in the north and rain in the south, there may be a transition zone in which warm air overlies cold air, and falling rain freezes as it hits cold air and cold roads, trees, and power lines. The resulting ice storm could be devastating. An alteration of 1°C at ground level may make the difference between thick ice and mere rain. Or the cold air may penetrate 5 miles farther south, bringing snow to a town that was predicted to have rain. Therefore, weather prediction will probably always have an element of uncertainty. Nonetheless, understanding the factors that affect multiyear cycles of weather has been largely a matter of better records (of ocean temperatures, global rainfall patterns, temperature) and of larger computers that can process the enormous amount of data.

Climate is the pattern that underlies weather cycles. It too has cycles, but the cycles are measured in millennia rather than decades. For example, the "Little Ice Age" lasted from the thirteenth to the eighteenth centuries,[13] interrupting a warming trend that began with the Holocene, which reached its peak between 7000 and 4000 B.C. Understanding the climate requires data for more years than we have records—more years even than there have been humans on earth. Fortunately, we have learned to decipher information present in natural deposits, notably tree rings and fossils. Tree rings have been used for some time to identify rainfall cycles. For example, tree ring data suggest that it was probably a drought that drove the cliff dwellers of the North American Southwest (New Mexico and Arizona) into the river valleys. In dry years, trees grow less and the growth rings of their trunks are small: counting these rings can take us back several thousand years, since trees need no longer be alive, only to have been alive during the period of interest. For "recent" times—back to the Rome of Pliny the Elder—we can check the tree ring record against recorded comments about weather, about crops grown, about drought or famine or plentiful harvests. But neither human records nor the oldest trees can record the cycles of warming and cooling that comprised the major Ice Ages. For such data, paleoecologists turn to fossil records, looking at the ratios of isotopes of carbon and oxygen. It is known that uptake of different isotopes depends to some extent on temperature. Alterations in isotope ratios of fossils from the same species several millennia apart can tell us something of the temperature variations in that time.

---

[13] To demonstrate the uncertainties of climatology, some sources say the Little Ice Age occurred from the sixteenth to the nineteenth centuries.

Current theories about the factors that drive climatic cycles focus on the Milankovitch cycles and on levels of CO_2 in the atmosphere (Figure 3.6). The Milankovitch cycles consist of:

- A 41,000-year cycle in the tilt of the earth on its axis, resulting in different distances between the earth's northern and southern hemispheres and the sun. The tilt ranges between 21.8 and 24.4° and affects the amount of the globe that is in the tropics (i.e., that part of the globe where the sun can be directly overhead) as well as the size of the area that has no daylight in winter and no night in summer. As the tropics increase in size, the differences between summer and winter are magnified.
- A 23,000-year cycle, during which the wobble of the earth's axis causes the poles to describe circles. The wobble alters the relationship between the seasons and the distance of the earth from the sun. At present this relationship is in phase with seasons in the southern hemisphere, so their seasons are aggravated while northern seasons are somewhat mitigated. In about 12,000 years, this *precession of the equinoxes* will reinforce northern seasonal variation but decrease differences between seasons in the southern hemisphere.

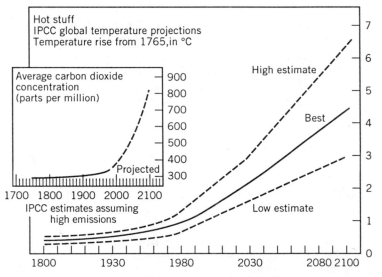

Figure 3.6 CO_2 levels in the atmosphere from the eighteenth to the twenty-first centuries. Projections beyond 1990 are based on computer models of an Intergovernmental Panel on Climate Change, using different estimates for the effects of cloud cover. (From *The Economist*, May 26, 1990, p. 93.)

· A 100,000-year cycle, during which the shape of the earth's orbit goes from its most nearly circular to its most extremely elliptical. Although the total amount of sunlight hitting the earth each year does not change during this cycle, the seasonal variation changes. This can affect seasonal occurrences such as monsoons.

Levels of CO_2 in the atmosphere affect global temperatures because CO_2 is transparent to (transmits) shortwave solar radiation but is opaque to (blocks passage of) infrared emission. Since radiation of heat from earth into space is by infrared radiation, increased levels of CO_2 serve to "trap" solar heat. Considerable evidence links past fluctuations in CO_2 to known temperature variations. During the most recent Ice Age, CO_2 made up approximately 0.018% of the atmosphere, or about one-half the present levels. It is thought that the oceans are a "sink" for CO_2—that is, CO_2 is removed from the atmosphere by being captured by the oceans.[14] The anthropogenic causes of increasing atmospheric levels of CO_2 are discussed below.

Several theories exist as to why the oceans should vary in the amount of CO_2 that they trap. One hypothesis is that the driving force is the amount of CO_2 drawn from the upper layers of the ocean into the deeps. This removal of CO_2 from the surface layers of the ocean causes increased movement of CO_2 from the overlying air into the water. This hypothesis depends on a subsidiary hypothesis, the functioning of the *North Atlantic conveyer*. The North Atlantic conveyer hypothesis argues that the cooling of the margins of the ocean by cold land winds causes cooled salt water to sink into the deeps off Greenland, generating a current that flows southward in the Atlantic deeps, then flows across the Indian Ocean, and finally, flows east across the southern Pacific. During the water's passage, detritus sinks into it from the aquatic biosphere. As the organisms decay, they produce CO_2. Therefore, CO_2 is created in the deeps. As long as the water returns to the surface in the Pacific, so does the CO_2. If for some reason the current stops, there would be less CO_2 in the surface water, because detritus that fell into the deep ocean would decompose but the CO_2 would stay in the deeps. To compensate, more CO_2 would be drawn out of the atmosphere. The earth would cool. Calculations of the nutrient levels in deep and shallow waters during the Little Ice Age suggest that the North Atlantic conveyor was then moving slowly, if at all.[15]

---

[14] On the other hand, a recent examination of the correlation between temperature and solar flares (sunspots) shows an excellent correlation between recent weather patterns and the length of the sunspot cycle. Some climatologists think this is a spurious effect; others think it may require reevaluation of the theories of the role of CO_2 in global warming.

[15] During the last ice age, it took approximately 2000 years for water to move from the North Atlantic to the South Pacific via the deep current. It is estimated that the transit now takes 1500 years.

CHANGING THE CLIMATE

As we have begun to understand the mechanisms underlying natural cycles of weather and climate, we have in the past decade also had to recognize the plausibility of anthropogenic climate changes. Three types of alterations are postulated, of which two are thought to be under way:

1. Global warming results from the accumulation of carbon dioxide and other heat-trapping gases in the atmosphere, largely as a result of the burning of fossil fuels.
2. Destruction of the ozone layer of the stratosphere by chlorofluorocarbons alters the amount of UV radiation striking the earth.
3. Catastrophic alteration of climate could also be caused by nuclear war (see Chapter 2). This possibility has receded somewhat since the end of the Cold War and the breakup of the Soviet Union, which occurred a few years after the concept of "nuclear winter" was first publicized. Nonetheless, most of these nuclear warheads still exist. Moreover, the proliferation of nuclear weapons among smaller nations means that détente between the superpowers does not necessarily suffice to prevent nuclear war.

The amount of carbon dioxide in the atmosphere is a product of two opposing processes: respiration and photosynthesis. Respiration, both by plants and by animals, produces CO_2. Photosynthesis of green plants uses CO_2 and generates oxygen. Atmospheric levels of CO_2 are also affected by inorganic processes such as volcanic eruptions, but over millennia such processes average out. There have been very great increases in consumption of fossil fuels over the past century, however. At the same time there has been massive deforestation worldwide as trees were cleared for fuel and for farming. More recently the damage done to forests by acid rain has also led to net decreases in photosynthesis. In 1800, 0.028% of the atmosphere consisted of CO_2. Today's levels have risen to 0.035%. Further increases, due to the burning of fossil fuels, are expected to produce the *greenhouse effect,* also known as *global warming.* While the magnitude of the greenhouse effect is under dispute, a minimum estimate is that it will increase mean global temperatures by 1.5°C within the next century, with greater temperature increases in the winter than in the summer (Figure 3.7). An upper estimate is for a 4.5°C rise in mean global temperatures within the next century, with consequent melting of polar ice caps, shifting of the zones that are suitable for agriculture, and significant increases in desertification of what are now temperate zones. The greenhouse effect will make the earth warmer, and will do so with unprecedented rapidity.[16]

[16] On the other hand, several indicators suggest that the earth is actually moving into a new ice age, which may be forestalled by the greenhouse effect.

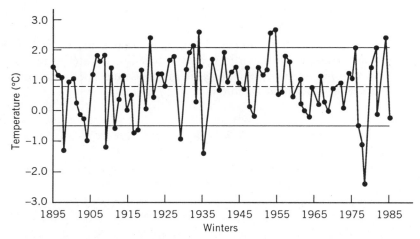

Figure 3.7 Mean winter temperatures in the United States, 1895–1985. Random variation makes it extremely difficult to distinguish trends. (From the *Bulletin of the American Meteorological Society* 65:1302, 1984.)

What effect will such a seemingly trifling rise in mean temperatures make? No one knows. The present is already a very warm period in the earth's climate cycle. The projected CO_2-induced global warming will add to that temperature maximum. The rapidity of the shift will probably make it difficult for slowly reproducing species such as trees to adapt. Extinctions are quite probable. Moreover, because CO_2-induced temperature increases will not originate in the relationship between earth and the sun, the greenhouse effect is unlikely to be seasonal, or to differ in direction between the northern and southern hemispheres, as do the effects of the Milankovitch cycles. To gauge the effects of a 4.5°C increase in mean temperature, consider that the English climate was only 1°C cooler during the Little Ice Age than before. This 1°C drop ruined the vineyards of medieval England, altered the pattern of fisheries, may have been the cause of famine in the fourteenth century, and may have been responsible for the failure of the Viking colonies in the New World. Glaciers grew. If a 1.5°C rise in the mean global temperature does nothing more than raise the ocean levels an estimated 20 to 140 cm before the year 2100, it will drown entire archipelagos and large parts of Florida, and will threaten most coastal cities. The estimated rise in ocean levels by 2025 is 20 cm, about 8 inches.

Effects on ecosystems will depend very much on individual "before" and "after" variables. Increased CO_2 increases photosynthesis, so if all else were held constant, increasingly luxuriant vegetation might be seen. If rainfall decreased in a given area, however, growth might be stunted. Plant species in some areas would suffer from decreased rainfall; others would suffer (or benefit) because an increasing portion of the year's precipitation will fall as rain, with fast runoff, rather than as snow, with slow runoff. Effects will certainly be uneven across the globe, perhaps even across fairly small geographic distances, because certain weather phenomena are essentially dichotomous. Frost kills some plants that survive (albeit sulkily) if the temperature is only 1°C higher: increases in mean temperature say rather little about dichotomous phenomena.[17] Increasing temperature would stress plants at the warm boundary of their range (e.g., the deciduous trees that need a cooling period between growth periods). Slow-growing species might die out if their present habitat becomes unsuitable too quickly for adaptation or colonization of newly suitable habitats. In the Arctic, conifers could move north into subalpine areas, while birch (*Betula* sp.) and willow (*Salix* sp.) could encroach on moss and lichen in the alpine zone. Alterations in the ratio of open areas for grazing relative to forest cannot readily be predicted, since they depend on interactions between temperature, rainfall, and even levels of grazing. Similarly, the effects on agricultural systems will depend not only on mean temperatures, but on the extremes; on rainfall and when it occurs; and on the quality of soils in the more northern areas that open up. Corn, for example, is very sensitive to drought and to very high temperatures during certain parts of its growing cycle, although it is normally quite heat tolerant. Organisms at the northern (cold) edge of their range will probably benefit from global warming. Nonetheless, significant global warming will be disastrous for natural communities. Even if single species can migrate, or are moved by humans, it is unlikely that all the necessary elements of stable plant–animal predator–prey systems will make the transition intact. Raising the temperature also speeds up the reproductive cycles of many insects and could lead to excessive damage from insects that now form a minor component of a community.[18]

Minimizing the degree of global warming may drastically affect global economics (Figures 3.8 and 3.9). To reduce CO_2 output will require strong curbs on all forms of combustion. Strict controls not only on burning of tropical forests, but on burning of gasoline, coal, and other carbon-based fuels dear to the industrial world will be required. Shifting from carbon-based to other power sources will be costly at best. Electric power can eventually be generated by use of wind, water, and sun, as well as by nuclear

---

[17] For example, the winter of 1991–1992 broke records for both high and low temperatures in various parts of the United States. In Illinois, sharp cold snaps in November and May damaged many bushes, whereas January was abnormally warm. Although *mean* temperatures were above average, a cold April and May meant that farmers and gardeners could not plant earlier than usual.

[18] On the other hand, there are indications that some caterpillars, such as the larva of the buckeye butterfly, grow more slowly in higher concentrations of CO_2. Alterations in survival resulting from the changed rate of growth could affect predation on plants as well as food supplies for birds that feed on such caterpillars. Adding the complication of increasingly fragmented habitats due to increasing human activity probably means that more species will be threatened than helped by increasing CO_2 levels.

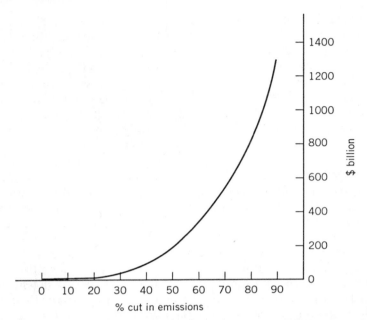

Figure 3.8 Annual cost to the world economy of cutting emissions of greenhouse gases, assuming gradual phase-in of efficient controls. Note that costs increase exponentially, whereas emissions decrease linearly. (From *The Economist*, July 7, 1990, p. 22.)

energy, but the costs of the changeover will be very high. In some cases, alternative technologies do not exist: How does one prevent cattle from generating methane as the bacteria in their gut break down cellulose? The collapse of entire industries could easily trigger worldwide depression. Even under the most optimistic scenario, the expectation is for sharp reductions in economic growth.

Given these economic consequences, there are many policymakers who question the necessity for strong action in the immediate future. Their objections to implementing draconic curbs on greenhouse gas production are multifaceted.

A fundamental problem for acceptance of the threat of global warming is the nature of the data. The *facts* are that atmospheric levels of carbon dioxide have increased over the past 150 years, that carbon dioxide is transparent to incoming sunlight but traps refracted heat by absorbing in the infrared range, and that atmospheric levels of several other gases that also trap heat are also increasing. Also generally accepted is the conclusion that the less developed countries, mostly in the southern hemisphere, will suffer most from adverse consequences. The greater hardships in the southern nations will be due to their lesser ability to compensate for natural disaster, not to greater climatological effect in that hemisphere.

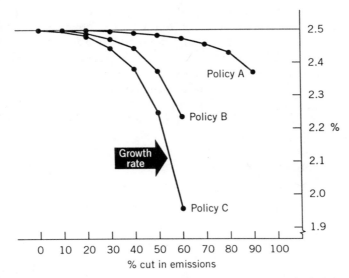

Figure 3.9 Effects on economic growth of more and less efficient methods of cutting greenhouse gas emissions. Two policies use market strategies. Policy A, least damaging to global economies, uses uniform international carbon taxes to allocate cuts efficiently, and phases the tax in slowly. Policy B uses carbon taxes as in policy A, but phases them in more rapidly, scrapping existing energy capital over two decades. Policy C uses inefficient regulatory mechanisms, phasing them in rapidly as in policy B. (From *The Economist*, July 7, 1990, p. 22.)

Less certain is the ability of existing "CO_2 sinks," such as the oceans and green plants, to absorb carbon dioxide. Also less certain is the current status of global warming. Weather is inherently chaotic, and there is no compelling evidence that global warming is already underway. Proponents of global warming argue that the recent spate of volcanic eruptions has so far delayed incontrovertible evidence of warming. Opponents note that the weather over the past 30 years has not been significantly warmer, that some of the most dramatic warming occurred before 1945 (and therefore before the rapid postwar increase in trace gas emissions), that polar temperatures do not show the predicted increases, and that confounding factors such as urbanization may be responsible for apparent increases in recorded temperatures.

The predictions of how fast the consequences of global warming will occur and how significant they will be are quite uncertain. These predictions are based on computer models that include numerous assumptions about weather outside the range for which data are available. As yet untestable assumptions are made about future levels of cloud cover, the time of day that warming trends will be most apparent, and mitigating effects of atmospheric pollutants that absorb incoming light.

The prognoses, if not the computer models, also include assumptions about interactions between increased temperature, changes in rainfall, and the effects of heat stress on plants and the resulting effects on crops and on natural ecosystems. In sum, opponents of immediate action to curb generation of greenhouse gases say that there are great uncertainties about the extent to which global warming will occur and that specific consequences of that warming are little more than speculation.[19] Proponents point out that by the time the effects of global warming are so obvious that they cannot be disputed, enormous damage will already have occurred, and that the economic consequences of unchecked global warming (of the order predicted by the models) are far worse than the consequences of slowing that process.

[19] Carefully stated climatological arguments against rapid and severe global warming are presented by P. J. Michaels and D. E. Stooksbury in their article, "Global warming: a reduced threat?" *Bulletin of the American Meteorological Society* 73: 1563–1577. Arguments against deleterious ecological effects of global warming, even if severe and rapid, are presented by J. H. Ausubel, "A second look at the impacts of climate change," *American Scientist* 79:210–220, 1991.

4

TERRESTRIAL AND HEALTH EFFECTS OF AIR POLLUTION

If the red slayer thinks he slays
Or if the slain thinks he is slain,
They know not well the subtle ways
I keep, and pass, and turn again.

—Ralph Waldo Emerson[1]

INTRODUCTION

At various times during the past 20 centuries, air pollution has been blamed for most of the world's ills. From the Rome of the Caesars and the Renaissance, to Restoration London, to 19th century New York, the wealthy abandoned cities during the summer, when pestilence arrived with the "miasmas." Night air was thought to carry disease, as was damp air. During the 14th century, people held bouquets of flowers, or posies, to their faces to ward off bubonic plague (which is actually transmitted by fleas carrying the plague bacillus, *Pasteurella pestis*.) The word malaria means "bad air," because miasmas were thought to cause this disease (malaria is actually transmitted by mosquitos carrying any of several species of parasites in the genus *Plasmodium*.) After the Industrial Revolution, as the germ theory not only identified the causes of major diseases like malaria, tuberculosis, and plague, but suggested ways to interrupt their transmission, soot and smoke from countless factory chimneys provided visible evidence of noxious air. After World War II, the concerted efforts of cities and industry to clean up pollution, coupled with the shift from coal to oil or gas for heating, led to improvements in visible urban air quality in most locations. The levels of

---

[1] From *Brahma*, by Ralph Waldo Emerson, in *A Treasury of Great Poems*, L. Untermeyer, ed., Stratford Press, New York, 1955, p. 790. This is also the poem quoted by Robert Oppenheimer on the occasion of the explosion of the first atomic bomb in New Mexico, 1945.

pollutants did not necessarily fall, however. Increasing numbers of automobiles generated pollutants that were usually invisible but nonetheless affected the lungs. When conditions were right, this pollution became visible as *photochemical smog*. The burgeoning petrochemical industry added organic vapors to the air, producing palpable pollution in areas where concentrations of chemical factories developed: along the Kanawha River in West Virginia, the lower Mississippi River in Louisiana, the Mexican border and the Gulf coast in Texas. The coal that is no longer used for household heating fuels massive power plants with chimneys so tall that the sulfur dioxide travels hundreds of miles before it is deposited as acid precipitation, and photochemical smog blurs vistas in our national parks. The oil and gas that are burned for home heating generate carbon dioxide that contributes to global warming. Once again air pollution can be blamed for a multitude of ills.

Most anthropogenic pollutants are released close to the earth's surface, and almost all (excluding only the pollution associated with supersonic aircraft and with space exploration) are released into the troposphere. Thus there is a tremendous overlap between the molecular species involved in stratospheric and tropospheric pollution. Nonetheless, differences exist. The CFCs are essentially inert in the troposphere, and anthropogenic particulates do not reach the stratosphere. Perhaps the biggest difference, however, is that in the troposphere there are direct interactions between pollutants and organisms, while the effects of stratospheric pollutants are mediated by their effects on solar radiation. Moreover, tropospheric interactions involve anthropogenic particulates as well as water and gases. Both particulates and gases can affect human health, and both affect ecosystems.

ECOLOGICAL EFFECTS OF AIR POLLUTION

The complexity of ecosystems, coupled with the multiplicity of tropospheric pollutants and the enormous complexity of their interactions, means that very few generalizations about their ecological effects can be proven. Analysis of ecological effects must often be repeated for specific pollutants at different sites, and certainly for different mixtures of pollutants in different ecosystems. A general consensus on the effects of *acid precipitation* exists, however, and will be presented. The equations for the interactions among the various gases have been shown in Chapter 3.

ACID RAIN

Acid rain is the common term for a phenomenon that has only been recognized in the last decade: the deposition of acidic particles or abnormally acidic rain, on land or water. Although environmental concern focuses on wilderness rather than on human artifacts, acid precipitation is not limited to wilderness areas.

Any source of acids or acid anhydrides is, in theory, a source of acid precipitation. In practice, the major acid producers are SO_2, SO_3, and NO_2. Volcanoes are the dominant natural source of these gases. Human sources include mobile sources such as automobiles, and stationary sources, notably power plants. Factories may contribute to some extent but are not considered major culprits. For the United States east of the Mississippi, power plants are the major source of acid-generating effluents (Figure 4.1). Coal-fired power plants are especially implicated because midwestern and some eastern coal fields contain up to 4% sulfur. In contrast, western coal and some southern coal have sulfur levels less than 0.5%.

NO_x are also acid-forming, but their fertilizing potential means that they can be utilized by plants rather than staying permanently in the soil. However, it has recently been suggested that precisely this fertilizing capacity may contribute to forest mortality. The nitrate precursors in air may act as foliar fertilizers, providing the option of increased growth to the tree. But since the soil is, at the same time, less fertile due to acidification, the roots cannot support the increased growth. The foliage suffers injury from the imbalance, and dies prematurely. This hypothesis fits the observation that conifers, which retain their needles more than one growing season, tend to be more sensitive to acid precipitation than are hardwoods, which shed their leaves each year. Moreover, spruce and fir, with 5-year cycles of needle-drop, are more sensitive than pines, which shed needles every second year.

There is also some inhibition of acid formation by the presence of other acids. Finally, the conversion of SO_2 or NO_x in the atmosphere requires oxidizing agents. Ozone, hydrogen peroxide, and hydroxyl radicals are among the more important oxidizing agents available, again emphasizing the interactions between the various pollutants in air.

Materials Damage

Damage to artifacts by air pollution is primarily a local phenomenon, in the sense that the effects of locally generated pollutants are far more severe than the effects of distant sources. This is in marked contrast with the effect of acid precipitation on wilderness, which is discussed below. Materials damaged include painted surfaces ($540,000,000 per year), steel (more than $1,160,000,000 per year in Europe alone), and stone. Effects on stone depend on the type, with European marble more reactive than the American variety. To illustrate, the effects of acid on sandstone are biphasic. First the calcite in the sandstone is converted to its anhydride, leading to a 28% increase in volume; subsequently, the anhydride is converted to gypsum, with a further 19% increase in volume. The volumetric changes themselves stress the matrix. Fabrics, automobiles and other human artifacts obviously suffer as well, but being less durable to begin with, their costs are shrugged off.

When the materials have a value other than the cost of replacement, damage estimates soar. Rome has budgeted $200,000 for a 5-year project to

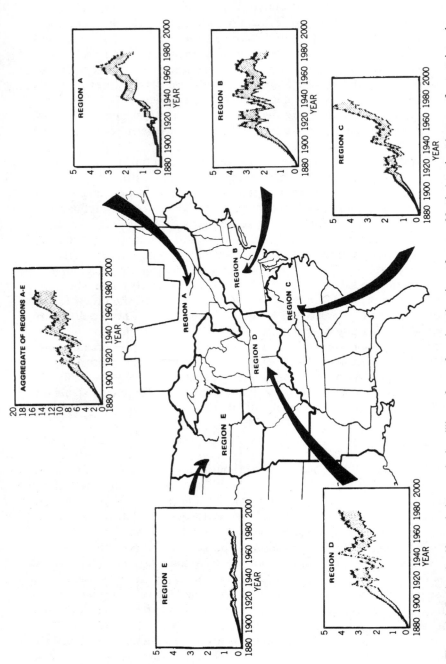

Figure 4.1 Regional sulfur emission trends, in millions of tons, for selected regions of eastern North America. A range of uncertainty, due to differences in the sulfur content of coal used within a region, is represented by shading in plots for each region. Note that emissions are increasing rapidly in the "Sunbelt," with its increasing populations, while emissions in the older industrial areas of the Northeast and Midwest are increasing more slowly or staying relatively constant. [From A. H. Johnson, *Environment* 28(4):9, 1986.]

repair ancient buildings; West Germany expects to spend $10,000,000 per year maintaining its cathedrals.[2] The air pollution in Athens is so bad that the friezes on the Parthenon have lost most of their detail: to appreciate the magnificent sculptures, it is advisable to go to the Parthenon Museum, or even to the British Museum in London, where the Elgin Marbles have been preserved by being indoors.

Environmental Effects

Crop damage depends very much on other environmental factors, especially the cation-exchange capacity of the soil. For ozone, damage to plants increases with increasing temperature and humidity but with decreasing light; increasing the soil water stress also decreases ozone damage. Such factor-by-factor analyses are quite unrealistic, since second-order and higher interactions among factors are to be expected. Moreover, all the factors noted above probably interact with other pollutants, with soil fertility, and with the presence or absence of specific pests, to affect yield. In the future, interactions with increasing levels of UV radiation will also have to be taken into account.

As with pest damage, it is important to distinguish *for the purpose of crops* between *injury,* which merely says that the plant is affected, and *damage,* which would be expressed in terms of decreased yields. Because crops tend to be annual, because they are grown on relatively fertile soil, and because they are usually selected for their good growth in a particular environment, the occurrence of crop damage would ordinarily require high levels of pollution. Stressed ecosystems, plants growing under marginal conditions, and perennials are all more likely to be affected than are the corn and soybeans of Illinois.

The question of acid rain has been and continues to be extremely political, for a number of reasons:

1. The political jurisdictions that generate the acid are not the same as those affected by acid precipitation. A senator from Ohio gets no credit, and considerable blame, if his vote eases pollution in upstate New York but causes unemployment in Cleveland.
2. Often, the political jurisdiction is another country. For example, Canada inherits a significant fraction of the SO_2 from U.S. power plants. In Europe, which comprises many smaller countries, this problem is aggravated. The richer countries of the EEC (e.g., Germany), and the more eastern countries that are heaviest recipients of airborne pollutants, want all upwind generators (e.g., France) to cut SO_2 genera-

---

[2] This estimate specifically *excludes* the costs of repairing the cathedral at Cologne (Köln). Cologne has a separate budget of somewhere between $1 and $20 million per year because it was completed in the nineteenth century with a particularly acid-sensitive limestone.

tion. The poorer countries (including France) say that they cannot afford the necessary pollution control systems. This controversy was finally partially resolved as part of the EC's preparations for unification in 1992. Some parts are still not resolved, frustrating the Common Market's policies on both the permissible types of gasoline and on speed limits.

3. Identification of the source of the pollutants is not easy when generators are hundreds of miles from damaged areas. Because air is a common resource, impossible to partition, it is extremely difficult to assign costs of damage. This problem is discussed in Chapter 13.

4. Nearby sources of acid produce more damage than does any *single* more distant source, making it easy to accuse the target area of primary causation. For example, more than 50% of the acid in the Adirondacks is generated in the industries of the New York Finger Lakes region. Nevertheless, a good deal of the rest comes from the Midwest. Who should clean up first? What if the New York sources are multiple mobile sources (cars) and the distant sources are a very few power plants? What if the Midwest is economically depressed and New York economically prosperous?

5. Damage is slow to appear, rather nonspecific, and often in remote areas. In many places, severe weather, pests, or natural fluctuations occur concurrently with acid precipitation. It is then difficult to distinguish between causes of forest decline and damage that is a consequence of, or coincidental to, the decline.

6. Human health effects were not attributed to acid precipitation until very recently, despite the long history of the role of air pollution in human respiratory illness and death. Therefore, the entire armamentarium of fear of death could not be invoked.

Initial reports of damage to lakes in the Adirondacks of New York State were dismissed as irrelevant and/or reparable by liming (increasing the pH). More recent surveys demonstrate damage over wide areas of the United States, including the mountains of New England and New York, large areas of the Southeast and the West, and the Boundary Waters of Minnesota. Canada is more worried than the United States, and reports considerable damage in eastern Canada. Damage to forest cannot be repaired with liming because the affected areas are too large and too inaccessible. It is implausible to apply the necessary amounts of lime, and the cost is prohibitive. Lake liming is now considered a possible technique for reversing acidification of some lakes, but this technique can succeed only if further input of acidity is prevented and if acidification has not led to the leaching of toxic metals from the substratum. It is thought that such leaching will make reintroduction of native fauna impossible in many cases.

In the last 10 years Europeans have identified widespread damage to the forests of Germany (eastern as well as western), Poland, and Czechoslovakia.

In the mid-1980s it was estimated that up to 50% of European forests were damaged; that estimate has now risen to 80%. Evidence is increasing that the soils in many forests are damaged sufficiently to preclude easy reforestation if and when the affected trees die. The resulting erosion on slopes may well have permanent consequences on the capacity of the soils to produce vegetation. Such deforestation, with inability to reverse the damage, has occurred in several areas of the world, notably Iceland and Spain. Although deforestation may be considered a "worst case" estimate, it must be considered when the relative costs of acid rain amelioration are discussed. Strict economic analyses that neglect the difficulties of re-establishing forests on denuded slopes are inappropriate models for the comparisons between the costs of cleaning up coal before it is burned versus use of localized repair techniques such as liming.

Mechanisms by Which Acid Precipitation Causes Forest Decline

When forest decline or *Waldsterben* was first identified in Germany, the initial assumption was that air pollution was the culprit. More precise explanations proved to be essential for political purposes, since remediation was likely to affect "culprit" industries unfavorably. Differences between species and between regions made it difficult to assign causes, while the longevity of trees, and the fact that older trees were disproportionately affected, made laboratory experiments difficult to design and easy to criticize.

After 10 years it is clear that no single factor can be blamed; forest decline is due to a combination of pollutants. Trees on marginal soils or under climatic stress are most severely affected, but species differ in sensitivity. The syndrome appears to be the result of the cumulative effects of alterations in nutrient balance due to leaching of toxic minerals from underlying rock strata and to the greater acidity of the soil, probably coupled with ozone toxicity to foliage. More recently, the suggestion has been made that the NO_x, previously assumed less harmful than SO_2 because it can be used as fertilizer, is actually harmful precisely because it is used as fertilizer. The theory is that the increased fertilizer available to foliage, coupled with decreased availability of nutrients to the roots, leads to unbalanced growth and consequent weakening of the tree.

While the loss of trees to air pollution is disastrous in terms of effects on local ecosystems—increased erosion from steep slopes, decreasing habitat for numerous species—it is particularly worrisome at this time, when the effects of pollutants are thinning the stratospheric ozone layer (see Chapter 3). Because of their many leaves, trees are efficient photosynthetic machines. An acre of trees uses far more carbon dioxide, and generates far more oxygen, than the same acre planted in grass, corn, or sunflowers. It is estimated that the *addition* of 100 million trees would remove eighteen million tons CO_2 from the air each year. Loss of the same number of trees will increase quantities of CO_2 correspondingly, aggravating both global warming and ozone loss.

HEALTH EFFECTS OF AIR POLLUTION

Any discussion of the health effects of air pollution must distinguish between rich and poor nations, and also between the rich and the poor in those nations. In the United States as a whole, relatively few air pollutants are present in high enough concentrations in outdoor air to be demonstrated health risks. Ozone has long been considered to be the only serious exception to this optimistic view. Recent data suggest that even low levels of very small particulates, less than 10 μm in diameter, may also be quite hazardous. Nonetheless, for the United States, the task of discussing the health effects of air pollutants is relatively easy. In the country as a whole (if one excludes occupational exposure, which generally affects a small number of people who are very heavily exposed to a particular toxicant), the major health effects of air pollution can be ascribed to ozone (and perhaps particulates), to indoor air pollution, and to smoking.

Even in the United States, however, there are exceptions to this minimalist view of the health effects of ambient air pollution. Individual communities and some parts of larger cities suffer from severe air pollution due to industrial activity. In the past, chimneys belching smoke symbolized progress and prosperity; to most of us today they symbolize pollution. But to the people living in the shadow of polluting industries, the significance of the chimneys is complex. Active factories mean jobs, not only for the employees of the factory, but for merchants in the town. Jobs mean that taxes are paid, so towns thrive and schools benefit. When a community is poor, it is easy to suggest a conflict between jobs and clean air. Moreover, the conflict is genuine if the factory competes with factories in states or countries with less stringent pollution controls. It must be stressed that the health cost of such pollution-dependent jobs is often recognized by its victims: the poor are not stupid, only powerless.[3]

The extent to which industrial pollutants are present in outdoor air is even more severe in poorer countries. In Bulgaria, dust in the town of Srednogorie contains extremely high levels of lead, and 70 times the permissible levels of arsenic, a result of air pollution from the copper smelter nearby. Extraordinarily high rates of birth defects and cancer occur in the town. In the city of Leipzig, in the former East Germany, dust and sulfur dioxide are the major air pollutants, with over 92% of the population thought to suffer from

---

[3] Most hazardous waste facilities, for example, are sited in depressed areas, especially when minorities make up a large part of the population. In fact, race is a better predictor of sites for hazardous waste facilities than is the socioeconomic status of inhabitants, according to a 1987 study by the United Church of Christ. And, during the 1990 presidential elections, one poll purportedly found that 28% of whites, but 48% of Latinos and African-Americans, considered the environment a key factor in deciding for whom to vote. (*Source:* G. Di Chiro's article, "Defining environmental justice," *Socialist Review,* vol. 22, #4, pp. 93–130, December 1992. The poll is attributed to a major, unfortunately unnamed, environmental organization.)

respiratory problems caused by the latter.[4] South of Leipzig, radioactive wastes were indiscriminately dumped in unstabilized piles, often without documentation, for over 30 years. In the industrial areas of Poland, air and water pollution are blamed for high cancer rates. Over 80% of the forests of the Czech and Slovak Republics have been damaged by acid rain; the health effects have not been assessed. In Hungary, industrial air pollution is blamed for the high incidence of respiratory diseases in children.

Thus, in many parts of the "Second World" of the former Soviet Union and its allies, outdoor air pollution secondary to heavy industry probably exceeds indoor air pollution as a source of health problems. This is particularly true because formaldehyde, a major component of residential indoor air pollution in the United States, is less prevalent where buildings are stone or concrete and permanent press fabrics less universal.

In the "Third World," consisting of the developing and underdeveloped nations, industrialization is less advanced than in eastern Europe and the former Soviet Union. Where heavy concentrations of factories exist, air pollution can be devastating. In the 1970s air pollution was considered a major contributor to a 50% infant mortality rate among the poor of the *favelas* in an industrial area of Brazil.[5] Nonetheless, although air pollution in and around factories often poses severe health hazards in Third World countries, agriculture is still their major industry, and a smaller proportion of their populations lives near factories and smelters. Therefore, indoor air pollution will account for a major part of the toll exacted by air pollution in large parts of the Third World. But the constituents of indoor air pollution in developing countries are less likely to be synthetic chemicals, and more likely to be smoke from the coal, wood, or cow dung fires that are used for cooking and for heating in small, poorly ventilated homes. Moreover, cigarette consumption is increasing in Third World countries, adding one more source of smoke. Finally, increasing numbers of automobiles add to the pollution in urban areas.

It is clear, then, that the type of air pollution one is exposed to depends on many factors. Only a few can be discussed in this book. Specific industrial pollutants are almost innumerable and best covered in specialized texts. Individuals *choose* to smoke, presumably because they derive benefits from the habit; thus smoking is somewhat outside the scope of this

---

[4] The extent of pollution problems in eastern Europe is illustrated by the German authorities' insistence that the air pollution in the Leipzig region is nonetheless not as serious as the water pollution. This applies also to Srednogorie, where the copper smelter dumped sludge containing 200 tons of arsenic into a river whose water was then used for irrigation. The major rivers of eastern Europe—the Danube, the Elbe, the Oder—are all badly contaminated.

[5] This particular valley has been cleaned up by concerted efforts of the local and national governments and the generating industries (many of which are branches of multinational corporations). Both air and water pollution have been significantly reduced, and uncontaminated drinking water has been supplied to the slums. Infant mortality is decreasing.

book and will not be discussed at length. The discussion that follows will therefore focus mostly on pollutants in indoor air and on radioactive isotopes.

All forms of air pollution (gases, particulates, or liquid droplets) can damage human health. Until very recently, the human health effects of air pollutants were the major focus of concern about air pollution. Conversely, inhalation ranks as one of the three major routes of entry into an organism (Figure 4.2). The other two are ingestion and absorption through the skin. Two types of health effects from air pollution must be considered: pulmonary and systemic. Pulmonary injury occurs when inhalation of a chemical injures

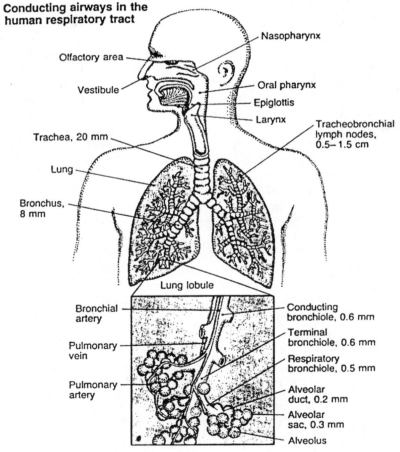

Figure 4.2 Human respiratory tract. Reprinted with permission from *Environmental Science and Technology* 17:170A, 1983. Copyright 1983 American Chemical Society.

the lungs themselves.[6] However, the lungs also provide a portal into the bloodstream. Since blood picks up oxygen in the lungs and carries it to every cell in the organism, an agent present in the lungs may affect any organ.

To be toxic to an organ, an agent must reach the organ, and the access of a chemical to the lungs depends first of all on its state. Gases readily penetrate to the lungs, as do vapors. For liquid droplets and for particulates, penetration depends on size. Particles and droplets best reach the lungs when they are less than 5.0 μm in diameter, because larger particles tend to be trapped by the cilia lining the respiratory tract. Once in the lung, an agent's toxicity depends on its reactivity and on physical parameters that are not well understood. A chemical may stay in the lung, damaging cells that it contacts. In such pulmonary toxicity, the lungs are the *target organ* for the toxicant. It is quite possible that a chemical that is toxic to all cells becomes strikingly toxic to the lungs (if administered by inhalation) because they are the first organ contacted and so receive the largest dose of the agent. For most chemicals that affect primarily the lungs, exposure by inhalation is the major, if not the only route of exposure. There are exceptions, however.[7]

Whatever they do to the lungs, most gases are absorbed into the bloodstream from the alveoli, and so enter the body. Particulates are either inert, in which case they may stay in the alveoli, or they dissolve and cross into lung tissue or into the bloodstream. In the latter case, many of the toxic effects are qualitatively similar to the effects of ingestion or dermal absorption. For example, inhaled lead (see Chapter 6), remains toxic to the brain and bones just like ingested lead. In the case of radioisotope inhalation, the lungs are a major target organ for some isotopes, but others (e.g., ^{131}I) are transported to other organs. Inert particles can also enter the bloodstream by phagocytosis.

Both lead and asbestos are examples of airborne particulates that cause major health effects. Asbestos is a pulmonary toxicant; lead is not. Lead's damage is chemical, and the airborne particulates are only one route of access to the organism. To cause lead poisoning, airborne lead particles must be dissolved and enter the body via the pulmonary circulation. Vapors of lead salts are more commonly implicated in respiratory lead poisoning, however,

---

[6] It is also possible for an inhaled substance to injure the nasal passages or bronchi. Cocaine, which constricts blood vessels, eventually destroys the nasal passages; certain chemicals cause nasal or laryngeal cancer. Nonetheless, the lungs are the primary target for inhaled toxicants.

[7] The most notorious example of lung damage after ingestion of a chemical is provided by the herbicide paraquat, 1,1'-dimethyl-4,4'-bipyridylium ion. In contrast to solvent-induced pneumonia, in which a chemical is inhaled while being swallowed or vomited, paraquat is absorbed from the stomach and actively transported into the lungs. It causes interstitial pneumonia and induces an irreversible fibrosis. Paraquat poisoning is rarely, if ever, occupational. It is most commonly ingested in suicide attempts, although documented examples of murder exist. A macabre aspect of paraquat toxicity is that because of the time required for fibrosis to develop, there is a lag of roughly a week between ingestion and death, but no treatment is available.

while particles are thought to enter through the gastrointestinal tract. Lead and its toxicity are discussed in Chapter 5.

In the case of asbestos, it is the particles themselves and their persistence in the body that are the problem. However, the etiology of disease is the same as for other particle-caused lung injury. Larger particles (those >5 μm in diameter) will be coughed up before they reach the alveoli. Small particles that settle in the lung reduce the elasticity of the alveoli, causing a *fibrosis* that decreases oxygen uptake by the lungs. Asbestos fibers also trigger the process of carcinogenesis. Asbestos-related diseases, consisting of asbestosis or asbestos-induced emphysema, lung cancer, and mesothelioma, are discussed more fully in the case history below. While asbestos fibrosis or *asbestosis* is the most notorious respiratory syndrome ascribed to particulate pollutants, one must neither forget nor minimize the other diseases caused by exposure of the lungs to dust: black lung among coal miners, brown lung among workers in cotton mills, and silicosis in tunnel cutters and perhaps among fiberglass applicators. Farmer's lung is thought to be an allergic response to molds found in silage. Each disease has a latency period that varies with the intensity of exposure, with duration of exposure, and with the individual.

Recently, evidence has been found that exposure to particulates is more harmful than previously thought, especially for children, the elderly, and those with pre-existing respiratory problems. In several studies, careful correlation of day-by-day mortality figures with the levels of particulates in the air of several cities suggests that mortality increases with surprisingly low levels of particulates. Small particulates, 10 μm or less in diameter, are implicated. It is suspected that the current 24-hour standard of 150 μm per cubic meter of air will need to be revised downward.

INDOOR AIR POLLUTION

In considering air pollution, one tends to think of cities, with miasmas of automobile exhaust; of factories belching smoke and noxious fumes into the air; of "them" polluting "our" breathing space. Ironically, the pollution we impose on ourselves, in our own houses, can be as severe as all but the most severe industrial pollution. The reason for the severity of indoor air pollution is, of course, that the space involved is small and the roof and walls inhibit the replacement of air. In effect, we generate pollution and do not let it dissipate, and then we stay in the polluted space for over a third of our lives.

Major sources of indoor air pollution are:

· Smoking by inhabitants. In addition to the damage that smoking does to the smoker's lungs, evidence is accumulating that other inhabitants also suffer. Nonsmoking wives of male smokers have a higher incidence of lung cancer than do nonsmoking women married to nonsmokers, and children living in households where one person smokes tend to have more respiratory illnesses than those living in nonsmoking households.

· Heaters, especially wood stoves and unvented kerosene heaters. The hydrocarbons as well as the soot increase risk of respiratory illness (acute or chronic), while carbon monoxide poses an acute risk of poisoning.

· Cooking, especially frying, which emits benzopyrenes into the air, as well as particles of soot and miscellaneous gases.

· Gas appliances, including furnaces and clothes driers, but especially gas stoves. Major pollutants are NO_x and CO_2. CO may be a problem if the appliances are using "coal gas" (no longer used in the United States) or are not adjusted properly. Children living in households using gas for cooking and heating may have more respiratory illness than do children living in "all-electric" homes.

· Furniture. If it includes plywood or particleboard, furniture can emit ("offgas") formaldehyde. The formaldehyde is used in the gluing process for both particleboard or pressed wood and plywood. It is exceedingly difficult, and often prohibitively expensive, to buy furniture that does not contain *any* plywood or particleboard. With care in manufacturing, the release of formaldehyde can be minimized. Painting particleboard, or coating it with vinyl or Formica, can also decrease emissions. However, for people sensitized to formaldehyde, minute quantities suffice to trigger severe allergic reactions.[8]

· Walls and insulation. Walls give off formaldehyde if they are made of plywood. Insulation made with urea–formaldehyde foam retains unreacted formaldehyde in the air spaces of the foam; this formaldehyde is released over the life of the insulation. As with furniture, care in production minimizes the amount of formaldehyde that is retained. Many houses and most mobile homes built since 1973 used urea–formaldehyde foam insulation, although some attempts have been made to ban it (see Chapter 12). Especially in the case of mobile homes, which have a small air volume relative to the surface areas, and therefore have relatively large amounts of plywood and insulation used in their construction, severe air pollution can result. Studies have shown that even carefully applied urea–formaldehyde foam insulation offgases significantly. In most cases the problem is severe primarily for those with formaldehyde allergies, but exposure can trigger such allergies.[9]

· Wall and pipe coverings. If walls or pipes are coated with friable asbestos, or if an asbestos coating is damaged, asbestos particles can flake off under conditions of ordinary use. The worst problems are seen in

[8] It is also possible for appliances, even toasters, to give off formaldehyde, if they contain plastics or glues that offgas.
[9] Formaldehyde is also a carcinogen. The case history of formaldehyde's regulation—or lack thereof—is given in Chapter 12.

some public buildings, where asbestos was sprayed on as a foam.[10]
Removal must be carried out with extreme caution. Undamaged asbestos—floor tiles or siding, for example—should be left in place as long as it remains intact. It should not be sawed or abraded.

· Sheets and clothing. New cloth and new sheets give off formaldehyde if they have been treated to be "permanent press." Cotton blended with synthetic fibers is particularly likely to be so treated. The emissions decrease sharply (to zero?) after one washing.

· Plastics in shower curtains, furniture, cars, and so on, frequently include polyvinyl chloride, which may offgas trapped vinyl chloride monomer. This is part of the "new car smell." The same slightly sweet, rather pleasant odor is also very noticeable with new shower curtains. Vinyl chloride causes liver cancer at occupational levels of exposure. Liver cancer remains rare in the U.S. population as a whole, however, arguing that environmental levels of vinyl chloride may be too low to be a problem.

· Household products such as:

Aerosol spray cans, which emit not only their product, but also the propellant, typically a hydrocarbon solvent. Among the solvents that were used in household aerosol spray cans between 1950 and 1980 were chlorofluorocarbons (CFCs), vinyl chloride, and methylene chloride. Today, many products advertise that they "contain no fluorocarbons," but the propellant is rarely identified. It is best to use manually pumped sprays.

Pesticides, which are used inside the house to kill fleas on pets (flea collars and flea powders), to eliminate or decrease flea or cockroach populations (crack and crevice sprays), and sometimes aerosols ("fogging"). Insecticides are also used in or on the foundations against termites, on or near the house against wasps, and inside or outside the house against ants. Seepage through foundation cracks or drift through open windows can contaminate the inside of the house. Herbicides and fungicides may be used on lawns: although it is theoretically possible that material would be carried in on shoes, this is not a serious source of indoor air pollution.[11]

---

[10] In many schools built in the 1960s, asbestos was sprayed onto the ceilings of gymnasiums. Such sprayed-on asbestos continually releases fine particles into the air. In addition, however, both students and teachers report that a popular lunchtime amusement in some schools consisted of throwing basketballs at the ceiling to see how much dust would come down. After lunch the janitor would sweep up the debris. The magnitude of the health damage will not be known for some decades. A less visible location for sprayed-on asbestos was on the outside of pipes running between floors of high-rise buildings. The spaces the pipes run through often serve as air-conditioning plenums. Thus workers are exposed to a steady barrage of largely invisible fibers.

[11] It may, however, be a significant exposure for pets. A recent epidemiological study of non-Hodgkin's lymphoma in dogs suggested that dogs whose owners used the herbicide 2,4-D on their lawns had a significantly higher risk of developing this cancer.

Household cleaners, which often contain solvents, chlorine, and/or ammonia. Perfumes are also present in most cleaning compounds and may trigger allergies in sensitized inhabitants.

Paints, many of which contain metals as pigments (cadmium, formerly lead) and which may contain mercury or other mildew-retarding agents. "Oil-based" paints also contain solvents. Lead and cadmium are primarily pigments and remain in the paint, becoming pollutants only when the paint is removed from the walls and the particles are inhaled or ingested. Mercury, which is used primarily to retard molding, is volatile, however, and steadily enters room air from the paint. In several recent studies, latex paints were found to be formulated with dangerously high levels of mercury. Paint removers formerly contained methylene chloride; newer chemical ingredients (e.g., xylene) may also pose hazards.

· Hobbies may be hazardous because of chemicals used. In many cases the chief risk is from ingestion, but soldering fumes contain lead. The use of solder in poorly ventilated areas is quite hazardous. Pottery glazes often contain metal salts, which can leach out if the pottery is not fired at high enough temperatures; this is rarely a source of air pollution. Artist's pigments contain mercury, cadmium, and lead, and careless use has on occasion resulted in poisoning; again, ingestion rather than inhalation is thought to be the major problem.

· Radon and radon daughters, usually from soil strata underlying the house, can also originate in water that flowed through uranium-rich rocks. This pollutant is dealt with in the next section.

RADIATION

The dangers of radiation have been argued since before the first atomic weapons, "Fat Man" and "Little Boy," were built in the New Mexico deserts. Although early investigators of radioactive materials were known to have died of radiation-associated illnesses, their exposures were high enough to cause acute symptoms, including burns (Table 4.1). That lower levels of radiation could be hazardous was not recognized until after World War II. Even now there is considerable controversy about the risks associated with low levels of exposure to radiation.

Certainly, the scientists working on the Manhattan Project at Los Alamos in the 1940s did not consider themselves at risk. In at least one case, plutonium was transported, inadequately shielded, in a briefcase carried by a courier. After the explosions at Hiroshima and Nagasaki, the horrors of high-level radiation exposure were obvious, but concerted efforts were made by the U.S. government to minimize the risks of less drastic exposures and to hide the degree to which Americans were exposed to radiation and radioactive materials as a result of nuclear weapons testing, accidental re-

Table 4.1 Definitions of Units of Radioactivity

Unit	Symbol	Defined
Of Decay:		
Curie	Ci	Rate of radon decay:
		3.7×10^{10} nuclear transformations per second
Bequerel	Bq	$1\ Ci = 3.7 \times 10^{10}\ Bq$
Of Dose:		
	rad	0.01 J/kg or 100 erg/g
	rem	The quantity of radiation that is equivalent *in biological damage of a specified sort* to 1 rad of 250 kVp of x-rays

Source: Adapted from C. Hohenemser, M. Deicher, A. Ernst, H. Hoffsaess, G. Lindner, and E. Recknagel, "Chernobyl: an early report," *Environment* 28(5):40, 1986.

leases, or by working as uranium miners. Considerations of "national security" were used to obfuscate the evidence of hazard that was available, to the extent that scientists falsified data derived from controlled studies. In addition, the data from Hiroshima and Nagasaki are open to several interpretations, as is typical in uncontrolled experiments.[12] The very horror of the event, and the social consequences attached to being known as a survivor (e.g., in obtaining insurance, or in marrying), have caused some to deny being there. Initial estimates of exposures have recently been shown to err on the high side: that is, the effects observed were actually due to lower levels of exposure than had been thought. This has required a complete recalibration of radiation-associated risks, demonstrating how much of our knowledge is based on these two populations.

Postwar enthusiasm for nuclear power was high. Optimists worried about a world in which energy would be so cheap that no one would have to work more than a few hours a week. Both nuclear fission using uranium as a fuel, and nuclear fusion, in which two hydrogen nuclei combine to form helium, were expected to be major sources of power within years. In the 1950s every science magazine predicted that fusion power was just around the corner. Controlled fusion has still not been achieved, but nuclear fission is indeed producing a significant fraction of our energy. The state of Illinois obtains more of its energy from nuclear power plants than from coal, even though it has large reserves of coal and none of uranium; and France obtains 80% of its electricity from nuclear power plants. Unfortunately, the promise of

---

[12] For example, in Nagasaki the epicenter of the explosion coincided with the largest enclave of Catholics in Japan. Because marriages between second cousins are favored in traditional Japanese culture, but not among Catholics, there may be an inverse correlation between inbreeding and radiation exposure, which could confound the genetic effects of exposure (see Chapter 10).

cheap energy has not materialized. Nuclear power remains more expensive than power from conventional sources, although its proponents claim that this is due to political, not engineering, constraints. In addition, a series of not-quite-disastrous accidents in the United States, coupled with the explosion of the Soviet reactor at Chernobyl (in what is now Ukraine), has made it difficult to site new nuclear power plants.

Most of the literature on nuclear power plants is biased. The Atomic Energy Agency (AEC), later renamed the Nuclear Regulatory Commission (NRC), was, until very recently, responsible both for developing and for supervising nuclear energy. Opponents remarked on the futility of having the fox guard the hen house; proponents countered with arguments of the NRC's scientific competence and of the opponents' ignorance.

The arguments for exploitation of nuclear power are

1. Nuclear power is intrinsically cheaper than fossil-fuel energy; only the constantly changing regulations and threat of litigation make it so expensive.

2. Uranium is plentiful in the United States, and existing reprocessing techniques make it unlikely that we will run short of fuel in the foreseeable future.

3. Nuclear power does not generate greenhouse gases because it does not depend on burning carbon; given the increasing probability of global warming, and our increasing dependence on unstable sources of foreign oil, we really have no choice but to use nuclear power.

4. Nuclear power is safer than coal or oil if the risks associated with mining, transport, and air pollution are considered.

5. Other forms of energy also generate toxic wastes and also damage the environment from which they are collected: for example, using solar power as our primary energy source would require covering half of Arizona with solar collectors.

6. Adequate technology exists for dealing with the very small volume of high-level wastes that are generated, while low-level waste is not a real problem.

The arguments against exploitation of nuclear energy are expressed by the negation of each of the above.

1. Nuclear power has always required subsidies to be competitive with fossil-fuel energy; even the French, who depend on it so heavily, cannot make it competitive. They hope to export nuclear-power-generated electricity, but it costs more to generate than the market will pay for the electricity. Moreover, the poor safety record of the industry makes it clear that more, not less, regulation is needed, even for existing plants.

2. Plutonium is so incredibly toxic, and its half-life of several thousand years makes its hazard so persistent, that reprocessing is insane, especially given the threat of nuclear terrorism.

3. It is true that nuclear power does not generate CO_2 or other greenhouse gases, but there are alternative strategies that could be used to minimize global warming. These strategies include use of solar, hydroelectric, and wind energy and emphasis on conservation through increasing efficiency of using power in both the private and commercial sectors.

4. Nuclear power is safer than coal or oil only if unrealistic assumptions about future events are made. The string of near-accidents that have already occurred, and the accidents at Three Mile Island and at Chernobyl, make it clear that the probability of a major accident have been understated by proponents.

5. We have enough trouble coping with wastes that persist for decades, as seen by the number of Superfund sites. It is not reasonable to think that we can monitor high-level nuclear wastes for millennia. Moreover, although the spent fuel is a very small quantity of waste, the plants themselves become "hot" during their lifetimes and will have to be dismantled and sequestered, or will have to be guarded on site, also for hundreds of years.

6. Proponents of nuclear power tend to assume that alternatives will be equally centralized: that is, they assume that hydrologic power depends on large dams, which are indeed environmentally disruptive, and that solar power would require large centralized installations of solar collectors "over half of Arizona." Opponents of nuclear power often recognize alternative strategies as well as alternative sources. If each building had a solar collector on its roof, a considerable fraction of household electrical power could be generated. Heating water is most amenable to this sort of small solar installation. Similar strategies can be applied to water-generated electricity: there are already some former mill ponds contributing to power generation in New England.[13]

---

[13] There are structural changes that would be required in our society for these options to work. For example, in New England, utilities considered small generators a nuisance. These "small generators" were mostly people who owned a millpond that generated most of their electricity. They used the commercial net for peak power use and sought to sell to the net the surplus they generated at other times. Court action was needed to force utilities to pay the generators of small amounts of power for their contribution to the net. Similarly, tax incentives and low-cost loans should be made available to individual households that install solar water heaters or hot-water heat coupled to a solar generator as well as to backup electric power. Such incentives were made available by the Carter Administration (1976–1980) after two major oil crises (1973 and 1978) made obvious the need for the United States to stop squandering petroleum. The incentives were rapidly phased out by the Reagan Administration (1981–1989) as an unnecessary reduction in tax revenues. Comparisons of the cost to taxpayers of large-scale tax-financed dams versus small loans for small-scale projects should be carried out.

7. If the Reagan Administration had not stopped subsidizing research and development of alternative sources of energy, we would have alternatives on line. Although good storage methods require a technological breakthrough, both solar and wind energy could be used to generate a considerable fraction of our energy. Energy efficiency would also minimize our dependence on foreign oil.

It is still reasonably accurate to summarize the status of the debate on nuclear power, and the attitudes of the opponents, in the words of Stobaugh and Yergin[14]:

"In the 1960s the nuclear advocates rushed ahead to attempt one of the most massive technological innovations in history—seeking to convert the world's electric generating system from fossil fuels to nuclear. In so doing, they confused dreams with reality. Utilities made their decisions to install billions of dollars of generating plants on the basis of costs *estimated* by the manufacturers without an adequate base of relevant experience. The proponents also ensured an increasing supply of dedicated enemies by the manner in which they dismissed the opposition, attacking them as revolutionaries, Luddites, and misguided children. In general, the advocates failed to conceive of nuclear power as an overall system, one that starts with the mining of uranium and ends with a satisfactory disposal of spent fuel. Meanwhile, nuclear critics fought a war of attrition in fish and game commissions, county commissioners' offices, and the federal courts. Wherever the critics went, they asked one question: "How safe is safe enough?""

To add that Stobaugh is a professor of business administration at the Harvard Graduate School of Business Administration suggests that not only radicals see problems in a nuclear future. That a book published 13 years ago is still relevant suggests that we are nowhere near a resolution in 1994.

CASE HISTORY: CHERNOBYL

In the spring of 1986, a Soviet nuclear reactor in the small town of Chernobyl in Ukraine[15] exploded and burned. First reports of the disaster came not from the USSR, but from Scandinavia: workers returning to a nuclear power plant on Monday, April 28, set off radiation monitors. Knowledge of prevailing winds plus intensive monitoring of airborne radioactivity, which signaled major increases in airborne levels of radioisotopes, permitted observers to identify the probable

---

[14] R. Stobaugh and D. Yergin, eds., *Energy Future: Report of the Energy Project at the Harvard Business School,* Random House, New York,1979, pp. 220–221.

[15] Although there is no Soviet Union any longer, the accident at Chernobyl occurred in the context of a highly centralized federation, the Union of Soviet Socialist Republics. Had the Soviet Republic of the Ukraine been an independent nation, Ukraine, in 1986, a different story might exist.

source of the radioactivity. Initial speculations included the possibility of an atmospheric bomb test, but the existence of the power plant at Chernobyl was known and an accident was considered the most likely cause almost at once. The accident was confirmed by the Soviets two days later.[16] The reactor burned until at least May 5, by which time approximately 49,000 people had been officially evacuated from the towns and farms near Chernobyl.

In the USSR itself, recognition of the disaster was delayed because the plant's operators initially thought the core was intact. For some hours they remained unconvinced that there was a serious radiation problem because the highest reading on their instruments was less than 0.1% of the actual levels. This conviction was disastrous, since it meant that appropriate measures were delayed for those hours. One victim of radiation poisoning was a man in Pripyat, who sunbathed on his balcony the day after the explosion. By nightfall he was vomiting uncontrollably. By June 1986, the toll from Chernobyl was reckoned as: 2 people dead in the accident, one of burns and one from falling debris; 25 dead of acute radiation poisoning and 1000 ill with radiation sickness; 100,000 officially evacuated from the area around Chernobyl and the new town of Pripyat. Tons of foodstuffs in Europe—especially Poland—were destroyed. People in eastern Europe were warned to avoid produce. After the 1986 harvest, farmers in southern Germany were thought to have enough radioactive hay and grain in their barns to qualify as radioactive waste storage sites. No one has ever explained what happened to that hay. Fortunately many of the most dangerous radioisotopes are short-lived.

Until the breakup of the Soviet Union, information on environmental contamination from the explosion at Chernobyl came mainly from western Europe. Monitoring data are not uniform across the globe, and there is considerable evidence that even for atmospheric levels, a high degree of variability exists. Ground-level data suggest that modeling by distance is not possible, and that the occurrence of rain is critical. Peak *air* levels of radionuclides reached 100,000 times background levels in Poland, which is 400 to 800 km from Chernobyl, and as high as 10,000 times background levels in Scotland, 2000 km distant (Figure 4.3). Peak *ground* levels of fallout varied greatly: peak levels occurred with rain, reaching 30 to 40 times background levels at Konstanz, 1500 km from the reactor. In the absence of rain, ground levels at 1500 km were only 1 to 2 times background. From the available data it is estimated that air levels in the immediate vicinity of Chernobyl reached 1 million times background levels, or 2000 Bq/m^3, and that ground levels after rainfall reached 200 to 400 times background levels.

The entry of radioactivity into the human food chain was highly variable: 1500 km away from Chernobyl, ^{131}I levels reached 500 Bq per liter of milk (compared to essentially zero levels as background) where cattle were left outside. Clearly, variability is increased not only by the variation in ground-level contamination, but by the degree to which local authorities monitored and warned citizens and by the degree to which warnings were heeded. For example, in Konstanz

---

[16] Soviet experts have since admitted that the Chernobyl plant had released significant amounts of radioactivity into the air on earlier occasions. At those times the prevailing winds were from the west, so these releases were not identified in western Europe. Effects on downwind populations are unknown and may not have been monitored.

itself, authorities kept ^{131}I levels in milk below 100 Bq/L; in a nearby area with similar rainfall, milk containing 600 Bq/L ^{131}I was marketed. Exposure of two children living a very few miles apart could therefore be radically different.

There was no asparagus harvested in Germany in 1986 because the season coincided with the accident at Chernobyl; residues were too high. As a result, the 1987 crop was especially bountiful. This is perhaps a triviality, but it illustrates the capricious nature of the effects of a disaster. What other crops or foodstuffs were destroyed during the 1986 growing season because of persisting radiation from Chernobyl is not clear, nor is the degree to which monitoring continued in 1987 known.

Kiev, with a population of 2,500,000, is the major city closest to the reactor. It is located 130 km south of Chernobyl. In Kiev, thousands of wells were closed, and 58 new deep wells dug, due to contamination. During the summer of 1986, streets were vacuumed to remove radioactive dust; that autumn, the leaves that fell from the trees were deposited in a radioactive waste dump.

Like all major disasters, Chernobyl can be analyzed from several viewpoints:

1. The safety of nuclear power plants has again been called into question. Although the design of the Chernobyl reactor is considered obsolete by experts in the United States and western Europe, at least one U.S. reactor (Hanford, Washing-

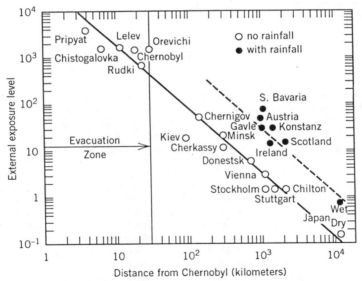

Note; The exposure, E, is measured as a multiple of the natural background, assumed to be 0.01 millirems per hour, and is proportional to the distance, r, from Chernobyl, or $E = (6.9 \times 10^4) \, r^{-1.4}$

Figure 4.3 External exposure levels at sites with and without rain as measured 7 to 10 days after the Chernobyl accident in regions affected by the main fallout cloud. (From C. Hohenemser and O. Renn, *Environment* 30(3):6, 1988.]

ton) of the same general design was in use.[17] Moreover, a relatively persuasive argument could be made that waiting until an accident proves the design of other reactors unsafe is foolhardy. The urgency of the question was underlined by riots and demonstrations in western Europe—especially in what was then the Federal Republic of Germany (western Germany)—against continued expansion of nuclear power. Data that must be considered include the graphite-moderated design of the Chernobyl plant, which is considered inherently less safe than the water-moderated design used in all but one reactor in the United States.[18]

Conclusion: insufficient data. Care must be taken to consider the level of risk posed by burning fossil fuels. These risks include acid deposition, contribution to global warning, mining accidents, and the black lung disease of coal miners. However, the disposal of radioactive wastes must also be considered in judging the costs of nuclear power.

2. The effects of even a small nuclear war on human health and activity were emphasized. Besides the fallowing of thousands of hectares of Ukraine, the explosion forced the destruction of milk and crops all over Europe; more losses were incurred because exports from presumably affected nations (e.g., Poland) were avoided by importers. The estimated cost includes an expected excess of "thousands" of cancer cases over the next 30 to 40 years.

---

[17] The inherent problem posed by the RBMK design and the concomitant evasion of the safety regulations designed to prevent Soviet RBMK reactors from going critical are described by R. Wilson, "A visit to Chernobyl," *Science* 236:1635–1640, 1987. A much longer narrative description of the accident and its aftermath are presented by the biologist Zhores Medvedev in *The Legacy of Chernobyl*, W. W. Norton and Co., New York, 1990. Medvedev's book benefits from the greater number of documents available in 1990, and also from the author's experience in (as he says) "the way pseudo-facts and misinformation are created in the Soviet Union to serve political and institutional interests".

[18] Several other reactors in the USSR use this same design, including one actually in the city of Leningrad (now renamed St. Petersburg). According to Zhores Medvedev, the Russian acronym RBMK stands for "reactor high power boiling channel type." This design was considered obsolete by nuclear experts in both America and Europe, where boiling water reactors (BWR) or pressurized water reactors (PWR) are the norm. However, the RBMK reactor in Hanford, Washington as well as several British RBMK reactors, were still in use in 1986. Before the accident at Chernobyl, however, the major problem of Soviet reactors was thought to be the lack of a containment building strong enough to prevent radioactive releases in a core meltdown. Core meltdowns, the most dreaded accidents of the nuclear industry, occur if cooling fails for a long enough time that the nuclear fuel becomes hot enough to melt. (Uranium melts at 1132°C; the zirconium that is used for nuclear fuel cladding melts above 1800°C). The concrete containment dome of the Three Mile Island reactor in Pennsylvania remained intact during the 1979 accident (which included partial melting of the nuclear fuel core) and thus prevented most releases of radiation outside the plant. In both RBMK and PWR reactors, failure of the cooling system allows heat to build up in the "core" (nuclear fuel), leading eventually to meltdown. But in the PWR reactors, water also moderates the energy of fast neutrons, with the result that loss of water slows the nuclear reaction. In RBMK reactors, graphite serves as the moderator, and the chain reaction continued after cooling failed at Chernobyl. Therefore, a major difference between the accidents at Chernobyl and Three Mile Island is that at Chernobyl there was an actual nuclear explosion. No containment structure would have withstood it.

Conclusion: It is argued by many that the Soviet push toward a reduction in nuclear warheads, which was marked in the months after Chernobyl, was due to their recognition of the intolerable results of any nuclear war, and that this recognition was the direct result of the costs imposed by Chernobyl.[19,20]

3. Human health effects can be modeled, using the measured levels of radioactivity found at different times to construct exposure curves for the population (Table 4.2). Input should include airborne radioisotopes as well as residues on plants that are ingested or transported to livestock (and thence to meat and milk). Exposure models must include the half-lives of different radioisotopes and must differentiate between doses to specific target organs. For example, it is expected that the thyroid cancers due to ^{131}I will have a low death rate, since these cancers are readily identified and removed. In contrast, cancers due to Cs (^{134}Cs or ^{137}Cs) would have an expected 50% mortality.

Conclusions: Depending on the models used and the levels of exposure estimated—and the latter levels differ depending on local weather as well as on distance from Chernobyl—a significant excess of cancer, ranging from thousands to tens of thousands of additional cases, is expected in the 300 to 400 million people exposed in 15 countries.[21]

4. Measurement of radioisotopes at different stations at successive times can be used to model air currents. As with microcosms, measurements are facilitated by tracers—radioisotopes, for example—but it is ordinarily not permissible to release enough radioactive material into the air to follow transcontinental air currents. Chernobyl did.

Table 4.2 Annual Risk of Death: Konstanz, Germany (Population 100,000)

Cause of Death	Deaths/Year
All	1100
All cancers	250
All smoking-related deaths	200
Traffic accidents	10–15
Cancers due to Chernobyl	1.6–2.9
Cancers due to atmospheric nuclear testing	1.5–1.6

Source: Adapted from C. Hohenemser, M. Deicher, A. Ernst, H. Hoffsaess, G. Lindner, and E. Recknagel, "Chernobyl: an early report," *Environment* 28(5):40, 1986.

[19] The calculation begins: if Chernobyl released as much radiation as an *x*-kiloton nuclear warhead, and if a small nuclear war would involve detonation of 10 to 100 *x*-kiloton nuclear warheads. . . .
[20] It is also argued that the accident at Chernobyl was directly responsible for the collapse of the Soviet Union and/or for the decision by Gorbachev to proceed toward economic reform.
[21] Assuming that the cancer incidence in Europe is similar to that in the United States, the background incidence is that 1 person in 3 will get cancer and 1 in 4 will die of cancer. So the "10,000 cases" comes on top of a base level of 100,000,000 cases and represents a 0.01% increase in cancer incidence. In effect, unless the cases are strongly clustered geographically, or consist of otherwise rare cancers, it will be difficult to prove their existence.

Conclusions: Very small variations in weather can lead to large variations in radioactive fallout; moreover, the action of local officials in the first days after the accident were crucial in minimizing exposure of cattle (and therefore of milk). Rainfall provided the critical difference—as much as 15-fold within 100 km—between heavy and light fallout *at ground level.*

RADON AS AN INDOOR AIR POLLUTANT

Radon, a decay product of the uranium-238 series (Table 4.3), has recently been shown to be a serious contaminant of surprisingly many houses in the United States. "Normal" soil levels of uranium are 1 to 4 ppm, but some localities have far higher levels. In Florida, areas lying above phosphate deposits have 50 to 150 ppm. Colorado has high-grade ores containing 1000 to 5000 ppm, and Canadian and African uranium ores contain 10,000 to 40,000 ppm.

Just as the first warnings of Chernobyl came from tripping of monitors at a nuclear power plant, so did warnings of radon pollution in Pennsylvania. In December 1985, a worker at a nuclear power plant set off radiation alarms on his way *into* the plant. Subsequent investigation showed that the radon levels in the worker's home were 13.5 WL (working levels, with 1 WL equivalent to 173 hours' exposure to 100 picocurie of radon and radon daughters per liter of air). In effect, the house was 13 times as radioactive as would be permitted for occupational exposure (Table 4.4).

It is now known that the source of the radioactivity in this house—and the somewhat lesser but still high levels in other houses in the area—are due to the "Reading prong," an area of high levels of uranium underlying portions of Pennsylvania, New York, and New Jersey. Similar high levels of uranium are found in areas of Maine. In Colorado the problem is compounded because tailings from the uranium mines were mixed into the cement of

Table 4.3 Important Elements in the Radon Decay Sequence

Element	Symbol	Half-life	Comment
Uranium	^{238}U	4.5×10^9 years	Decay to radon includes thorium, proctinium, and radium
Radium-226	^{226}Ra	1620 years	
Radon	^{222}Rn	3.8 days	Inert; gas
Polonium-218	^{218}Po	3 minutes	
Lead-214	^{214}Pb	27 minutes	
Bismuth-214	^{214}Bi	20 minutes	
Polonium-214	^{214}Po	<1 millisecond	

Source: Adapted from M. Eisenbud, *Environment* 26(10):11, 1984.

Table 4.4 Airborne Radon Standards for Buildings

Houses built on uranium mine wastes in the United States	<3 pCi/L
Phosphate mining regions in Florida	4 pCi/L: remedial action required 2 pCi/L: reduction to a reasonably feasible level required
Uranium mining regions in Canada	30 pCi/L: prompt remedial action required 4 pCi/L: remedial action required 2 pCi/L: investigation recommended
Union of Concerned Scientists	>5 pCi/L: remedial action indicated 2–5 pCi/L: remedial action suggested

Source: B. Hileman, "Indoor air pollution," *Environmental Science and Technology* 17:471A, 1983.

many house foundations and were also used in roadbeds. In northern Illinois, several municipal water supplies have been identified that are illegally radioactive (Table 4.5).

The major source of human exposure from these uranium deposits occurs as a result of the decay of ^{238}U to radium, which decays to the inert (noble) gas radon. Radium has a long half-life and is not itself terribly hazardous, but it is soluble in water. Radon has a short half-life but, being inert, is still less hazardous than the "radon daughters" that are formed as it undergoes nuclear decay. However, being a gas, radon can seep through soil and rocks and through porous or cracked foundations. Exposure to radon and radon daughters therefore occurs as the last step of a series of events: presence of uranium in soil or bedrock, followed by migration of radium and/or radon into the house via water or air, followed by further decay of the radon to highly radioactive (short-half-life) elements that are also chemically reactive.

Table 4.5 Indoor Radon Levels in the United States

Location	Number of Measurements	Highest Reading (pCi/L)	Percent >4 pCi/L
Northern California	80	7.4	15
Midwest	64	7.4	20
South	304	2.7	0
Northeast	133	77	20
New York	413	50	15
Pennsylvania	249	91	42
Maine	427	133	21
Other U.S.	826	—	—

Source: B. Hileman, "Indoor air pollution," *Environmental Science and Technology* 17:471A, 1983.

Because radon is a gas, and therefore radon and radon daughters are most likely to be inhaled—alone or as particles adsorbed to dust—the worst danger from radon is lung cancer. Radon is water soluble, so water flowing through uranium-rich soils is also a source of exposure, especially when water is heated and then released in a shower, since gases are less soluble in hot than in cold water. In some houses in Maine, risk estimates are that 1 person in 10 will develop lung cancer purely as a consequence of home radon exposure. This is approximately the same risk that attaches to smoking one pack of cigarettes per day for 20 years.

In Colorado, the tailings of the Grand Junction uranium mill were used as fill under the concrete slabs and around foundations of houses, as heat sinks in fireplaces, and in concrete and mortar. Remedial action has included surveys of 36,000 houses, with removal of the radioactive material from over 600 homes, schools, businesses, and churches at a cost of $23,000,000. Before removal was settled on as the solution, unsuccessful methods tried included epoxy sealants, air-cleaning equipment, and active ventilation systems. Many of these methods are being used (with reported success) in the houses on the Reading prong. The difference is that in Colorado, the radioisotopes were in the foundations of the house; in the Reading prong, the radon gas only seeps through the foundations. In Florida, high levels of uranium, and therefore of radon, are found in association with phosphates, which are mined for use as fertilizer. The use of these phosphates, containing approximately 100 to 150 ppm uranium, does enrich the uranium content of recipient soils, but this is not thought to pose a human health risk. The major risk is again the seepage of radon into the houses. In Illinois, parts of which lie above uranium-containing rock strata, considerable variation is found between counties (Figure 4.4).

Notwithstanding the environmental exposure to radon and radon daughters, the major source of radiation exposure for the majority of Americans remains the medical (diagnostic) x-ray. However, the number of houses that have now been *shown* to have excessive radon levels, and the area of the United States *known* to overlie granites of the kind that tend to have uranium in them, suggest that a significant fraction of lung cancer deaths may be due to radon (Figure 4.5). The highest estimate is that 10% of lung cancer deaths, or most of the lung cancer deaths not associated with smoking, are due to radon.

CASE HISTORY: ASBESTOS

Although the cumulative effects of outdoor smog and indoor air pollution have probably caused more morbidity than any single particulate, the notoriety of asbestos-induced illness is worthy of special note. Asbestos is an example of a valuable substance that became a major cause of occupational respiratory disease, because of carelessness and overuse. It is one of the few industrial products that has been shown to cause cancer in nonoccupational exposures.

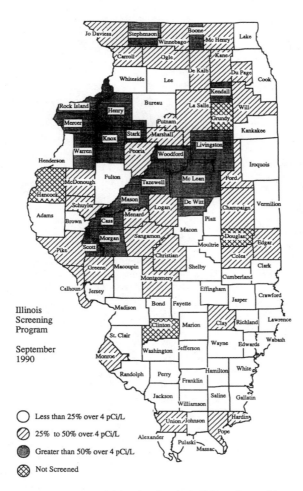

Figure 4.4 Example of the distribution of radon in buildings in a midwestern state. In most counties at least 20 houses were sampled, although Jersey County is represented by only eight samples. An accompanying letter warned both of inadequacies in the sampling and of uncertainties in extrapolating the data to health effects. (From the State of Illinois Department of Nuclear Safety, summer 1991.)

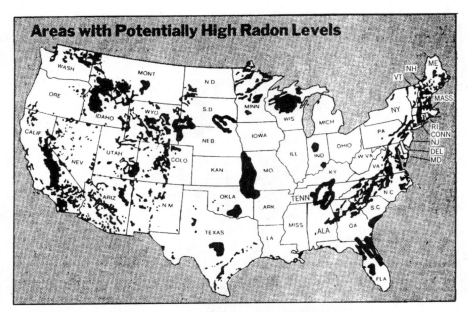

Figure 4.5 Location of areas with potentially high radon levels due to the uranium content of the underlying soils. EPA warns that soil permeability, type of construction, and ventilation patterns inside a house strongly affect indoor radon levels. (From the *New York Times*, August 16, 1986.)

Chemistry and History.   Asbestos is the name given to a group of fibrous mineral silicates, including both *serpentines* and *amphiboles*. Serpentines, such as chrysotile, are relatively pure magnesium silicate ($MgSiO_4$). They account for 90 to 95% of asbestos production. Amphiboles, such as amosite and crocidolite, contain cations other than Mg. There is some argument about the relative hazard of the two different types, since some researchers claim that the differences in fiber type are associated with differences in the incidence of asbestos-caused lung damage.

Asbestos is an amazing substance. It is fire resistant, can be woven, and is virtually indestructible. Asbestos is exquisitely fibrous: even under the magnification of the electron microscope, the fibers can still be seen. This is its value, and its bane.

It has been claimed that not only did the Romans use asbestos for royal funerary cloths, but that the pulmonary toxicity of asbestos was known to Pliny the Elder.[22]

[22] Several sources mention Pliny, including Kruus and Valeriote in *Controversial Chemicals* (see Appendix), who say that "Pliny referred to asbestos as 'the funeral cloth of kings'" (pp. 16–17). I cannot find this quote in Pliny's *Natural History*. Book XXXVI does refer, however, to asbestos: "Amiantus, which looks like alum, is quite indestructible by fire . . . and affords protection against all spells, especially those of the Magi." According to Rackham's (Oxford University Press) translation, *amiantus* is asbestos.

Certainly, the Romans did not use asbestos widely, and knowledge of its characteristics probably disappeared with the end of the Roman empire. Relatively few
uses for asbestos existed until the twentieth century. In 1920, only an estimated
20,000 tons of asbestos was produced worldwide; by 1970, this had increased to
4,300,000 tons worldwide, with 1,700,000 tons mined in the province of Quebec
alone. Most of Quebec's output was processed in the United States; most of the
rest of the world's production, 2.5 million tons per year at peak production,
originates in the countries that made up the Soviet Union. Asbestos is also produced in South Africa.

Uses. Given that most asbestos has been mined and used in this century, its
ubiquity is remarkable. The major use of asbestos, reckoned in tons, was in
cement. Other uses included, but were not limited to, asphalt, wallboard, pipes,
textiles, insulation, food and beverage processing, plastics, shingles and siding,
and filters (for beer as well as for gas masks). In the home asbestos was present in
spackling (plastering) compounds, in potholders and hot pads, and in the insulation
around pipes. Although minor in terms of tonnage, firemen's clothing and theater
curtains were probably the uses that most dramatically evoked the versatility of
asbestos: a woven, fireproof mineral. For visibility, there was the fluff of asbestos
that was sprayed onto ceilings and walls in the 1950s and 1960s as simultaneous
decorator material and insulation.

Release to the Environment. As can be imagined, the multiple uses of asbestos
insured that it entered the human as well as the external environment. The processes releasing the largest volumes of asbestos into the environment are its
mining and the processing of ore. Traditionally, the ore residues, or "tailings,"
contaminated with varying quantities of asbestos, were left in piles near the mine.
That was accepted practice for any mining residue. These piles were open to
the air and could be played on by children living nearby. Another source of
nonoccupational exposure was due to the dust generated during the manufacture
of asbestos-containing products. This dust not only blew through the factories
but often coated adjacent towns with snowlike powder.

Even after manufacture, many opportunities exist for asbestos release. During
application of blown-on asbestos insulation, high levels of airborne particles are
released. Disposal of worn-out asbestos-containing material often resulted in airborne particles. In addition, certain products release asbestos during use. The
most common and most hazardous of these products are asbestos brake linings.
As usual, occupational exposure is much higher than environmental exposure,
but even the latter can be very high and has been associated with illness.[23]

Asbestos releases can be divided into two types: point source releases and
diffuse (nonpoint) sources of asbestos pollution. Point sources include mine tailings, released into air or water; waste from construction sites or from demolition,
released into landfills and into water from runoff from dumps; asbestos dust

---

[23] A *Nova* program on asbestos presents the case history of a boy who helped his car-mechanic
father blow the dust out of brake linings. The father was still healthy; the son, whose exposure
was limited to weekends, died of mesothelioma.

released into air from factories or from manipulating asbestos-containing products (pipe, wallboard, insulation). The taconite iron ore mined in Minnesota contains some asbestos. Reserve Mining Company deposited these tailings in Lake Superior, resulting in contamination of the water supply of other cities on the lake. Nonpoint sources of contamination include brake linings, wall coverings, air ducts, asbestos-lined water pipes, and other asbestos-containing materials. With the exception of inhaled asbestos, health hazards from exposure are not proven (but see Table 4.6, which summarizes data on the risks for various cancers following occupational exposure to asbestos). The inhalation of asbestos leads to lung damage that is expressed as asbestosis, lung cancer, and/or mesothelioma, which is a cancer of the lining of the pleural cavity.

Toxic Effects.   In 1918, both U.S. and Canadian companies stop selling life insurance to asbestos workers. Since insurance companies use actuarial risks to determine excluded classes, and since actuarial risks are based on past experience, there must already have been statistically significant data demonstrating the lethal effects of asbestos. In the 1920s, *asbestosis* was described medically as a pneumoconiosis, similar to black lung, silicosis, berylliosis, and so on. Symptoms include the loss of elasticity of lung tissue, leading to shortness of breath. The loss of elasticity is progressive, irreversible, and dose dependent; severe cases are fatal. In this, asbestosis resembles all the other varieties of emphysema, which may result from any of several pollutants, including smoking. Asbestosis is characterized by frequent pleural abnormalities, which are absent in other pneumoconioses.[24] In addition, asbestos fibers can be identified in the lungs and pleura at autopsy. The latency period for symptomatic asbestosis is ordinarily 20 years from *onset* of exposure, although latency shortens with severity of exposure.

In the 1930s, the suspicion that there was an increase in the incidence of lung cancer in asbestos workers who survive asbestosis was first raised. Note that the latency for lung cancer is somewhat longer than that for asbestosis, so if exposure levels are too high, death from asbestosis occurs before lung cancer is seen. Similarly, in Italy, even asbestosis is said to be rare because tuberculosis kills asbestos workers before asbestosis appears. In the 1950s, the epidemiologist Richard Doll demonstrated a 10-fold increase in lung cancer among asbestos workers. Irving Selikoff, at Mount Sinai in New York City, used union records to study asbestos workers (Table 4.6). The lung cancer risk was heavily associated with smoking but was also increased for nonsmoking asbestos workers. Mesothelioma, an otherwise extremely rare cancer of the pleura, occurred at increased rates in asbestos workers, and did not show synergism with smoking.

Until mid-1986, OSHA (the Occupational Safety and Health Administration) allowed air levels of 2 million fibers per cubic meter (2×10^6 fibers/m^3) for workplace exposure. Given that an average adult inhales about 1 cubic meter of air each hour (1 m^3/hr), this standard means that 16 million fibers could be inhaled per 8-hour day. Since the test used to determine fiber levels uses optical micros-

---

[24] The pleura is the lining of the pulmonary cavity, on the outside of the lungs and the inside of the ribs. For asbestos to affect the pleura, it must somehow penetrate the alveolar sacs.

Table 4.6 Causes of Death among Asbestos Workers

Cause of All Deaths	Number of Deaths, 1943–1973		Observed Expected
	Expected	Observed	
Total deaths, all cases	300.65	444	1.48
Cancer			
All sites	51.26	198	3.86
Lung	11.86	89	7.62
Pleural mesothelioma	0	10	—
Peritoneal mesothelioma	0	25	—
Stomach	5.10	18	3.43
Colon/rectum	7.50	22	2.93
Asbestosis	0	37	—
All other deaths	249.39	209	0.84

Source: Adapted from S. Epstein, *The Politics of Cancer,* Sierra Club Books, San Francisco, 1978, p. 86.

copy, only fibers >5 μm are counted. Smaller fibers, especially those between 0.5 and 5.0 μm in size, are known to be both more numerous and more hazardous than the counted fibers. Moreover, the standard of 2×10^6 fibers per cubic meter does not even protect against asbestosis. Using data from occupationally exposed populations, it has been shown that exposures at levels of the pre-1986 standard lead to

20% lung cancer deaths

10% gastrointestinal cancer deaths

10% mesothelioma deaths

10% "other" cancer deaths

10% asbestosis

among asbestos workers. Note that even among people who smoke a pack a day for 20 years, only 10% get lung cancer. Because the latency period for the new exposure levels is not yet over, we do not yet know what the resulting disease frequencies will be.

To some extent the question has become moot. When it was demonstrated that the major asbestos companies had been aware of the hazards of asbestos for years, a flood of lawsuits drove the major U.S. producer, Johns Manville, into protective bankruptcy. In mid-1986, as part of a phasing out of asbestos use, allowable workplace levels were reduced by 90% (i.e., to one-tenth the previous levels). Most uses of asbestos in the United States have now been banned. No really good substitutes exist for some uses, and it is becoming clear that other mineral fibers, notably spun glass (fiberglass, rock wool), carry some of the same risks as asbestos.[25] Inhaling foreign substances is hard on lungs. Coal produces black lung, cotton mills cause brown lung, farmers get a lethal allergic syndrome called farmer's lung, and tunnel cutters were dying of silicosis before fiberglass was invented.

Epidemiologically, families of asbestos workers, residents of towns near asbestos dumps, and others exposed to asbestos at high levels (even if only briefly and/or in childhood) are known to be at increased risk of lung cancer and of mesothelioma. Environmental exposures of unknown significance include the use of patching and spackling compounds; drinking water from asbestos-lined pipes; the presence of asbestos fibers from insulation in airflow passages, ceilings, and walls; asbestos fibers released into air from brake lining wear; and drinking water containing asbestos fibers (notably, but not exclusively, water from Lake Superior, from Reserve Mining's dumping of asbestos-contaminated taconite tailings). Many people have, of course, had significant exposure to asbestos without realizing it: mechanics who repair brakes, hobbyists who polish the mineral serpentine, homeowners who patch their own plaster, and gawkers at construction sites where asbestos was being sprayed. The full toll will not be known before 2020.

---

[25] However, there is little evidence that spun glass (fiberglass) is carcinogenic, and the substance has been in use long enough that the absence of evidence is probably evidence of absence. It is suggested that spun glass may be less hazardous than asbestos because the fibers are not crystalline (i.e., that surface characteristics matter). Alternatively, it has been suggested that glass dissolves with time, and so is less "irritating." These are only speculations at present. Moreover, it is still advisable to wear a dust mask while applying fiberglass insulation, since silicosis is also a form of emphysema.

5

WATER POLLUTION, PERSISTENCE, AND BIOACCUMULATION

Sweet, sweet, sweet, O Pan!
  Piercing sweet by the river!
. . . The sun on the hill forgot to die,
And the lilies revived, and the dragon-fly
  Came back to dream on the river.
. . . The true gods sigh for the cost and pain,
For the reed which grows nevermore again
  As a reed with the reeds in the river.

—Elizabeth Barrett Browning[1]

INTRODUCTION

Water pollution has proven to be a very serious and very visible form of environmental contamination. Some incidents are given nationwide publicity, as when the Cuyahoga River "burned" in 1969, when Lake Erie was declared dead in the 1960s, and when the acidification of lakes in the Northeast first became known. More often, however, water pollution causes anxiety when people find out that their tap water not only smells bad but is unsafe to drink or even to wash in. Anger follows the realization that the stream one waded in as a child is now too vile to walk near, and that signs warn against eating the fish that still swim in a favorite lake. Too rarely, local opinion is mobilized because an industry threatens the water supply, even though no damage has yet occurred.

---

[1] *A Musical Instrument* by Elizabeth Barrett Browning, from *The Brownings: Letters and Poetry*, C. Ricks, ed., Doubleday Books, New York, 1970.

HISTORY

Water pollution is hardly a new problem. One of the earliest public health successes was the 1849 case of the Broad Street Pump in London. A physician, John Snow, recognized that a cluster of cholera cases shared a single common experience: all the victims got their water from a single pump. He removed the pump handle and ended the epidemic. This first successful use of epidemiology to control disease actually preceded the germ theory of Pasteur. Once the germ theory was validated, protection and purification of water supplies decreased the incidence of typhoid, cholera, and infantile and summer diarrhea to the vanishing point in developed countries.[2] Almost certainly clean water, even more than antibiotics, is responsible for the increased life expectancy in Western countries. It is so normal, and we are all so used to clean water coming out of every tap, that we are in danger of forgetting both how recent and how fragile the victory over waterborne diseases is. Moreover, secure in the efficacy of water treatment methods against *bacteria,* we have neglected to protect the sources of that water from *chemical* contamination.

To a large extent public health measures to protect drinking water have been, and continue to be, aimed at the infectious diseases of past centuries. Although it would be disastrous to decrease precautions against waterborne diseases, these measures are no longer enough. In the nineteenth century, population density in the United States was low enough that nearby rivers supplied water to much of the population, while those living near polluted rivers could dig individual wells. Increasing population density led to large-scale reservoir development, epitomized by the magnificent aqueducts that carry water from the Catskill mountains to New York and from northern California to Los Angeles. Elaborate precautions preserve the purity of the source reservoirs. Swimming, boating, and even fishing are carefully controlled. In the Catskills, private ownership of land is subject to severe restraints above certain altitudes in order to maintain the necessary watersheds to supply New York City. In California, elaborate systems of canals, viaducts, and regulatory bodies supply 30,000,000 people and the fields that produce half the fruits and vegetables of U.S. agriculture. For many towns in the water-rich midwestern farm belt, on the other hand, reservoirs are relatively shallow lakes surrounded by fields, and towns all over the United States rely on small well fields to serve residents. Rural residents depend on individual wells. All of these water sources are subject to contamination

---

[2] In the Third World waterborne diseases still take their toll. The seventh global cholera epidemic began in Africa in the 1960s and has now reached Peru. Even though dehydration, the lethal symptom of cholera, can be controlled, 10% of cholera victims in Africa die. Childhood diarrhea, especially in malnourished children, is a leading cause of death in sub-Saharan Africa and in any country torn by war, such as Iraq.

from agricultural, industrial, and private uses of the surrounding or overlying land.

State and county health officials monitor both private and public wells for coliform bacteria, which are accepted indicators of biological contamination. In agricultural areas, warnings are sent to pediatricians when fertilizers leaching into reservoirs raise nitrate levels high enough to cause methemoglobinemia in infants. Because of the enormous numbers of chemicals involved chemical contamination is more difficult to identify. Chemical assays are rarely undertaken unless there is a major spill, or chemical taste or odor is present in tap water. All too often the occurrence of a cluster of disease cases is the trigger that leads a population to question its water supply. All too often it is local residents who must not only identify the cluster, but badger local and state officials to acknowledge its existence.[3]

In the past few years there have been enough documented cases of chemical pollution of wells, aquifers, and rivers to demonstrate the need for more wide-ranging monitoring of drinking water sources. As can be seen in Figure 5.1, water can be contaminated by airborne pollutants that are carried in rain, by soil-deposited pollutants that are washed into rivers, lakes, and/or groundwater, and by direct discharge into bodies of water. This has always been the case. Natural pollutants—whether bacterial or chemical—have always been released into air, soil and water, and have always found their way into water. The tanneries of the Middle Ages produced wastes as surely as do the tanneries of today, and released them with far fewer controls. In previous ages, however, only mineral wastes were persistent. It was perhaps not always known that one should not take drinking water from the stream just downstream of a neighbor's privy, but taste alone warned one not to take drinking water from the stream directly below the tannery. It was also a commonplace that rivers cleaned themselves. Far enough downstream of the factory and the privy, the water was clean again. Until the 1950s, it was thought that 10 miles sufficed to clean a river.

As chemical production increased after World War II, the same principles were applied to the new organic chemicals as to their precursors. Rivers clean themselves; the solution to pollution is dilution; if you don't see, smell, or taste it, it can't hurt you. In addition, because the new miracle chemicals—plastics, pesticides, solvents, and household cleaners—were not as acutely poisonous as older materials, people stopped worrying about chemicals as poisons. The generation born since World War II has never known the toxic chemicals that our ancestors dealt with: the corrosive lye used to produce home-made soap, coal gas (CO) used in heaters and ovens, so that a leak was lethal; nicotine extracted from tobacco for use as an

---

[3] Case histories are presented in numerous popular books, including *Hazardous Waste in America*, S. S. Epstein, L. O. Brown, and C. Pope, Sierra Club Books, San Francisco, 1982; *The Politics of Cancer*, S. S. Epstein, Sierra Club Books, San Francisco 1978; *Malignant Neglect*, J. H. Highland and R. H. Boyle, Vintage Books, New York, 1980.

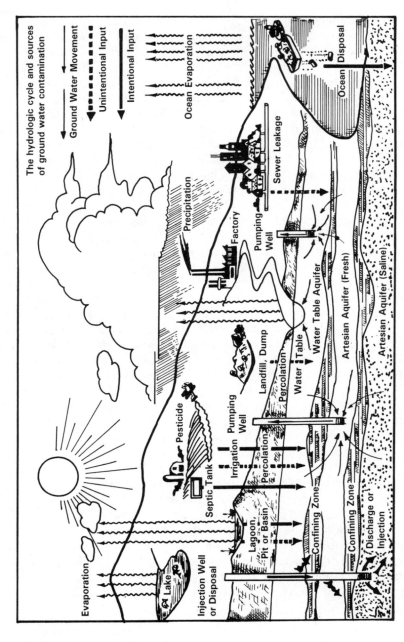

Figure 5.1 Hydrological cycle. (From *EPA Journal* 12(7):16–17, 1986.)

insecticide. The generation that developed DDT had grown up with arsenical weed killers toxic enough to be a mainstay of real as well as fictional murderers. An insecticide that could be sprayed on the skin was obviously harmless.

DEGRADATION AND PERSISTENCE

The major routes of *environmental degradation* are hydrolysis, photodegradation, and biodegradation. Hydrolysis occurs only in the presence of water: one reason why substances last longer in dry climates. Photodegradation occurs primarily in the presence of UV light, and biodegradation is primarily by microorganisms, although scavengers such as beetles, vultures, and hyenas also contribute. The need of all living organisms for water is the other reason that substances last longer in dry climates: fewer microorganisms mean that there will be less biodegradation. *Persistence* is the result of resistance of a substance to attack by these mechanisms. Because water is an integral part of both the atmosphere and of living organisms, persistence is very directly and inversely related to water solubility. As a first approximation, water solubility is inversely related to lipid solubility, so persistence tends to be positively correlated with lipid solubility. The reactivity of a chemical is also a factor, of course. Finally, the concentration of a chemical may alter its persistence. Many chemicals can be degraded reasonably well in dilute solutions or applications but are very persistent at high levels. Such increased persistence of concentrated chemicals is especially true when soil bacteria break down a chemical.

The modern "age of chemicals" that began after World War II was almost immediately marked by widespread environmental pollution, to a considerable extent because the new organic compounds seemed to be chemically inert (PCBs, plastics) or astonishingly less toxic than their predecessors (DDT). The types of effects that occurred can best be described by examples.

UNFORESEEN EFFECTS OF PESTICIDES

Persistence and Bioaccumulation: DDD

Clear Lake is a popular tourist area as well as a major breeding ground for the Western Grebe (*Aechmophorus occidentalis*), a particularly graceful, large water bird. It is also the home of the Clear Lake gnat, an insect that is quite harmless but so abundant that it is a major nuisance. Clear Lake was treated to control the gnats in 1949, 1954, and in 1957 at progressively higher levels of DDD (Figure 5.2), an insecticidal analog of DDT. The results provide an excellent example of the interactions among species in an ecosystem.

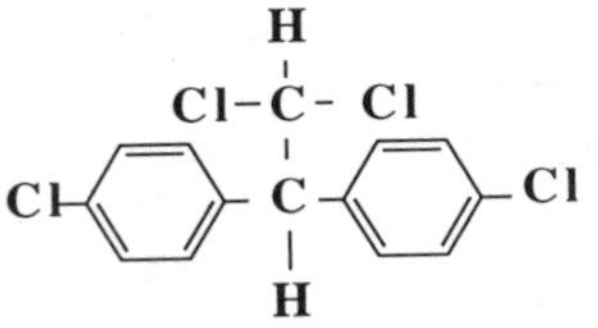

$$Cl-\underset{\underset{H}{|}}{\overset{\overset{H}{|}}{C}}-Cl$$

Cl—⟨benzene ring⟩—C—⟨benzene ring⟩—Cl

Figure 5.2 Structure of DDD.

After the 1949 spraying, water levels of DDD reached 14.2 ppb. No adverse effects were noted. DDD is a persistent insecticide, and gnat levels were suppressed for some years, but gradually increased again at Clear Lake. In 1954, DDD was again sprayed. Water levels of DDD reached 20 ppb. Although higher than after the first spraying, this concentration is still trivial in terms of animal toxicity or even of insecticidal activity. There were grebe deaths noted after the 1954 spraying, but no illness was found to account for the deaths, and no connection to the DDD spraying was made. In 1957 gnat levels had again increased to the point that spraying was considered advisable (note the decreasing intervals between sprayings, which suggest that the gnat population was developing resistance to DDD). This third spraying led to massive grebe deaths. Before 1949, the Clear Lake grebe population had consisted of approximately 1000 pairs of breeding birds; by 1960 it had decreased to 30 pairs, and even these had produced no young since 1957. Again, no illnesses were found. When assays for DDD were carried out, water levels remained in the range of parts per billion. Levels in organisms were startlingly higher. DDD levels were 5 ppm in plankton, 40 to 300 ppm in plant-eating fish, and as high as 2500 ppm in the fat of one carnivorous fish. Grebes had up to 1600 ppm DDD in their fat. These data illustrate the concept of *bioaccumulation*: fat-soluble compounds (which are relatively water-insoluble compounds) that are not readily degraded will accumulate in fat and are found at increasing levels in organisms as one progresses up a food chain (Figure 5.3). The kinetics of bioaccumulation are discussed in Chapter 7.

The saga of the Clear Lake grebes was repeated many times in the 1950s and early 1960s. The insecticide was not always DDD or even its close analog DDT. Several other *organochlorine insecticides* had been developed to compete with DDT, and each was lauded for its long persistence in the field, which meant that farmers had to treat less often, saving time and money. Aldrin, dieldrin, chlordane, and heptachlor were the most heavily used of these compounds. In time, each was shown to bioaccumulate, and each was eventually banned, as described in the case history on the cyclodiene insecticides (see below). Unfortunately, the lesson that was learned from the experience with DDD was not that *either* undue persistence *or* bioaccumulation was hazardous, but only that *bioaccumulation* was. Not until several organochlorine insecticides had been banned were factors other than bioaccumulation recognized as undesirable. The following cases illustrate

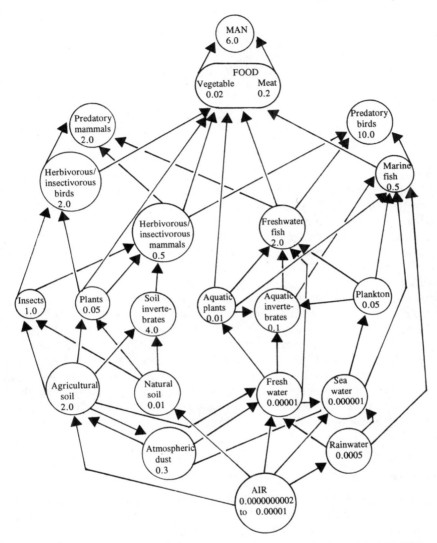

Figure 5.3 Typical transport and food chain concentrations in ppm of DDT (and DDD and DDE) residues. (Redrawn from C. A. Edwards, *Persistent Pesticides in the Environment,* CRC Press, Boca Raton, Fla., 1973, p. 1.)

several patterns of toxic effects observed as a result of the indiscriminate use of persistent pesticides that marked the post–World War II decades.

Persistence and Acute Toxicity: Endrin

Endrin (Figure 5.4) is an organochlorine insecticide, belonging to the family of insecticides designated the *cyclodienes* because they are generated by the

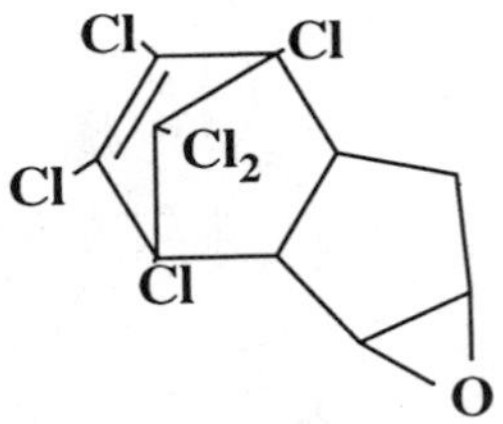

Figure 5.4 Structure of endrin.

Diels–Alder condensation reaction. Endrin is very closely related to aldrin and dieldrin and somewhat less closely to chlordane and heptachlor. Endrin is both persistent and bioaccumulative, but unlike DDD it is also very acutely toxic to vertebrates. It is so toxic to mice that it has been used as a rodenticide. Its toxicity to fish is so great that numerous fish kills occurred whenever endrin was used or produced near bodies of water. Endrin use east of the Mississippi was banned because it was considered impossible to keep its levels in waterways low enough to prevent fish kills. To get that ban, however, EPA had to compromise: endrin use west of the Mississippi was allowed, since it could be argued that there were fewer waterways in the high plains and fishing was less important. In 1981, endrin was used west of the Mississippi against a severe infestation of grasshoppers, which threatened the wheat harvest of the Great Plains. After widespread spraying, endrin bioaccumulated in waterfowl, leading to a ban on hunting of ducks and geese in Dakota, as well as causing numerous bird deaths. Public health authorities calculated that a person could be poisoned by eating normal portions of meat from birds that had eaten contaminated insects, even though the birds had not died.

Persistence without Bioaccumulation: Aldicarb

Aldicarb (Figure 5.5) is a carbamate pesticide produced under the brand name Temik. It was banned in Wisconsin potato fields after contaminating groundwater.[4] Compared to the organochlorine insecticide DDD, aldicarb is quite water-soluble (water solubility >1ppm). It shows little bioaccumulation in laboratory studies, but it is highly toxic to mammals as well as to insects. After it appeared in the water in Wisconsin, it was also found in the groundwater of Long Island, where it was also used on potato fields. When wells on Long Island were shown to contain aldicarb at levels as high as 0.5 ppm (which is the LD_{50} to fish), aldicarb use on Long Island was banned. In Florida, aldicarb was used in orange groves as a soil treatment against nema-

---

[4] L. McWilliams, "Groundwater pollution in Wisconsin: a bumper crop yields growing problems," *Environment* 26(4):25, 1984.

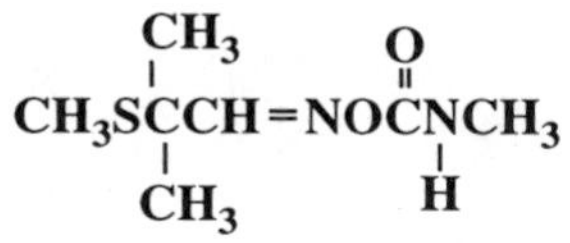

$$CH_3 \quad O$$
$$CH_3SCCH = NOCNCH_3$$
$$CH_3 \qquad H$$

Figure 5.5 Structure of aldicarb.

todes. Massive groundwater contamination occurred after heavy use in 1983–1984. In all three states, aldicarb had been used according to the label[5] but had been applied to soils that were sandy and leached readily. In 1985, aldicarb contaminated watermelons in California, leading to illness for over 1000 people and to the destruction of a sizable fraction of the state's watermelon crop. This last incident was probably due to illegal use of aldicarb on watermelon fields rather than to unintentional environmental contamination. It must be stressed, however, that the contamination was not direct—aldicarb was not sprayed *on* the watermelons. Only the soil was treated, since aldicarb is a systemic insecticide which is picked up by plants and carried into many parts of the plant, including the fruit. Aldicarb did not bioaccumulate in the plants, either: it was present in plant water at levels no higher than its concentration in soil water. Nonetheless, 1000 people became ill—a rather dramatic illustration of aldicarb's toxicity.

Aldicarb pollution results primarily from its persistence. Even though its relatively high water solubility prevents bioaccumulation, its long persistence underground provides ample opportunity to reach, and contaminate, groundwater. Because of its toxicity to vertebrates, even low levels of contamination suffice to cause health effects.

Acute Toxicity without Undue Persistence: Fenthion

Fenthion (Figure 5.6) is an organophosphate insecticide that is neither terribly persistent like endrin nor as toxic to mammals as aldicarb. It does not bioaccumulate like mirex or DDD, and is favored because it is not extremely toxic to mammals. It is, however, very toxic to birds and is licensed as an avicide. When fenthion was used to poison starlings in central Illinois in

---

[5] A *pesticide label* is a legal document that identifies permissible uses of a product. Using a pesticide on other crops or for uses other than permitted on the label is illegal. Unfortunately, even deliberate violation of the law is only a misdemeanor, which has led to flagrant abuses of federal and state pesticide laws. In 1992, over 80 restaurants in Chicago and its suburbs were sprayed with parathion-methyl by an unlicensed applicator. Parathion-methyl has never been labeled for indoor use; the applicator had been warned at least once and had then disappeared, together with his 55-gallon drum of parathion-methyl. There was evidence that he had crossed into Indiana and was spraying restaurants there. Lawyers for the city of Chicago thought that the egregiousness of the violation *might* actually lead to a jail term. But the maximum term in jail for a misdemeanor is 364 days, even under federal law. Neither lawyers nor city health officials thought the penalty would deter the applicator.

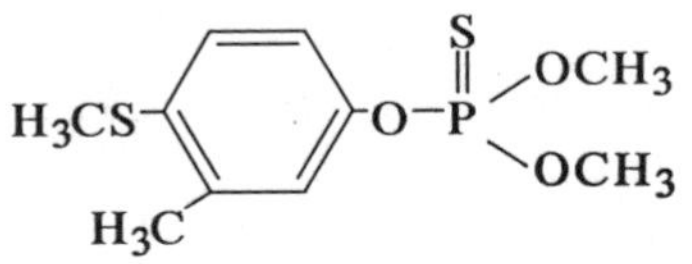

Figure 5.6 Structure of fenthion.

1986, owls, falcons, and possibly some predatory mammals died from eating the carcasses of fenthion-poisoned birds. The problem with fenthion is that it is extremely toxic both to the seed-eating pest species (starlings, blackbirds) at which it was directed and to the highly desirable predatory species (owls, hawks) that prey on the pest species. Moreover, in avicidal applications fenthion is applied to roosting surfaces, and absorbed through the skin of the feet. Thus it is possible for predators to be exposed to the unabsorbed toxicant on the bird's feet, as well as to the absorbed toxicant. It is not known whether unabsorbed material played a role in this episode of secondary poisoning.

Pesticides as Hazardous Wastes

Even when pesticides have supposedly been disposed of, they can still cause problems. Of the many hazardous waste sites that have attracted media attention in the past decade, most have involved waste from pesticide production. Like many of the problems with which our society is now grappling, this was also foreshadowed within a decade of the first use of DDT. The Rocky Mountain Arsenal near Denver, Colorado, produced chemical warfare agents until 1949; it was then sold to a chemical company, which continued to manufacture toxicants on the site. In 1951 it became apparent that groundwater was contaminated for a radius of several miles. In 1959, tests showed arsenic, phosphonates (nerve gas and/or pesticide derivatives), chlorides, and chlorates in the groundwater. The herbicide 2,4-D was supposedly not manufactured at this arsenal, but nevertheless appeared in the groundwater, apparently generated in situ. Contamination of groundwater by the various chemicals led to contamination of shallow wells used for irrigation, to illness in livestock, and to reductions in crop yields. The U.S. government filed suit against the chemical company, claiming that much of the pollution is due to pesticide production rather than to chemical warfare agents. The company contested this, and a settlement was eventually reached. Cleanup had not been completed in 1992.

Summary

Since the Clear Lake study of 1960 the examples of bioaccumulation and of persistence of chemicals have multiplied: DDD, PCBs, aldrin/dieldrin, chlordane/heptachlor, dioxins, Kepone. Contamination by persistent chemicals must be suspected not only in areas of use, but also where the chemicals were manufactured and where the by-products of their manufacture are dumped. Thus the concern about mirex contamination in Mississippi, where it was most heavily used as an insecticide, is mirrored by the concern of Canadian health authorities, who still find measurable residues of mirex in fish near factories where mirex was synthesized as a flame retardant. PCBs contaminate numerous waterways (see the PCB case history) and seriously damaged reproduction of mink on ranches along the Great Lakes. The closely related PBBs (see the PBB case history) were dispersed throughout Michigan via an accident that was followed by the determined efforts of the responsible organizations to minimize their liability.

The frequency and the irreversibility of such incidents of environmental pollution make it clear that we cannot depend on postmarketing experience to discover the environmental hazards associated with chemicals produced by the kiloton. A "quick test" or short-term assay is as badly needed for environmental pollution as for carcinogenesis, and for the same reason: the latency period between release of the chemical and identification of the hazard is so long, and the cost of repairing the harm so high (if it prove possible at all), that prevention is the only possible solution. Such a short-term assay can be found in the *microcosm*.

PREDICTING ENVIRONMENTAL BEHAVIOR: THE METCALF MODEL ECOSYSTEM

Microcosms, also called model ecosystems, can be designed to evaluate the toxicity, transport, or persistence of chemicals within a simplified ecosystem. The system can be terrestrial and/or aquatic, it may or may not measure volatility, and it may be either static or dynamic. Its cost, obviously, varies with its complexity. The degree of complexity, in turn, varies with the purpose of the microcosm. Model ecosystems in use today sometimes consist of "mesocosms," often large outdoor pools or even ponds, which are used to investigate the dynamics of ecological systems, and are correspondingly complex. The first microcosms, in contrast, were designed to evaluate the fate and transport of toxicants, especially of pesticides, in soil and water, and to do so quickly, cheaply, and repeatably. This first microcosm was the Metcalf model ecosystem, called the *terrestrial aquatic laboratory model ecosystem* by its originator, R. L. Metcalf.

The Metcalf model ecosystem (Figure 5.7) simulates a farm pond surrounded by fields under cultivation. It consists of a 10-gallon aquarium containing a sloping shelf of washed quartz sand as its "field" and a "lake" of 7 liters of standard reference water. The water is aerated and provides mineral nutrition for plankton, algae, snails, mosquito larvae, and fish in the aquatic phase and for sorghum plants in the terrestrial (sand) phase. The entire system is kept in an environmental growth chamber at 26.5°C with a

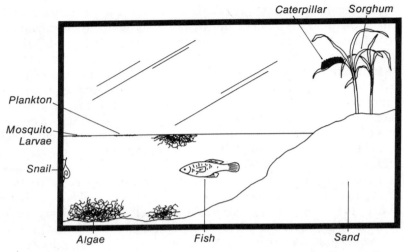

Figure 5.7 Metcalf model ecosystem: a terrestrial–aquatic model ecosystem modeling a pond surrounded by cultivated fields. (Courtesy of the Institute for Environmental Studies, University of Illinois.)

12-hour diurnal cycle of 50,000 foot-candles of fluorescent light to mimic summer light and temperatures.

The chemical to be tested is preferably, but not necessarily, radiolabeled[6] to simplify monitoring. The chemical is applied to the sand or to the sorghum plants using 1 to 5 mg of test chemical per experiment. This is equivalent to a realistic level of 0.2 to 2.0 lb/acre for pesticides and can be altered when higher or lower levels of a compound are known to be applied. Ten last-instar saltmarsh caterpillars (*Estigmene acrea*) are introduced to eat the treated sorghum. The caterpillars and their excretory products contaminate the lake portion of the ecosystem. The radiolabeled products enter the several aquatic food chains:

plankton → daphnia → fish
algae → snail

The movement of radiolabeled compounds from plants to water is measured by counting the radioactivity of duplicate water samples at specified intervals. Twenty-six days after treating the ecosystem, 300 mosquito larvae are added. Three days later, 50 larvae and 50 *Daphnia* (also called "water

---

[6] The radiolabel consists of incorporation of a radioactive isotope, either ^{14}C or ^{3}H, into the structure of the chemical being evaluated. The isotope emits particles that can be counted in a scintillation counter in order to quantify the amount of the chemical present in each component of the system. In the absence of radiolabeled compounds, gas chromatography/mass spectrometry can be used to identify the chemical, but this is far more tedious.

fleas") are removed for analysis, and the food chains are completed by adding 3 mosquito fish (*Gambusia affinis*). Three days after the addition of fish the system is terminated (day 32). All organisms are weighed, homogenized, and extracted with an organic solvent to obtain the lipid-soluble radio-labeled products. The water from the system is also extracted and analyzed. Aliquots of all samples are counted for total radioactivity by scintillation. Then the extracts are concentrated to a few milliliters and applied to thin-layer chromatography (TLC) plates. TLC is carried out with solvents appropriate for the chemical being tested, and the chromatograms are placed against x-ray film and exposed for 1 to 2 months to determine the regions containing the radiolabeled products. These areas of the TLC plates are then scraped into scintillation vials and the amounts of the individual degradation products determined by scintillation counting. It is possible to extend the analysis to the identity of key degradation products if the work is warranted by their importance.

The *ecological magnification* (EM) and *biodegradation index* (BI) of the parent compound (and of degradation products) can be calculated as:

$$\text{EM} = \frac{\text{concentration of parent compound in the organism}}{\text{concentration of parent compound in water}}$$

and the degradative index is

$$\text{BI} = \frac{\text{concentration of polar metabolites}}{\text{concentration of nonpolar metabolites}}$$

The Metcalf model ecosystem was not designed to identify the toxicity of chemicals to particular species or particular food webs. Its purpose is to identify the environmental fate, transport, and persistence of environmental pollutants, particularly of pesticides. The major weakness of the Metcalf model ecosystem is that it is static, and so does not adequately model stream and other flow-through systems. For example, a fish living in a DDT-treated stream is exposed not only to a fixed amount of DDT from spray "fallout" but also to increments due to freshly contaminated water that flows downstream. The total DDT uptake may be much higher than it would be for fish in a pond, with a fixed amount of water and no additional inflow of DDT-contaminated water. The Metcalf model ecosystem models the pond. Once the ecological characteristics of a chemical are known, however, it is relatively easy to translate "pond" behavior into "stream" behavior.

Against this minor weakness are set the advantages of the Metcalf ecosystem: it is quick, cheap, and reproducible. Since its design in 1965, over 150 chemicals have been assessed in this system. For many of these chemicals, real-world data on persistence and biodegradation are available, often under both static and flow-through conditions. Therefore, remarkably accurate

**1 Ecological Magnification of Selected Organic Chemicals
in Two Organisms**

Chemical	EM (Fish)	EM (Snails)
Carbaryl	0.32	3.5
Mirex	219	1,597
Kepone	118	637
DDT	13,550	5,725
Aldicarb	42	—
PCBs	83.40	
Trichloro (2,5,2′)	—	—
Tetrachloro (2,5,2′,5′)	753	—
Pentachloro (2,5,2′,4′,5′)	12,152	—
Methidathion (insecticide)	51.30	1.94

Source: Data from R. L. Metcalf and J. R. Sanborn, *Illinois Natural History Survey Bulletin* 31:379–436, 1975; R. L. Metcalf et al., *Archives of Environmental Contamination and Toxicology* 3:151–165, 1975; and B. M. Francis and R. L. Metcalf, *IES Research Report 9*, 1981.

predictions can be made with respect to the fate of new compounds by comparing data to ecosystem and to field data for the other 150+ chemicals that have been evaluated. Moreover, the system is reliable even when metabolites of parent compounds are evaluated. For example, desbromoleptophos, which is a photodegradation product of the organophosphorus ester insecticide leptophos, had EM = 3888 in fish when calculated in the leptophos ecosystem, and EM = 3477 when evaluated in its own system. Examples of EM values for several chemicals are shown in Table 5.1.

CASE HISTORY: POLYCHLORINATED BIPHENYLS

There are 210 possible isomers that are considered to be "polychlorinated biphenyls" (PCBs; Figure 5.8) if one includes (as the law does) monochlorinated biphenyls. The bulk of commercial products consists of about 50 isomers, however. PCBs were first produced in 1929 as Aroclors. Commercial mixtures of polychlorinated biphenyls were designated by four digit numbers, of which the first pair [12] designated the category of PCBs, and the second pair of digits designated the approximate percentage of chlorine: for example, Aroclor 1242 has 42% chlorine, equivalent to the composition $C_{12}H_7Cl_3$ and is a liquid with a boiling point of 380°C,

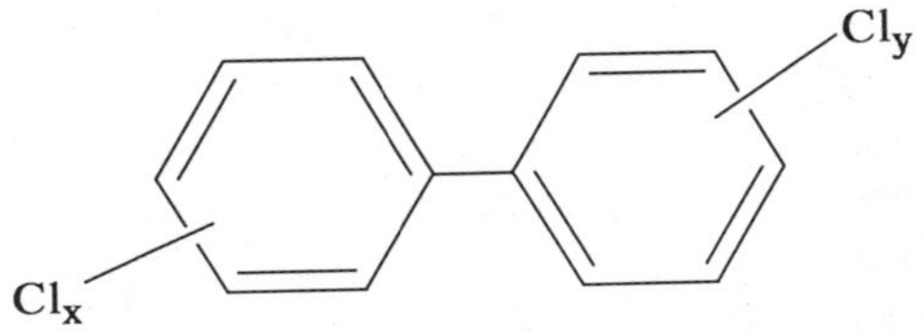

Figure 5.8 General structure of PCBs.

while Aroclor 1260 is 60% Cl, equivalent to $C_{12}H_4Cl_6$ and is a solid. Individual PCB isomers can differ sharply in their persistence, with the monochlorinated compounds generally being quite biodegradable. PCBs with three or more chlorines on each phenyl ring are generally quite persistent, highly lipophilic (soluble to 10% in fat), very water insoluble (ca. 10 to 50 ppb), and extremely stable in the environment. Commercial names for PCBs include Aroclor (United States), Phenoclor (France), Fenclor (Italy), Sovol (Russia), Kaneclor (Japan), and Clophen (Germany).

PCBs were developed as high-boiling-point, stable heat-transfer fluids for use in transformers and heat exchangers. Subsequent uses included hydraulic fluids, plasticizers, carbonless carbon paper, inks, lubricants, waxes, cutting oils, and adhesives. Of U.S. production from 1930 to 1970, approximately 50% was used in capacitors and transformers and 20% as plasticizers. In 1970, at the peak of production, U.S. production reached 85×10^6 lb. World production that year was 200×10^6 lb. Cumulative U.S. production is thought to have exceeded 1×10^9 lb. After 1970, PCB use was restricted to closed systems such as capacitors. Increasing evidence of their environmental bioaccumulation led to suspension of all uses by 1977. Nonetheless, because of the long life span of PCB-containing capacitors, there are still many units containing PCBs in use.[7]

Release to the Environment. An estimated 55×10^6 lb of Aroclor was lost to the environment each year that PCBs were used. Of this, 80% was released to the atmosphere during the burning of paper, plastic, or paint, and 20% was released to fresh and coastal water by leaks, disposal of industrial wastes, leaching, and atmospheric fallout. This made PCBs the first industrial (nonpesticidal) products to be identified as ubiquitous environmental pollutants.

Biological Effects. PCB residues in human fat averaged 1 to 2 ppm; in milk, 0.5 ppm, mostly in the butterfat. Fish in Lake Michigan contained as much as 10 to 15 ppm. Bioaccumulation is widespread, with PCB levels in shark liver reaching 218 ppm, while brown pelican fat can contain 266 ppm. In Lake Superior, the bioaccumulation from water to fish was 10^6, or 1 million-fold.

Precise information on the effects of PCBs on organisms, or even their degradation and bioaccumulation, is hard to obtain because different mixtures have quite different characteristics. When chickens were fed Aroclor 1242, 11% accumulated in fat; if Aroclor 1254 was fed, 30% accumulated; and if Aroclor 1260 was fed, 51% accumulated in fat. Moreover, within each mixture, the more highly chlorinated isomers are selectively sequestered in the fat as the less chlorinated isomers are hydroxylated and excreted. In aquatic ecosystems, the most highly chlorinated congeners are relatively immobile in sediments, while the least chlorinated conge-

---

[7] In 1991, a laboratory at a major university was badly contaminated when students were assigned to remove leaking fluids from a transformer. The university had bought the units used and had asked the vendor if there were PCBs in them. The vendor apparently assumed that there were not. Prudence dictates that in the absence of definite information about date of manufacture, the presence of PCBs in transformers should be assumed. Enormous cleanup costs and the contamination of students, who may face possibly serious long-term health effects, could have been avoided.

ners can be dechlorinated by microorganisms. Fish therefore tend to accumulate PCBs with 4 to 6 chlorines. Unfortunately, some of the most toxic isomers fall into that range.[8]

In birds, *chick edema disease* is caused by the presence of 10 to 20 ppm PCBs in poultry feed. This syndrome includes enlarged livers, kidney necrosis, decreased egg production, decreased hatchability of eggs, and developmental defects in those chicks that hatch. The malformations include crossed or short beaks and misshapen limbs. There have been repeated episodes of poisoning in poultry over the decades of PCB use, and further losses to farmers have resulted from the destruction of flocks to prevent sale of contaminated meat. A 1978 episode required destruction of hundreds of thousands of chickens, as well as destruction of eggs from contaminated laying hens. The source of that episode was fish meal, contaminated in a fire in a Puerto Rico warehouse, which was shipped to Texas, Idaho, and Arkansas. The chicken feed contained up to 62 ppm of PCBs (0.2 ppm is the legal maximum).[9] What are now recognized as PCB-induced malformations have been seen in birds on the Great Lakes and in eastern shorebirds since the late 1960s, and are still occurring in Great Lakes birds.

The mammalian species with the greatest sensitivity to PCBs is mink, in which dietary levels of 5 ppm cause complete reproductive failure. In contrast, 100 ppm in the diet of rats causes only moderately decreased survival of offspring, and 10 to 20 ppm of PCBs in the diet of farm animals results in liver damage and reproductive toxicity. Rats are sensitive to other aspects of PCB-induced toxicity, however, since even 0.5 ppm in the diet will cause microsomal induction in the liver.

In humans, the most obvious symptom of PCB exposure is chloracne, a severe form of dermatitis that may persist for years. At least 10 deaths due to liver necrosis and cirrhosis are known. The Japanese Yusho disease affected 600 people who ate rice oil that had become contaminated with PCBs. Symptoms included chloracne, melanosis, edema of the eyes, swelling and stiffening of the limbs, as well as headache and hearing difficulties. Fully half the victims still had symptoms after 3 years. Children born to Yusho mothers had a number of birth defects. The human effects of PCBs are not well delineated because in both the Japanese and the Taiwanese episodes of Yusho, some of the symptoms may have been due to heat-engendered chlorinated dibenzofurans. PCBs themselves are thought to be carcinogenic, probably acting as promoters (see Chapter 11).

Environmental Distribution. PCBs are ubiquitous contaminants of the environment because they were widely used for decades and because their manufacture included routine releases into the environment. Releases were essentially unregulated because PCBs were considered to be so inert. River systems where PCBs were manufactured are badly contaminated. Sources of river pollution include the

---

[8] Toxicity of PCBs is usually measured by their ability to cause death (acute toxicity), chloracne (clinical toxicity), or cancer (chronic toxicity). These endpoints correlate reasonably well with the fit of an isomer at the aryl-hydrocarbon (Ah) receptor, and there is some effort by regulatory agencies to consider that interaction with this receptor totally explains PCB toxicity. Recent studies indicate, however, that some PCBs are potent neurotoxic agents and that the neurotoxic effects are not correlated with fit on the Ah receptor.

[9] *Champaign–Urbana News Gazette*, April 12, 1978.

GE plant in Bloomington, Indiana, and several GE plants along the Hudson River in New York. Landfills often contain poorly sequestered PCBs, while spills have contaminated too many sites to track. Finally, buildings with transformers using PCBs as the heat transfer medium have had to be abandoned after relatively minor fires because of the production of dibenzofurans and dibenzodioxins when PCBs are heated.

It is possible—although expensive—to remove PCBs from the remaining closed sources, especially transformers, but cleaning up the existing environmental contamination is extremely complex as well as unimaginably expensive. Most environmental residues are soil associated. Soil is bulky, and the contamination is often only in parts per million. Moreover, soil can be burned only at tremendous cost in fuel. Because of the costs, and because there is literally no place to put contaminated soil (should it be placed in a landfill?), soil-associated PCBs will in most cases be left in place unless or until they are detected in leachate or nearby streams. This is especially true in landfills not known to be leaking, in smaller lakes and streams, and in many areas of the Great Lakes. The hazard posed by PCB residues in rivers and the Great Lakes will be declared solved by publishing advisories that warn about contaminated fish and that urge fishermen not to eat too much of their catch. For navigable waterways, neglect is not an option. Rivers and harbors need to be dredged periodically to keep channels clear for ships. Dredging churns up contaminated sediments, once again releasing buried PCBs to the water. Moreover, the dredged material has to be taken elsewhere, and this is difficult if the sediment is known to be contaminated with PCBs. For the Hudson River, it is estimated that merely dredging shipping channels and disposing of the sediments would cost over $3 million. Cleaning up *all* the contaminated sediments, however, would cost over $200 million. Since the partial cleanup involved in dredging channels and disposing of the sediments would have to be repeated at intervals, one can say that the eventual cost of the cleanup—assuming there were no inflation—would obviously exceed $200 million. And the Hudson River is neither the only nor even the worst example of PCB contamination.

CASE HISTORY: POLYBROMINATED BIPHENYLS

Polybrominated biphenyls (PBBs; Figure 5.9) are a mixture of highly brominated biphenyls, used mostly as a flame-retardant and fire-proofing material for plastics, synthetic fibers, and electrical products. The technical product is a solid with a melting point between 200 and 250°C, designated $C_{12}H_2Br_8$ (octabromobiphenyl). The average composition is

heptabromobiphenyl	1%
octabromobiphenyl	33%
nonabromobiphenyl	60%
decabromobiphenyl	6%

PBBs are very resistant to attack by light, hydrolysis, air, or enzymes. They are also very insoluble in water (0.028 to 0.030 ppm at 25°C) and have a very high

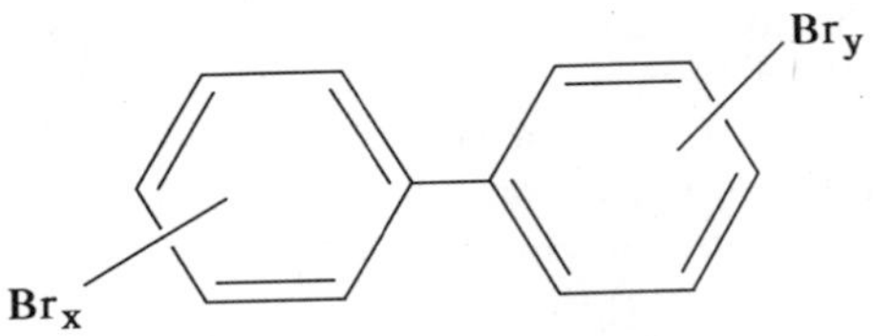

Figure 5.9 General structure of PBBs.

fat solubility, indicated by the high octanol–water partition coefficient of 340,000.[10] Identification of PBBs is readily made by electron-capture gas–liquid chromatography at high temperature and with short columns and neutron activation analysis, but only if their presence is suspected.

Production and Release. PBBs were produced from 1970 on by Michigan Chemical Company, St. Louis, Michigan, under the name Firemaster; annual production was approximately 5×10^6 lb. Uses included incorporation into typewriter, calculator, and business machine housings (50%); into radio and television parts, thermostats, electric shavers, and hand tools (30%); and into thermoplastics (15%). The remainder was used in other types of electrical equipment. In the ordinary way, release to the environment would have occurred slowly, as these items were discarded and the plastics disintegrated. The lifespan of small equipment is estimated to average 5 to 10 years, but the persistence of the plastic might be considerably longer, especially in landfills. Environmental contamination by PBBs was much greater than these uses suggest, however, because the manufacturer released an estimated 0.25 lb/day into the Pine River, which flows near the St. Louis plant, during ordinary manufacturing processes.

Extent of Environmental Contamination. Nonetheless, production of PBBs was a small-scale affair if compared to production of PCBs, and in their ordinary use PBBs might never have come to toxicologists' attention. There was, however, an accident. In 1973, several hundred pounds of Firemaster was substituted for Nutrimaster, a dairy feed supplement produced in the same plant. Nutrimaster was distributed to farmers throughout Michigan by the Michigan Farm Bureau. The substitution was not recognized for some time. When authorities realized that there was something wrong with the feed, they remixed the "bad batches" with clean feed ("the solution to pollution is dilution"), aggravating the long-term problem. Eventually, more than 15,900 cattle, 2700 swine, 1200 sheep, and 1,500,000 domestic fowl were destroyed and buried in a special landfill at Kalkaskia, Michigan. It is estimated that in 1980, 98% of the population of Michigan's lower peninsula carried PBBs in their fat. Body fat levels in the obviously affected Michigan animals averaged 110 to 2480 ppm, and milk levels in dairy cattle were 44 to 900 ppm. In 1974, FDA established permitted levels of PBB in meat and milk fat as 0.3 ppm; levels in eggs and feed were to be no higher than 0.05 ppm.

---

[10] J. M. Norris et al., "Toxicology of octabromobiphenyl and decabromobiphenyl oxide," *Environmental Health Perspectives* 11:153–161, 1975.

Biological Effects. The toxicology of the PBBs is not entirely clear but probably parallels that of PCBs. Acute toxicity is low: the rat LD_{50} value of the commercial mixture is greater than 2000 mg/kg, but high lipophilicity leads to extreme levels of bioaccumulation. PBBs are readily absorbed through the skin. Liver enlargement is seen in rats fed 10 to 100 ppm in the diet; reproductive effects are seen in rats when 100 to 1000 ppm are included in the diet. Human effects of PBBs are disputed. Both carcinogenicity and immune suppression were suggested by initial observations in exposed people and animals, but could not be proved to the satisfaction of lawyers, due to confounding factors in the epidemiological data. Symptoms seen in animals that ate contaminated feed in Michigan included weight loss, decreased milk production, loss of hair, grotesquely abnormal hoof growth, hematomas, reabsorption or spontaneous abortion of embryos, stillbirths, and neonatal deaths. Some of the symptoms seen in affected animals are probably unique to PBBs or may be due to unidentified dibenzofuran contaminants. Many of the effects that were identified by farmers and by the clinicians who worked with them were initially dismissed by toxicologists. These effects included psychological effects (depression, irritability, sleep disturbances) and increased susceptibility to infectious diseases. More recent evidence about the animal toxicology of PCBs confirms the plausibility of such effects. PCBs, and the more toxic dioxins, are now recognized as immunosuppressants. There is no reason to suppose that PBBs do not have similar effects.

Unlike the PCBs, PBBs did not become ubiquitous contaminants of the environment. Their more specialized uses and localized manufacture precluded the kind of pervasive, low-level contamination that has occurred with PCBs. The mixup and subsequent scandal prevented the exploitation of their indubitably useful properties to increase production. As a result, the major environmental damage occurred during manufacture, since the manufacturer neither prevented releases to the environment (the Pine River) nor protected employees against skin absorption. The inclusion of Firemaster in dairy feed was a genuine accident and allows only the conclusion that accidents are inevitable, especially when the hazards of chemicals are minimized.[11] The subsequent refusal of the Michigan Farm Bureau and officials to admit any error aggravated the consequences tremendously. Unfortunately, such "coverups" are probably as inevitable as accidents, since any organization attempts to minimize its liability. Finally, because officials did not identify the nature of the problem in the feed even when they finally admitted that there was something wrong with the feed, they remixed it with clean feed, thus spreading the PBBs into the entire Michigan agricultural pipeline. The high bioaccumulation potential of the PBBs made such dilution a disaster.

CASE HISTORY: CYCLODIENE INSECTICIDES

The most heavily used cyclodiene insecticides are aldrin and its epoxide dieldrin, heptachlor, and chlordane (Figure 5.10). Compounds with substantial but

---

[11] Among the factors conducive to a mixup were the similarity of the bags used to contain Firemaster and Nutrimaster; the similarity of the trade names; and the insistence by management that Firemaster was harmless.

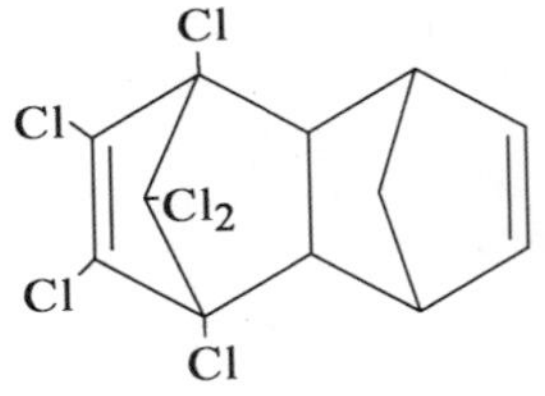

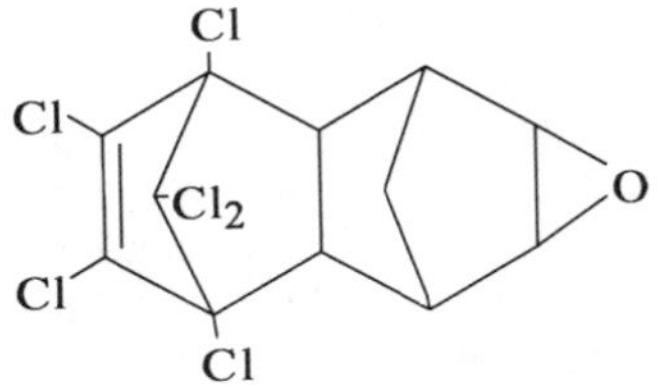

Figure 5.10 Structures of the cyclodiene insecticides aldrin and its epoxide, dieldrin. The structures of some other cyclodiene insecticides are shown in Figure 5.4 (endrin) and Figure 5.11 (mirex and chlordecone).

significantly smaller production included the closely related endrin and the somewhat different chlordecone and mirex. During the 1960s, production of cyclodiene insecticides reached 20×10^6 lb/yr. In Illinois alone, over 80×10^6 lb of aldrin, heptachlor, and chlordane were used as soil insecticides. In 1974, agricultural uses of aldrin and dieldrin were banned by U.S. EPA, although their use was still allowed for termite control. Within a few years, heptachlor use had been restricted to control of aphids on pineapple, endrin use was allowed only west of the Mississippi, and chlordane was permitted only for termite control. In the late 1980s, even these uses were under attack. Thus these compounds seem to be only of historical interest, at least in the United States. Nevertheless, their history is that of the modern age of pesticides and to avoid a repetition, should be remembered.

Uses.   The heaviest use of the cyclodiene insecticides was in control of soil-inhabiting insects, notably aldrin and heptachlor against the corn root worm larvae (*Diabrotica longicornis, D. virgifera, D. undecempunctata*), black cutworms, white grubs in lawns, and Japanese beetles; chlordane against termites and ants; and mirex against ants. Dieldrin was also used in mothproofing of both clothes and rugs, and for control of fruit tree insects.

Biological Effects.   The acute toxicity (LD_{50} to rats) of the cyclodiene insecticides is quite variable, ranging from 3 mg/kg for endrin to 335 mg/kg for chlordane (Table 5.2). All of the cyclodienes are neurotoxicants, causing violent convulsions in poisoned animals. Endrin is particularly toxic to fish and was implicated in numerous fish kills due to effluents released during its manufacture. Dieldrin, used in malaria and Chagas disease control, caused chronic poisoning in 2 to 40% of the workers applying 0.5 to 2.5% suspensions. Symptoms were indistinguishable from epilepsy and persisted for years. The variability in incidence of poisoning

Table 5.2 Acute Toxicity of Cyclodiene Insecticides in the Rat[a] (mg/kg)

Aldrin	40	Endrin	3
Dieldrin	46	Chlordane	335
Heptachlor	100	Mirex	235
Chlordecone	95		

Source: Courtesy of R. L. Metcalf.

[a] These values are from the World Health Organization Testing Programme. A range of LD_{50} for each cyclodiene, with sources, are given in many books, including the *NIOSH Registry of Toxic Substances*.

may well have been due to differences in clothing worn at different times and in different climates. When fed to sheep and monkeys, dieldrin caused behavioral abnormalities at chronic dosing levels of 0.1 ppm. Dieldrin and heptachlor epoxide are also highly carcinogenic. When mice were fed dieldrin at 0.1 ppm, 21% of males and 30% of females developed liver cancer. Human dietary levels of dieldrin during the 1970s were approximately 0.00008 ppm and could result in 20,000 to 200,000 cancers.

Persistence and Release to the Environment. The use of cyclodienes as insecticides assured their immediate release into the environment. Additional releases occurred during manufacture. Aldrin is rapidly converted to dieldrin, its epoxide, as is heptachlor to its epoxide, heptachlor epoxide. Persistence of these compounds in soil is measured in decades, and major losses from soil appear to be due to volatilization, leaching, or erosion rather than to degradation. In the body, dieldrin has a half-life of about 12 months. Dieldrin, present in Lake Michigan water at 0.000002 ppm (2 ppt) bioaccumulated to average levels of 0.26 ppm in lake trout. Both dieldrin and heptachlor epoxide are essentially universal contaminants of human fat, with average levels of 0.29 ppm dieldrin and 0.24 ppm heptachlor epoxide. In the late 1970s, human breast milk in east central Illinois contained dieldrin levels of 0.29 ppm, providing 7.5 times the WHO acceptable daily intake (ADI) for milk-fed infants.

The restrictions on cyclodiene use, beginning with the banning of dieldrin in 1974, were designed to minimize environmental and human exposure to these compounds. Several exemptions were permitted, however, notably the use of chlordane and aldrin as termiticides, the use of endrin west of the Mississippi, and the use of heptachlor to control aphids on pineapple in Hawaii. In each case, human exposure has resulted.

After most of its uses were banned in the United States, heptachlor use was still allowed in Hawaiian pineapple fields. Pineapple producers argued that heptachlor was needed to control aphids on pineapple. The mealybug or aphid is not itself harmful to the pineapple, but carries a disease, "mealybug wilt," that withers pineapple roots. The aphid could be quite easily controlled, except that it is protected from its natural predators by ants, which harvest the "honeydew" produced by the aphids. Ants also carry the aphid to new plants. To protect the pineapple crop, a series of insecticides was used to kill the ants. Heptachlor was the latest of these. Rigorous restrictions on the interval between application and harvest were designed by Hawaiian agronomists to prevent heptachlor residues from reaching the pineapple fruit. Meanwhile, university biologists very diligently sought a use for the tops of the pineapple (which are inedible) and eventually found that these could be turned into silage. They developed a mixture of pineapple tops and other plants that provided a nutritious diet for both beef and dairy cattle. The agreement was that this "chop" would not be made from pineapple tops within 1 year of spraying. Several lawsuits are now in progress to decide whose error led to the feeding of freshly sprayed pineapple tops to the Hawaiian dairy cattle. It is certain that feeding this "chop" to dairy cattle caused considerable contamination of milk, which was not recognized for months (or perhaps years) and was then hushed up as "harmless." As in Michigan after the PBB contamination, numerous cattle had to be destroyed, and massive human contamination occurred.

Aldrin, used as a termiticide, was overapplied and so thoroughly contaminated a house on Long Island, New York, that the owners had the house bulldozed. The owners did not consider that they could sell it in its contaminated state, since they themselves could not live in it without becoming ill. This episode of the 1980s is matched by less well publicized episodes in Illinois due to chlordane. The usual pattern is that the termite control results in contamination of heating ducts and/or cinder-block foundations, making removal of the chemical impossible. On occasion, contamination of dirt floors in crawl spaces or old-fashioned cellars has occurred. The cyclodiene persistence does the rest.[12]

CASE HISTORY: MIREX AND CHLORDECONE [KEPONE]

Other persistent pesticides related to the cyclodienes are mirex and chlordecone (Figure 5.11). Mirex is a cyclodiene insecticide used primarily in southern states against the imported fire ant (*Solenopsis invictus*). Most residents, and wildlife, in these states have body burdens of mirex, even though application rates were in grams per acre rather than pounds per acre. In the Great Lakes, where mirex was manufactured and where the manufacturers released effluents, fish have significant burdens of mirex in fat. It is interesting that although the major focus of regulators has been on the toxicity of the *insecticide* mirex, 90% of the amount produced was used as a flame retardant. Mirex is the least degradable of the organochlorine insecticides. Only a few degradative products are found. One is chlordecone (see below); another, photomirex, results from photolysis and is as long-lived as mirex itself. The half-life of mirex in mammals is estimated to be 10 years; in small mammals, this means that the half-life is longer than the life span of the animal.

---

[12] In treating a house for termites, it is important to block access of the insects into the house. Termites nest outside, using wood as food but not living in their lunch box. Thus a good termite treatment is one that lays a chemical barrier between the wood inside the house and the termite nest outside. Ideally, this is done without letting the insecticide into the house. In practice, applicators tend to treat the outside of the foundations *until the insecticide seeps through cracks in the wall,* since they then know that they have applied enough. Since most exterminators guarantee the elimination of termites, and since re-treating a house means a significant cost in materials and labor, exterminators have little incentive to treat properly. If this procedure contaminates a crawl space from which a forced-air furnace draws its air, or if heating ducts are breached, or if cinder-block walls "breathe" the insecticide into playrooms or below-grade living areas, health problems can easily result.

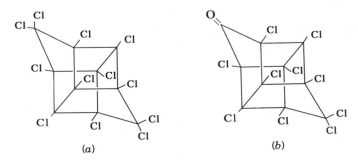

Figure 5.11 Structures of mirex and chlordecone.

Mirex is not terribly acutely toxic but is suspected of causing cancer. It causes birth defects (congenital cataract) in some species.

A close analog of mirex is chlordecone. This chemical, trade-named Kepone, is somewhat less persistent than mirex. It nonetheless figured in a massive industrial pollution in which the James River of Virginia was heavily polluted, and eventual pollution of the Chesapeake Bay is feared. Chlordecone was never a massively used insecticide in the United States. Developed by Allied Chemical Corporation in the 1950s, most chlordecone was exported to other countries. In 1974, Allied licensed a small company, Life Sciences Inc., to synthesize chlordecone. Although nominally an independent company, Life Sciences was founded and managed by former Allied employees, consisted of a single plant in Hopewell, Virginia, that produced nothing but chlordecone, and sold all of the chlordecone it produced to Allied Chemical Corporation. The courts eventually held Allied responsible for the pollution generated by Life Sciences.

In 1975 a worker sought medical help for symptoms, including tremor, irritability, and memory loss, and commented that others in the factory had the same symptoms. Chlordecone was identified in his blood. The Life Sciences factory was ordered closed within the month, but chlordecone remained a problem 10 years later. Life Sciences' chlordecone factory in Hopewell was on the banks of the James River and released chlordecone-containing effluents into the river both directly and via the Hopewell sewage system. The James River flows into the southern part of the Chesapeake Bay, one of the richest wetlands in this country and a major source of shellfish and fish. In 1976, fish in Chesapeake Bay contained from 0.02 to 0.09 ppm chlordecone, and fish caught off the coast of New Jersey contained 0.02 to 0.03 ppm. It is known that the sediments of the James River are contaminated and that leaching of chlordecone from the sediments continues, while the contaminated sediments may be carried toward the bay. Dredging to remove contaminated sediments was considered. Not only was the $7 billion cost of dredging prohibitive, but the probability that dredging would actually increase the amount of chlordecone reaching the bay made it inadvisable.

The human contamination with chlordecone included both workers in the factory and others in Hopewell, although clinical symptoms were seen only in workers. These symptoms included both reduced fertility and neurological symptoms. Many of the workers were essentially incapacitated by tremors. Because chlordecone has a prolonged half-life in animals, recovery was slow. Long-term effects have not been seen yet, but chlordecone is a carcinogen when fed to rats or mice at 10 ppm in the diet.

SELENIUM

Although most of the dramatic examples of water pollution involve synthetic chemicals, there are many natural sources of water "pollution." Water that contains dissolved salts is known as "hard" water if the salts precipitate during heating or leave a scum when soap is added. A major industry exists to make such water more pleasant to use. In the western United States ground water is often quite alkaline, and therefore bitter in taste, or even toxic. Near hot springs, water may have a pronounced odor of rotten eggs because sulfides are present; some hot springs were formerly valued for their radioactivity, the result of dissolved radon (see Chapter 4). Streams flowing over highly organic soils may contain high levels of phenolic compounds, lending a brown tint and unpleasant taste to the water. Such contamination of water with natural chemicals often occurs in the absence of human activity. In other cases, human manipulation of waterways leads to movement of minerals into water. One example of this phenomenon is the contamination of a wildlife refuge in California by selenium leaching from irrigated fields.

Selenium, with an atomic weight of 79, occurs naturally in soils in many parts of the world. Selenium can occur as the element, as selenate (+6), selenite (+4), and as the selenide (−2). Thus it has both metallic and nonmetallic properties and is a metalloid rather than a metal. Selenites and selenium itself are almost insoluble in water, as are selenides of heavy metals such as arsenic. Selenates, in contrast, are quite soluble.

Industrially, selenium is used in many electric components, because it can produce an electric current when light shines on it. This makes it extremely valuable in photoelectric dells, light meters, and photocopiers. It is also used in glassmaking, either to remove greenish tints from glass or to produce the deep red color favored for traffic signals.

Selenium is widely but unevenly distributed in the earth's crust. The levels of selenium in soils are the primary determinant of animal and human exposure, because it is taken up by plants in proportion to its presence in soil or water. Under most circumstances, water levels tend to be low and are less important predictors of human exposure than food levels. The environmental toxicity of selenium poses an interesting story for two reasons: first, the classification of selenium has changed, over the past decades, from that of a primarily toxic element with possible benefits at low levels, to that of a beneficial element with serious toxic effects in oversupply. Selenium is a constituent of the enzyme glutathione peroxidase, which functions to protect cells from the damaging effects of oxidative reactions. There is considerable evidence that selenium has anticarcinogenic properties, probably due to its mediation of antioxidant effects. Selenium derivatives act as antimutagens and/or anticarcinogens in several experimental systems, and epidemiological evidence correlates high levels of soil selenium with decreased incidences of several human cancers. In addition, the selenium-deficient Keshan area of China has an extremely high incidence of cardiomyopathy that is successfully mitigated with selenium supplementation. At high levels, however, selenium is extremely toxic to the liver.

With the exception of occupational exposure, there is no widespread concern about the human health effects of selenium exposure. It is possible that people living in areas with high soil levels of selenium, and eating food grown locally, might ingest toxic quantities, but there is little evidence that this occurs. Selenium toxicity in livestock, on the other hand, is a well-known phenomenon.[13] In the United States, most people are estimated to have an intake of 60 to 170 μg/day, and 50 to 200 μg/day, corresponding to between 0.03 and 3.0 ppm in the diet, is considered safe. Interactions of selenium with protein and other trace nutrients may alter these limits. Perhaps because selenium is derived from a wide variety of foodstuffs, including meat, poultry, grains, seafood, vegetables, fruits, and milk, very few cases of human selenosis or deficiency have been reported.[14] On the other hand, a case of selenium contamination in California illustrates the diverse ways in which human activity can contaminate the environment.

The Kesterson Reservoir and National Wildlife Refuge consists of a series of 12 drainage ponds (the reservoir) and an associated marshland (the wildlife refuge) in the highly agricultural San Joaquin Valley of California. The drainage ponds collect the runoff from the San Luis Drain, which contains irrigation runoff from the irrigated fields in the San Joaquin Valley. The creation of Kesterson was intended to solve several problems. First, California has destroyed or developed almost 95% of its wetlands in the last century, resulting in a dearth of marshes hospitable to the migrating waterfowl of the Pacific flyway. The presence of such wetlands is essential for survival of migrating water fowl. At the same time, ongoing irrigation of the fields of the San Joaquin Valley was complicated by shallow groundwater lying above an impervious clay ("hardpan"). The importation of large quantities of irrigation water into the topsoil above the groundwater led, over the 100 years since irrigation began, to a rise in the groundwater table that threatened to drown the roots of crops. The solution to this agricultural problem was a series of drains, some on the surface and some subterranean, and it seemed logical to use the drain water for optimum management of water levels in

---

[13] Two forms of the disease exist: "alkali disease" and "blind staggers," reported from the western United States and from New Zealand, respectively. Neither can be cured: the only available treatment is to remove the stock from selenium-rich forage.

[14] A good summary of the human toxicity of selenium, and the human risks associated with the contamination at Kesterson, is provided by A. M. Fan, S. A. Brook, R. R. Neutra, and D. M. Epstein in "Selenium and human health implications in California's San Joaquin Valley," *The Journal of Environmental Toxicology and Health* 23:539–559, 1988.

the wildlife marshes. Then the water should have drained into San Francisco Bay (via Suisun Bay, the conjunction of the Sacramento and San Jose rivers). But other drains had been built as the San Luis Drain was planned, and their completion led to increasing concern about the ecological effects of agricultural runoff on the ecology of California's coastal waters. By the time the last segment of the San Luis Drain should have been dug, a combination of financial constraints and ecological concerns blocked its completion. The agricultural water drained into Kesterson Reservoir and evaporated in the 12 ponds laid out for that purpose. The soils of the San Joaquin Valley developed from marine shales and are high in several salts, including selenium salts. The absence of an outlet led to salination—the process by which salt builds up in soil as a result of irrigation water evaporating from the fields, leaving its burden of dissolved salts behind.

At its extreme, salination produces salt water, as it has in the oceans, in the Great Salt Lake in Utah, and in the Dead Sea in Asia Minor. Not enough time has elapsed for overt saltiness at Kesterson, but salts, especially selenium, built up in the water. The selenium was taken up by aquatic organisms, which were eaten by fish. The fish also took up selenium directly from the water. The large population of birds living on and around the ponds ate fish, or aquatic plants, as did the mammals in the same area. Between 1983 and 1986, scientists identified selenium as the cause of extremely high rates of egg and hatchling loss among birds nesting near Kesterson. Grebes, ducks, and stilts, which spend only part of the year at the reservoir, were affected; coots, which live at Kesterson throughout the year, were most severely affected. Embryos often had structural abnormalities, and almost no hatchlings were seen, suggesting that few of the hatchlings were normal enough to survive. Levels of selenium in water were 300 ppm, while levels in vegetation were 10 to 100 times as high as in marshes that did not receive runoff water. Tissues from birds living on the reservoir also contained extraordinarily high levels of selenium. Laboratory studies confirmed the role of selenium in causing the observed effects, including both prehatching mortality and the particular malformations seen. Interestingly, small mammals living on the pond shores continued to reproduce normally. It is not entirely clear whether mammalian reproductive processes are less sensitive to selenium toxicosis or whether land organisms were simply less heavily exposed.

The tragedy of the contamination at Kesterson is that there is no cure, and very little can be done to prevent continued exposure of migratory birds. The people living in the vicinity have been supplied with bottled water for drinking and cooking, because of high selenium levels in groundwater. Hunters have been advised to limit their consumption of birds from Kesterson, and advisories for pregnant women are published. But the only immediate protection for the migratory birds, is a "shooing operation" to frighten them away from this one marshy oasis among the unhospitable tilled fields. Long-term solutions, including flushing the drainage ponds and their underlying soils, have become prohibitively expensive in California's decade of drought, tax rebellion, and recession. As at Minamata, Love Canal, the James River, and the Hudson River, there is no way to clean up contaminants that are spread through the soil and the water. *The only solution is never to repeat the pollution anywhere.*[15]

SUMMARY

Forty-five years after the start of the post–World War II boom in chemical use, one can safely say that the manufacture of persistent chemicals leads to their dispersal in the environment. Such dispersal may be intentional, as with pesticides, or it may be accidental. It may occur in a single burst as did the PBB release in Michigan, or by slow trickles from legal waste pipes, as did the release of PCBs into the Hudson River. Release can occur during manufacture, use, misuse, or disposal, but it occurs. Moreover, however the release occurs, persistent chemicals seem to find their way into the human environment and ultimately into the human dietary.

Chlordecone and PBBs demonstrate the effects of accidental releases: effects are relatively local but may be very severe, and time or mismanagement may turn a local contamination into one with much wider ramifications. Certainly, both the people of Michigan and the fish of the Chesapeake Bay carry the marks of the "local" spills. Such incidents are often identified when a small group of people or animals suffer recognizable clinical symptoms, giving rise to a "cluster" of disease cases.

PCBs typify the effects of slow releases over time: low levels in the water are deemed harmless until one or more species in the ecosystem suffers incontrovertible damage. Generally, however, all organisms over large areas are contaminated, but the incidence of clinical symptoms is low, both because of the low degree of acute toxicity of the pollutant, and because of the low levels of exposure. Ecological damage from slow releases can be extremely severe, since recognition of the problem comes only after whole ecosystems are irreversibly contaminated. Very often it is reproduction that is visibly affected. This sensitivity of reproductive processes to xenobiotics is logical, since successful reproduction not only requires survival of the organism at all stages of its life cycle (and some stages may be more sensitive than others to a given toxicant) but also because reproduction requires the coordinated functioning of numerous organs, and the coordinated presence of the right levels of numerous hormones, in both sexes.

Not the least of the lessons to be learned from these case histories is that organizations responsible for contamination tend to deny both its occurrence

[15] Miyamoto, 1975 in P. Kruus and I. M. Valeriote, eds., *Controversial Chemicals,* Multiscience Publications, Montreal, Quebec, Canada, 1979.

Table 5.3 Insecticides Whose Use in the United States Has Been Banned, and the Basis for Their Banning

Pesticide	Year	Reason
Aramite	1955	Carcinogen
Aminotriazole	1959	Carcinogen
Alkylmercury fungicides	1970	Chronic toxicity
DDT and DDD	1971	Bioaccumulation
Strychnine and sodium fluoroacetate	1972	Acute toxicity
Aldrin and dieldrin	1974	Carcinogens, bioaccumulation
Heptachlor and chlordane	1976	Carcinogens, bioaccumulation
Strobane	1976	Carcinogen
Mirex	1976	Bioaccumulation
Leptophos	1976	Delayed neurotoxicity
DBCP	1977	Male sterility
Chloranil	1977	Carcinogen
Chlordimeform	1977	Bladder carcinogen
Chlordecone (Kepone)	1978	Neurotoxicity
BHC	1978	Off flavors
2,4,5-T and silvex	1979	Dioxin contamination
Endrin[a]	1979	Fish toxicity
Chlorobenzilate	1979	Carcinogen, male sterilant
Pyriminil	1980	Pancreatic poison
Nitrofen[b]	1980	Teratogen
Aldicarb	1981	Water pollutant
EDB	1983	Carcinogen
Toxaphene	1983	Carcinogen, fish toxicity
Dicofol	1986	Bioaccumulation
Dinoseb	1986	Teratogen
Alachlor	1987	Carcinogen, water pollutant
Chlordane	1987	Indoor air pollutant
Parathion-ethyl[c]	1991	Acute toxicity

Source: Courtesy of R. L. Metcalf.
[a] Banned east of the Mississippi only.
[b] Technically, nitrofen was *withdrawn* by the manufacturer.
[c] Most uses banned by U.S. EPA.

and its consequences. Such denial is equally vehement whether the organization is a company incorporated to make a profit or a government agency charged with protecting citizens. Both in Hawaii and in Michigan, the authorities sought to limit damage to their agency rather than their constituents. One of the results of such actions has been the banning of a long list of insecticides (Table 5.3).

6

HAZARDOUS WASTE DISPOSAL

Off the side of the side road was another fifteen foot cliff, and at the bottom of the cliff was another pile of garbage. We decided that one big pile was better than two little piles, and rather than bring that one up, we decided to throw ours down. And that's what we did. . . .

—Arlo Guthrie[1]

INTRODUCTION

Garbage and the problem of its disposal have been an inevitable accompaniment of civilization. Hunter-gatherers can dispose of their waste along the way, scattered rather than heaped, leaving no record for the future, but even they tend to leave heaps of bones, and discarded and broken tools, that can be used to identify their life-styles. Settled clusters of humans put their wastes into piles, out of the way. It accumulates because some items, such as potsherds, do not decompose readily. Prehistoric middens were interspersed with dwellings and include both pottery shards and food residues. Our knowledge of the earliest civilizations comes in large part from investigating these garbage heaps. A people's diet can be reconstructed from the bones and seeds they discard; their art, from fragments of broken pottery; their technology, from weapons and tools broken or blunted by use. The larger the city, the more effort must be put into waste disposal or yesterday's debris will choke today's commerce. In Rome, an entire section of the city is built on a hill composed of pottery broken over millennia.

Historically, there have been two ways of dealing with pollution: solid waste was usually removed from sight, either in a pit or on a heap. Liquid waste was either dumped directly onto soil or diluted into the nearest creek or river. The Romans built sewers to match their aqueducts: neither feat was repeated in Europe until the dawn of modern times. Well into the nineteenth century, most waste disposal was managed by each household

[1] From *Alice's Restaurant*, written and recorded by Arlo Guthrie, 1967 for Reprise Records division of Warner, Inc., from the compact disk.

individually. Even city houses had a yard, and each yard had both a privy and a kitchen midden. The privy received human excrement; the midden received kitchen wastes and ashes. Liquid household waste other than urine could be thrown into the street, and in many towns the upper story of houses projected over the lower story so that slops could be thrown directly into the street from upper windows. In larger cities, provision was made for the periodic removal of human excrement, first by cart, later by sewer pipes. This human excrement ("night soil") was used as fertilizer in many parts of Europe and England well into the twentieth century and continues to be used as such in China today, albeit with rigorous sanitary precautions to prevent the breeding of flies or transmission of disease.

The *Encyclopaedia Britannica* of 1910 warns of the dangers of disease transmission if excrement is used as fertilizer, but the same article notes wistfully that disposal of sewage into the sea leaves the land poorer. Initially, the use of sewage pipes also led to a tremendous increase in waterborne illnesses, since these pipes simply led to the nearest river, which was used by the next town downstream as the source of drinking water. It was an easy extension to include other waste liquids in the sewer pipes. By the beginning of the twentieth century, flush toilets, cesspools, and sewage disposal systems were common in cities. Garbage haulers took solid wastes to dump sites at the edge of the city. Privies and compost heaps remained the rule in rural areas, where each farm would also have its own dump site for irreducible trash: cans, bottles, old couches.

Ironically, the indiscriminate mixing of food wastes with essentially durable wastes is a recent phenomenon. For most societies throughout most of human history, recycling has been necessary for survival. Fats were turned into candles or soap; food wastes were fed to dogs or pigs. Not only were rags turned into paper, but some historians postulate that society was only able to benefit from the printing press because increasing prosperity made rags discardable. A hundred years before Gutenberg, rags were too valuable to become paper. It was left to the twentieth century to discard the prudent dictum: "waste not, want not."

LANDFILLING

How do we dispose of garbage today? What is a modern landfill? Do we know how to construct a safe depository for our garbage? If not, are there alternatives? And perhaps most critically, do we have enough room for all the garbage we generate? If not, what are the alternatives, and what risks are associated with such alternatives?

This preoccupation with garbage *per se,* as opposed to concern over the disposal of excrement and prevention of water pollution, is less than 30 years old. Even the modern age of sewage disposal can only be said to date from the 1840s, when Dr. John Snow removed the handle of the Broad Street Pump in London and so ended a cholera epidemic. In the next 50 years the acceptance of the germ theory as the primary cause of contagious human illness completed the connection between contaminated water and disease, and the entire panoply of sanitary engineering focused on providing clean water and removing dirty water. To this emphasis on clean water are credited many of the public health advances of the twentieth century, including tremendous decreases in infant mortality. It can easily be argued that sanitation, more than antibiotics, revolutionized health care. The major aspect of that sanitation was removing pathogens from drinking water.

Nonetheless, sanitary systems of the nineteenth and early twentieth centuries still depended on the biological degradation of wastes by aquatic organisms. Each city along a river dumped its wastes into the river. Cities often took—and still take—their drinking water from the same river. They are, of course, careful to withdraw water upstream from their own sewage outflow. In the past, time and the river were expected to decontaminate the water for the cities downstream. Today, the leftovers from upstream cities are dealt with by water treatment: sedimentation, clarification, and chlorination or other disinfecting techniques. As long as the only waste entering the river is sewage, these measures suffice. It is said that the water in New Orleans, which comes from the Mississippi, has been drunk six times before it reaches the faucets of New Orleans.

Industrial liquid wastes proved harder to detoxify. Because many industries use large amounts of water in their processes, and because many synthetic processes (and therefore their waste streams) are primarily aqueous, river banks and lake shores are favorite sites for industrial plants. Especially obnoxious examples of industrial waste include the highly acid, reeking effluent from paper pulp mills, and the long-lasting froth put out by detergent manufacturers in the 1950s. Less obvious pollutants were not controlled quickly. Adjacent waterways served as sewers for wastes containing cadmium and mercury in Japan, lead and mercury in the United States, and mining wastes everywhere. The municipal sewage of Milwaukee is so heavy in cadmium that the sludge cannot be used as a soil amendment. GE legally trickled PCBs into the Hudson River for almost 40 years before the damage was noticed. On the banks of the James River in Virginia, Life Science Inc. disposed of Kepone (chlordecone) wastes into the municipal sewage system of the town of Hopewell. The insecticide could not, in any case, be degraded by the bacteria used in the sewage treatment plant, but it actually poisoned these bacteria, so that the town released both Kepone and effectively untreated sewage into the James River.

In the United States, with its wide-open spaces and frontier mentality, solid wastes other than excrement were not dealt with systematically until the second half of this century. Even in the 1950s, the typical American city dump consisted of a pile of refuse, ranging from discarded household machinery to coffee grounds, open to the air and to rodents, often catching fire or set afire by vandals, and mined by enterprising scavengers. Solid wastes from households and from industry were dumped into the same pits. In rural areas, each household took care of its own trash, including discarded cars. In semirural areas, people from more built-up areas dumped old furniture and other trash along the roadsides.

With increasing fastidiousness and increasing population density, the odors and frequent smoldering fires of the "town dump" came to be considered public nuisances. Trash was put into pits and covered at intervals. Nevertheless, open dumps were familiar sights into the 1950s, even in relatively urban areas. Cities along the ocean dumped their garbage at sea well into the 1980s. In small towns and rural areas, dumps existed primarily for large unsightly objects such as couches and old hot water heaters; individuals still burned, buried, or dumped most wastes on their own land. Most laws dealing with waste disposal were concerned with transmission of disease by water contamination or by vermin. In the 1960s, the accumulation of unsightly trash, and a growing awareness of the permanence of industrial wastes, led to increasingly strict control of all kinds of garbage. First municipal dumps were covered daily, and localities began regulating who could dump garbage within their boundaries and where they could dump it. States began to think about distinctions between *municipal* waste—what you and I put in our garbage cans at home and at the office—and *hazardous* waste—some of the things industry puts out[2] (Figure 6.1). Gradually, in the process of identifying

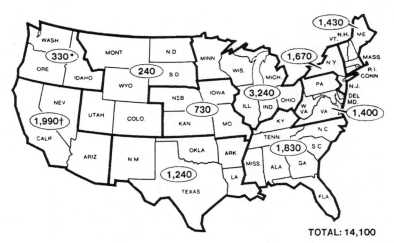

TOTAL: 14,100

Figure 6.1 EPA's estimate of the number of U.S. facilities producing hazardous wastes in 1981, by region. *The northwestern region includes Alaska. †The western region includes Hawaii. (From *The New York Times*, August 31, 1983.)

[2] This is, of course, an artificial distinction. A large fraction of industrial waste is the same sort of trash as households generate: paper, dirty water, and plant and animal matter. And households can dump some nasty items: pesticides, photographic chemicals, and acids. For an entertaining and informative account of our garbage habits, read *Rubbish!* by W. Rathje and C. Murphy, HarperCollins, New York, 1992.

what was going into garbage dumps, governments at all levels began to realize that they also had to worry about *how much* of everything there was. By 1990, garbage disposal had become a very important political issue, and there is no evidence that it will become less so in the next decade.

Controversies can be roughly grouped into two categories. First, there are debates between states over who will accept particularly noxious wastes. PCBs and radioactive waste are the most controversial, followed by anything classified as "hazardous". States that have until now accepted certain types of waste are refusing to accept them from states that do not have their own hazardous waste landfills. Midwestern states worry about becoming New Jersey's landfill. The western states do not wish to be the "national sacrifice area" for radioactive material from all over.

Second, there are bitter debates within states over where new disposal sites will be located. Any place selected objects immediately. Even if the local authorities want the facility—some jobs are always generated—opposition inevitably develops. Citizens' groups put pressure on their legislators, who typically respond by including a referendum as part of the siting process. No matter how good the site is in the view of the experts, a determined opposition can then prevent the facility from being built.[3] These debates impinge on national policy as well. Several proposed consortia for pooling radioactive and hazardous waste disposal between states have broken down, because the recipient state cannot persuade any community that has an acceptable site for such a landfill to allow it to be sited. The many toxic waste sites resulting from earlier careless disposal provide fuel for landfill opponents in every state (Fig. 6.2).

Despite the genuine paralysis over siting new landfills, there has been progress in landfill management. After some years of dispute over the best technologies, landfills are again becoming standardized. In place of open heaps of rubbish, a modern landfill consists of mounds, area fill, or dike containment. If the method used is *dike containment,* trenches are dug into the ground, with intertrench width of 1 or 1.5 ft for every foot of depth. Rarely are the trenches more than 20 ft deep; most often they are 6 to 10 ft deep. Their width, however, ranges from 2 to 100 ft. As an alternative to trenches, *area fill* involves deposition of waste above ground level, covered on all sides with dirt and/or synthetic barriers. This method is particularly useful in locations where the water table is high. It also provides a means

---

[3] The validity of the experts' opinion usually becomes the first target of the opposition. Geological and hydrological data are disputed; aquifers are discovered; soil scientists are pitted against each other. Old laws may be dusted off. Nor is the expert opinion always vindicated. In the case of a dump at Wilsonville, Illinois, which U.S. EPA proposed to expand as a repository for all PCB-containing wastes from the eastern half of the United States, experts were adamant that the site would not leak for at least 100 years. The courts nonetheless decided that the town could not be forced to allow such a repository. Two years later, the existing landfill was found to be releasing leachate—leaking—into a small stream. A vein of sand, missed in surveying, was present in the perfect soil.

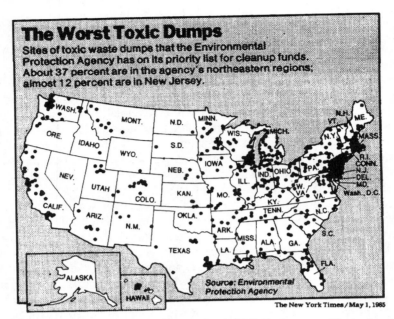

Figure 6.2 The worst toxic dumps, as identified by the U.S. Environmental Protection Agency before 1985. (From *The New York Times,* April 1, 1985.)

of extending the useful life of a trenched landfill: trenches are simply filled to several feet above ground level. *Mounds* are closest to the old rubbish heap and can be used only for household or municipal waste.

The precise method used by a community depends on the average solid content of waste, on the intensity of land use that is required, and on the need to avoid absorption of leachate. New York City, for example, uses mounds in Freshkills Landfill on Staten Island, since these can be piled far higher and do not require the intertrench distance of dike containment. Because the garbage is above ground, it is not subject to leaching by groundwater, although precautions must be taken to avoid rainwater acting as leachate.

When the waste includes hazardous materials, far different precautions must be taken. Much hazardous waste is still stored above ground, but this is generally represented as temporary storage. *Landfilling* of hazardous wastes almost inevitably means that the waste is placed in soil trenches and dike containment.

Hazardous Waste Landfills

Site selection is the most critical aspect of hazardous waste landfill design and function. The ideal site overlies very sparse groundwater, preferably protected by a layer of bedrock. There should be little surface water, the site should not drain into a river or lake, and should be outside the 100-year floodplain of any nearby rivers.[4] Clay is less permeable to leachate and is the preferred soil. Pure clay is rare, however, and there should be considerable concern over the possible existence of sand pockets or veins of sand underneath the landfill.

Until the 1980s, it was assumed that the permeability of clay to leachate from the landfill could be modeled by the leaching of water. This is a reasonable first approximation, since in the past most leachate was indeed primarily aquatic, due to percolation of surface and/or groundwater into older landfills. When laboratory data demonstrated that clay is up to 100 times more permeable to organic than to aqueous leachate, *synthetic* liners became obligatory quite rapidly, although 10 years earlier they were banned because of concern that they would be breached (Figure 6.3). The possibility of puncture now seems to be ignored. Moreover, a soil permeability of approximately 3.1 cm/yr seems acceptable at present.

It is now recognized that the interactions between leachate and soil are actually very complex and depend both on the nature of the soil[5] and on the leachate. Moreover, gas control is absolutely required for landfills, including municipal landfills or "town dumps." The anaerobic decomposition of carbon-containing material in municipal landfills can lead to a buildup of methane, which can result in fires or explosions. The methane can migrate laterally through soil, especially where a dense cover overlies the landfill. In several cases, manhole covers have "popped" in nearby towns as methane pressure from municipal landfills increased. In more serious incidents, houses have literally exploded due to methane from "mere" municipal landfills. The possibilities of disasters when materials escape from hazardous waste sites are limited only by the nature of the waste they contain. In Louisiana, hexachlorobenzene escaping from a landfill by sublimation resulted in contamination of animals, especially cattle, for 100 miles. In 1992, a town in Guadalajara, Mexico, was devastated by the explosion of combustible waste that leaked into the sewers over a period of a week.

Covers are required for landfills. These are of two kinds. Daily covers serve to minimize smells, blowing of debris, and access of vermin to the garbage, and also prevent the entry of water into the landfill. The final cover is constructed very much like the liner of the trench and includes both natural and synthetic material. The integrity of the permanent cover is essential to maintenance of the landfill after closure, since any water that enters a modern landfill sits in the lined area, turning it into a stewpot in which wastes can mix, combine, and react with each other.

The deleterious effects of water on maintenance of landfills is particularly acute in hazardous waste landfills, which now segregate waste by type in

---

[4] The 100-year floodplain is that area that is *statistically* expected to be flooded *no more than once in a century*. There is no guarantee that it will not be exceeded twice in two years. . . . .
[5] For example, the plasticity of soil can range from crumbling to flowing. Sandy soils crumble, whereas certain clay soils, when saturated, become slurries. The rate at which liquids move through soil will be affected by the plasticity.

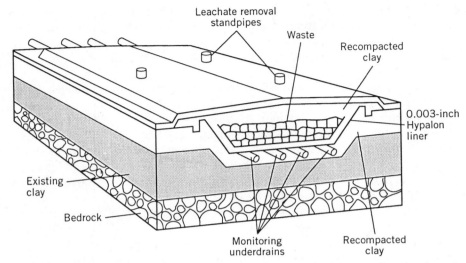

Figure 6.3 Construction of a modern hazardous waste landfill. Containment of wastes depends on the integrity of the synthetic liner (e.g., Hypalon), and further protection is offered by recompacted clay, location on clay soil, and bedrock. (From *Chemical & Engineering News*, March 8, 1982, p. 16).

order to prevent undesirable reactions (e.g., between acids and bases). *Leachate collection* is therefore required for all new hazardous waste landfills (see Figure 6.4). This monitoring of leachate below the liners also provides data on the integrity of the landfill liner, whether it is synthetic or only clay.

Disposal of the collected leachate proves to be somewhat unsatisfactory. Current methods include redeposition of the leachate on or in a landfill. Alternatively, the leachate can be concentrated and disposed of as are all other hazardous wastes. Disposal via municipal water treatment plants is often tried, but toxic wastes tend to kill the microorganisms in the sewage treatment plant. In several cases the result has been discharge of untreated sewage plus leachate into streams. Finally, some leachates can be disposed of by "landfarming," that is, by dispersal over land at low levels. Land farming is inappropriate for all but biologically degradable materials, however, and many of these could also be cycled through sewage treatment plants. Therefore, land farming does not represent a new option except for leachates that are known to be both biodegradable and toxic to the bacteria in sewage treatment plants: surely a small category.

Permit Requirements for Landfills

The exact pattern of regulations varies from state to state. To illustrate some of the complexities facing waste disposal regulators, we will use Illinois as an example. Obviously, other states may have acted earlier or later, but the general pattern has been similar. The state of Illinois first regulated landfilling in 1965. The first regulations simply consisted of registration of solid waste disposal sites; no justification was required. In 1970, a permit system was instituted, and now a *developmental permit* is required before construction begins, followed by a *permit to operate,* which is not granted until an inspection has demonstrated that the site design has been met.

Granting of the developmental permit for a hazardous waste site theoretically depends on demonstrating the geological suitability of the site for (hazardous) waste disposal, but the majority of hazardous waste sites currently in operation were actually grandfathered in, and self-selected on the basis of their owner's willingness to comply with permit requirements. Thus many sites are geologically or hydrologically unsuitable (Figure 6.4). In Illinois, for example, one of seven sites accepting hazardous wastes in the early 1980s was known to have flooded in the 1950s (sites should be off the 100-year floodplain), while another had been considered adequate only for small amounts of household refuse at the time it was licensed for general waste disposal.

Categorization of waste by hazard began in 1974 in Illinois. Initial efforts were often minimally informative (e.g., one entry reads "contaminated soil,"

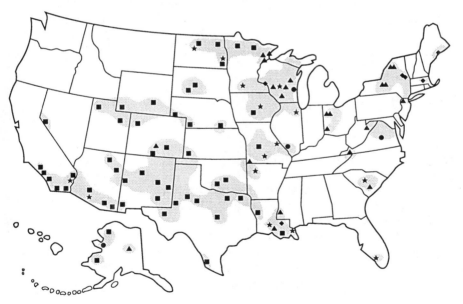

Figure 6.4 Drinking-water problems due to contaminated groundwater. Gray areas show areas of reported pollution; white areas, although not necessarily problem-free, are not considered to have major problems. Sources of pollution include industrial chemicals other than chlorinated hydrocarbons (●), heavy metals (▲), chlorinated hydrocarbons (◆), coliform and other bacteria (*), saline water (■), and general municipal and industrial wastes (◠). (From *The New York Times,* August 13, 1981.)

with no further explanation; another lists the waste as "toxic racehorses." Currently, chemical analyses of new waste streams are required, as are permits for new waste streams. In the near future, all hazardous wastes will be banned from landfilling except when it is not "technologically feasible or economically reasonable" to dispose of the material otherwise. It is anticipated that monumental court battles will be waged over this terminology, as producers, site owners, local residents, and lobbyists argue over what is "economically reasonable" and/or "technologically feasible." If encapsulation is available for $500 a ton, is that economically reasonable? Or can one continue to use steel drums at a fraction of the price?

In the meantime, federal laws are also sharply restricting what can be put into landfills. During the Reagan Administration, the federal EPA under its Administrator, Ann Gorsuch-Burford, and the head of the waste-disposal division, Rita Lavelle, rescinded restrictions on the disposal of liquids in hazardous waste landfills. Within a year, public pressure had forced a reversal by the EPA. Today, soil disposal of liquids that are designated as hazardous is sharply curtailed. In the description of alternative methods for hazardous waste disposal, it must be remembered, however, that *there will always be wastes that are disposed of in the soil.* Among the items that simply cannot go anywhere else are contaminated soil (e.g., soil from the Michigan PBB incident), ash from incineration, and construction debris. In the case of low-level contamination, incineration may be too costly (oil is a nonrenewable resource); in the case of elemental or radioactive wastes, burning is not destructive.

MINIMIZING HAZARDS OF LANDFILLED MATERIALS

Immobilizing Wastes

Techniques that consist of immobilizing waste in an inert matrix include cement-based processes and pozzolonic processes, which mix wastes with fine-grained silicaceous materials and water to form a concretelike mixture. Thermoplastic and polymer techniques mix wastes with plastics at high temperatures or with prepolymers and catalysts, in order to capture the wastes in a matrix for further encapsulation. Microencapsulation involves pressing or bonding the wastes together and coating them with an inert material. Although some forms of encapsulated material might be recycled into other uses, encapsulation is most generally a prelude to landfilling.

Innovative Biological Techniques

Biological seeding refers to the use of specially developed strains of microbes that are capable of digesting a specific waste. This technique is most suitable for fats, greases, aromatic hydrocarbons, pulp mill wastes, and so on, because such materials are potential carbon sources for microorganisms, and microbes that prefer them can often be selected. Seeding has been used in a few oil spills and holds some promise for soil decontamination, although there is a tendency for both the oil and the microbes to disperse in aquatic environments. It is harder to use biological seeding for halogenated wastes, which are only marginally biodegradable. There is a bacterium which, in the laboratory, in the absence of alternatives, will use the herbicide 2,4,5-T (2,4,5-trichlorophenoxyacetic acid) as a carbon source; its performance in the "wild," where nonchlorinated sources of carbon are available, is not known. Such bacteria may degrade low concentrations of moderately halogenated compounds, as data from PCB-contaminated sediments suggest.

Fluidized-bed bioreactors provide surfaces of solid material that allow microbial development in solutions in which the organic contaminants serving as food for the microbes are too dilute to support growth in the fluid itself. The fluid is circulated through the inert bed material and organics are gradually removed. This technique is still in the testing stages.

ALTERNATIVES TO LANDFILLING

Minimizing Quantities of Wastes

One category of alternatives are methods used in conjunction with landfilling, which serve to minimize either the quantities of the waste streams and/or the hazard associated with them. Pretreatment and recycling of reusable components of a waste stream result in a reduction of waste stream volume as well as, in some cases, permitting reclamation of valuable substances. The simplest form of pretreatment is the removal of water from dilute aqueous wastes. A more difficult case is the recycling of silver salts in photographic processes, which is worth it because of the high cost of silver. Solvents are, in many cases, easily recovered for reuse.

Process alteration to minimize the production of hard-to-handle or toxic waste streams is especially applicable during construction of new factories. At that time, more economical and efficient production processes can be designed into the manufacturing process. If factories have to be retrofitted, expenses rise. There may also be considerable cost attached to idle time while the factory is being redesigned. Nonetheless, even retrofitting can be worth it. In the case of vinyl chloride, producers were forced to cut emission by 90% when vinyl chloride was found to cause liver cancer in workers. Despite carefully marshalled arguments by the industry that the new rules would drive half the producers out of business, OSHA persisted. One plant manufacturing vinyl chloride did close. The remainder found that by recycling vinyl chloride that had formerly been emitted into the air, they increased their profits despite the cost of enclosures.

Pretreatment methodologies will obviously be as diverse as the waste streams they minimize. Some alternatives include flocculation, sedimentation, or screening to remove solid particles from water; flotation to remove suspended oils and greases; filtration; centrifugation; chemical separation of compounds in solution, including precipitation, neutralization, chemical oxidation, or reduction; chemical dechlorination (an emerging technology); separation of components in solution by dialysis, reverse osmosis, or membrane separation methods; or by ultrafiltration, electrodialysis, or phase exchange processes. Solutions can also be detoxified by adsorption of the pollutant on resins or activated charcoal. In many cases, pretreatment generates a new category of waste in place of the usually dilute liquid it cleans up. As long as the new waste is either less voluminous or solid instead of liquid, it is easier to handle and easier to dispose of. But since it is still necessary to dispose of the contaminated material, waste stream reduction is rarely a panacea. Stabilizing or encapsulating wastes to prevent leaching is used in conjunction with landfilling but may be used to permit a "hazardous waste" to be categorized as "nonhazardous."

Deep Well Injection

This technique consists of inserting wastes into geological formations that are believed to be effectively (geologically) isolated from aquifers. The waste is injected into a layer of porous rock several thousand meters underground. Overlying formations should consist of fracture-resistant rocks, such as shale or salt. The wastes must be deposited into a pipe, not directly into the bore hole, to avoid movement of the waste into intermediate strata. The use of deep-well injection for hazardous waste disposal is inherently limited because there are not many sites with the requisite geology, and space within those sites is limited by the pore volume of the deposition layer. Figure 5.1 (page 96) illustrates the siting of an injection well relative to rock strata, aquifers, and other wells.

Deep-well injection is routinely used to dispose of brine that is pumped up with the oil in oil wells. The technique has also been used to dispose of hazardous waste, with occasionally disastrous consequences. At the Rocky Mountain Arsenal near Denver, Colorado, wastes were injected 3600 meters deep into fractured metamorphic rock. Earthquakes occurred in the region between 1962 to 1965, apparently because the liquid increased pressure on the fractures and initiated movement. In California, the earthquakes from fluid injection into an oil field probably contributed to failure of a dam, while gas buildup in a disposal well in the eastern United States caused a "blow out" that released millions of liters of toxic wastes into a lake. Additional problems associated with deep-well injection are dissolution of the deposition stratum, which allows the wastes to migrate and which may cause subsidence of overlying terrain, and excess precipitation of solids in the well bore, which clogs the delivery pipe. The dramatic failures of the technique underline the possibility of more stealthy risks, such as failure of the well liner, which would allow deposited wastes to leach into intermediary strata (including aquifers). Several states, including Illinois, have banned construction of new injection wells for hazardous wastes.

Waste Exchanges

One industry's waste or by-product may very well be identical to another company's raw material. Waste exchanges seek to match such industries. In effect, waste exchanges are one facet of waste-reduction methodology, together with process alteration and pretreatment. Despite the intuitive appeal of waste exchange, it must be noted that several very unpleasant episodes of pollution have originated in "recycling" efforts, particularly of solvents. Also, the tendency of companies to guard their production methods is an obstacle to the information exchange needed for successful operation of waste exchanges. Finally, the threat of lawsuits over impurities that may remain in the exchanged material makes many companies hesitate to sell their waste streams. Nonetheless, waste exchanges are in use in some instances.

Incineration

Incineration is often thought of as an alternative to landfilling, and can, to a considerable extent, eliminate landfilling of hazardous waste. It is also often used in conjunction with the methods above, especially with dewatering. However, incineration of all but the most innocuous substances requires "scrubbing" of the effluent gases to remove hazardous emissions. Therefore, the residue from incineration includes not only the ash (which should of course be captured, not released into the air) but also the scrubber sludge or dust. Incineration is particularly good for concentrated organic compounds which generate heat as they burn (and so minimize the needed input of fuel): organic solvents, for example. Burns above 1000°C will also destroy dioxins.[6] Incineration is not effective for elemental or radioactive wastes, or for very aqueous pollutants.

The major problem with incineration is that it is difficult even for experts to monitor the process with absolute accuracy. Brief temperature fluctuations or partial failures of the scrubbing systems may release extremely hazardous effluents. Similar charges can be made against landfills, but leaks from landfills are in soil and water, which can be more easily monitored and more easily controlled. Once an incinerator has released its effluent, it is irretrievable. Moreover, the tremendous volume of material handled by incinerators means

---

[6] 2,3,7,8-Tetrachlorodibenzodioxin, a by-product of trichlorophenol production, is the most toxic synthetic chemical known. Its presence in the soil of Times Beach, Missouri, led to evacuation of the town and its purchase *in toto* by the U.S. government. The exact toxicity of dioxin to humans is disputed, as is its carcinogenicity, but is certainly measured in μg/kg rather than in mg/kg (see Chapter 7). The panic evoked by its presumed presence makes it a highly political toxicant.

that even a miniscule release, on the order of parts per billion, can amount to large quantities of material being deposited over time. As a result, many health officials are much more concerned about incinerators than about other methods of disposal. Plans to build new incinerators invariably meet with vehement opposition from people living nearby.

Incineration is the best developed alternative to landfilling, and the most versatile. Incineration of municipal waste is common and is increasingly combined with generation of heat. There are also incinerators designed to burn hazardous waste, although it is extremely difficult to site new incinerators for this purpose. Most of the hazardous waste incinerators that are in operation on land, including one at the southern edge of Chicago, have been on their present site since well before 1980 and have merely been expanded or modified to burn hazardous waste as the regulations tightened. Incineration of hazardous waste also has intrinsic limits. High combustion temperatures mean that a high cost in fuel is involved unless the hazardous wastes are themselves combustible. In addition, special pollution controls are required to prevent creation of hazardous air pollution by the exhaust plume. Ocean incineration holds some promise. The *Vulcanus* is a ship that burns hazardous waste on the ocean. It was used to dispose of the Agent Orange[7] from Viet Nam, with a small island in the Pacific serving as the base of operations.

After incineration, most waste streams leave behind some ash, which must usually be landfilled as hazardous waste. If the waste is such that scrubbers must be used to catch acids, there may also be considerable waste from the scrubbers—which often consist of salts that absorb and neutralize the acids.

Several techniques exist for incineration. Each is designed for particular kinds of waste, although some designs are quite versatile. *Rotary kilns* are versatile and can be used to burn solids, liquids, and gases. Postburn cleanup involves removing particulates from the chimneys. An additional problem is that installation costs are very high. *Multiple hearths* can handle sludge as well as solids. They are fuel efficient, which is an advantage in burning low-heat substances such as sludges. On the other hand, maintenance costs may be high. *Liquid injection* is the most common form of incineration but can be used only for liquids. The liquid wastes are sprayed through nozzles to speed vaporization and thus combustion. Wastes can be blended before combustion to minimize fuel addition and maximize the efficiency of the burn. These systems are frequently used as after-burners for the gases from other incineration systems. Residence time of the waste in the burn ranges from 0.5 to 2 seconds, and the temperatures range from 1300 to 3000°F.

---

[7] Agent Orange was the code for a mixture of two herbicides, 2,4-D and 2,4,5-T (2,4-dichloro- and 2,4,5-trichlorophenoxyacetic acid, respectively). Since 2,4,5-T is inevitably contaminated with dioxin, disposal of this mixture in the United States became a highly political and emotional issue, even though 2,4,5-T had been used for 30 years on pastures, lawns, and along rights-of-way without any questions being asked.

Fluidized-bed incineration uses a bed of granular material, often sand, over a perforated metal plate. Air passes through the plate, keeping the bed of sand "tumbling." The sand is first heated to 840 to 1600°F. Then the wastes are injected into the bed. The resulting extremely rapid heat transfer from the large mass of sand leads to rapid and complete combustion. Compactness and small capacity recommend these incinerators for small installations. Also, selection of the proper "bed" can allow for scrubbing of selected gases from the exhaust by chemical reactions, which suggests that fluidized-bed incinerators are most effective when dedicated to a small number of waste streams. These incinerators are still at the demonstration stage rather than being in commercial use.

Molten salt combustion is a small-scale, quite economical variant of fluidized-bed incineration. The lower temperatures used in molten salt combustion are the basis of its fuel economy, but also make it less attractive for some wastes, notably those that might generate dioxins or dibenzofurans. Ash and combustion products are retained in the salt, which must be removed periodically and landfilled as hazardous waste unless a means of reclamation exists. The gases emitted from the burn must also be treated. Molten salt combustion is probably most suitable for small installations. Liquids, solids, or gases can be injected from below the molten salt. Destruction of most organic compounds occurs at 1500 psi and 177 to 315°C, in less than 1 hour of residence time; these temperatures are not high enough for PCBs or other compounds that generate dibenzodioxins or dibenzofurans.

Cement kilns, which heat some constituents to 2600°F, could be designed to use hazardous combustibles instead of clean fossil fuels as part of their fuel supply. Barriers to such use are the increased particulate production resulting from burning wastes, which would probably require additional scrubbing of chimney effluents and the need to upgrade facilities not equipped to handle hazardous materials. Use of cement kilns is actually a form of co-incineration. Because of loopholes in existing regulations, some cement kilns are used to burn hazardous waste without scrubbers or monitoring of effluents.

Co-incineration uses hazardous combustibles as part of the fuel in boilers. This technology can be used in existing boilers simply by adding combustible wastes to the fuel to be burned. Alternatively and preferably, new, specially designed boilers can be built. Since startup costs using existing boilers are low, this method is dangerously attractive. Corrosive products may injure existing boilers that are not specifically designed to handle such materials. Moreover, assuring complete destruction of wastes is difficult unless boilers were designed to handle them. Boilers that were not built for co-incineration rarely have appropriate scrubbers, and monitoring is difficult.

As a technology for new boilers, co-incineration is more expensive but also more conducive to rational design and appropriate monitoring. In the 1980s many communities, as well as the U.S. EPA, hoped that co-incineration would solve two problems at once: energy generation and garbage disposal.

Plants were actually built in several cities. The technology did not live up to its promise, however. First, even municipal waste streams include enough toxic materials that there is considerable worry about effluents, and serious objections by neighbors. When incinerators are expected to burn hazardous waste, panic often occurs in the surrounding community. Second, costs of running the co-incinerators proved to be much higher than expected. Third, precisely those wastes that have the high heat content that is needed to make co-incinerators economical are the most easily recycled: paper and oil, for example.

Several forms of incineration offer promise for destroying difficult wastes in the future. *Pyrolysis* involves thermal destruction of complex hazardous organics in the absence of enough air for complete combustion. Degradation products can be reclaimed or burned by more conventional methods. In pilot models, it has been found that some products generated by pyrolysis are potent carcinogens; caution is warranted. *Plasma arc torches* produce temperatures up to 90,000°F. At this temperature, solids and liquids decompose to simpler, less hazardous forms. A major benefit of plasma arc torch incineration is that it does not produce toxic intermediates. Moreover, the units are energy efficient relative to the temperatures produced, and can be made portable. This technology is only in the development stage, however, and actual costs of using it for waste disposal are unknown.

CASE HISTORIES: LIFE ON THE EDGE OF A LANDFILL[8]

In the past 20 years, the contamination of homes, neighborhoods, and entire towns by hazardous wastes has become a staple of the evening news. The names of the affected communities have become familiar: Love Canal, Valley of the Drums, Stringfellow, Times Beach. In addition, each state has its own list of "Superfund sites," many of which are familiar only to state officials, to a few lawyers and toxicologists, and to the people who live there.

Experts tell us that hazardous waste landfills are nothing to worry about. They assure us that hazardous waste sites are very low on the list of things likely to shorten our lives. Automobiles are far more likely to kill us than is living near a dump. Passive smoking, radon in our basements, or even a high-fat diet are far more likely to cause cancer than are the chemicals we disposed of in soil and water, even when those chemicals leach from their graves.

If one looks at the epidemiological facts, the experts are undoubtedly right. Each year 5000 people die in automobile accidents in the United States. Even the most vociferous lobbyist does not claim that so many people suffer from cancer, reproductive dysfunction, or neurological problems because chemicals leached into their basements (Love Canal, Niagara Falls, New York), because fumes from settling ponds wafted across their neighborhood (Stringfellow Pits, California) or because dioxin-tainted oil was used to settle dust on their roads

---

[8] This was the title of a news story in the *Daily Illini*, the University of Illinois student newspaper, about the people living on a farm at the edge of an abandoned landfill in Danville, Illinois.

(Times Beach, Missouri). Unfortunately, knowing that most people do not suffer as they do is small comfort to the relatively few, but clustered, people who are affected.

Even in this book, we cannot take the space to detail their case histories, which has been done in newspaper stories and in several books dealing with the problem of hazardous wastes.[9] When one reads these case histories of notorious landfill sites, it becomes clear that the affected populations share certain characteristics. These include a lack of political power, due either to small numbers or to relative poverty, and a perception of powerlessness, due either to fear of unemployment or to ignorance. Simply put, hazardous waste landfills are not located in affluent neighborhoods. In the rare instances in which substantial houses are built on a forgotten dump site, remedial action is usually instituted soon after identification of the problem: either by public agencies responding to pressure, or by private companies responding to legal action. If the abandoned dump site is near very few houses, legislators and public agencies are less subject to pressure from constituents than if heavily populated areas such as Love Canal are involved. In poor communities, affected people may not realize that there are legal remedies or may not have the resources to pursue them. They may not have private doctors, but use clinics where they see different doctors on consecutive visits. They are unlikely to know how to find toxicologists. Politicians are less likely to listen. Media attention may be the only way the affected population can be heard, and only dramatic stories attract media attention.

For the same reasons, new hazardous waste landfills are invariably sited in rural areas. An additional benefit to such locations is that they are often areas of high unemployment. Increased prices for marginal land attract sellers, and at least some of the site's neighbors consider the benefits of jobs and a better tax base for the community to outweigh the risks. Cynically, one might say that absentee landlords and more distant "neighbors" will be more enthusiastic than adjacent farmers and next-door homeowners.

Moreover, a recent study[10] suggests that more people may be affected by hazardous wastes than is indicated by the notorious incidents. The study evaluated the relative risk of having a baby with a birth defect for women living within one mile of a hazardous waste site compared to women of the same age, socioeconomic status, and parity who did not live near such a site. Children with birth defects born in New York State in 1983 and 1984 were identified through the New York State Congenital Malformations Registry, and controls were selected from normal children born in the same census areas during the same peirod. A total of 917 hazardous waste sites was identified, and the distance between the sites and the homes of the mothers mapped to within 200 feet. There was a 12% increase in the overall incidents of birth defects among babies born to women living within

[9] For example, *Malignant Neglect,* J. H. Highland and R. H. Boyle, Vintage Books, New York, 1980; *The Politics of Cancer,* S. Epstein, Sierra Club Books, San Francisco, 1978; *Hazardous Waste in America,* S. S. Epstein, L. O. Brown, and C. Pope, Sierra Club Books, San Francisco, 1982.

[10] The information comes from an article by S. A. Geschwind, *Chance* 5:3–4, 1992, and is based on "Risk of congenital malformations associated with proximity to hazardous waste sites," by S. A. Geschwind, J. A. Stolwijk, M. A. Bracken, A. Fitzgerald, A. Stark, C. Olsen, and J. Melius, *The American Journal of Epidemiology* 135:1197–1207, 1992.

one mile of a hazardous waste site. The risk was even greater when the probability of maternal exposure rose because the nearby landfill was leaking and/or the site had been judged relatively hazardous. Most striking, however, was the finding that 30% of all pregnant women—cases as well as controls—lived within one mile of a hazardous waste site. This suggests that a large part of the American population is actually at risk of exposure to hazardous wastes and that the experts' complacency may be based on absence of evidence, rather than evidence of absence, of harm.

Despite common features, each hazardous waste site is unique. The mixture of chemicals differs between sites, as do the soil and water conditions that determine where and how quickly chemicals leach. Moreover, all of these factors vary by location within sites, and even with time at the same location. As a result, the symptoms of affected individuals are not uniform. One person may develop lymphatic cancer, another lung cancer; a third, neurological problems, a fourth, kidney failure. All of the problems may be due to drinking water from the same waste site, but differences in the leachate with time or region of the site lead to different illnesses. The diversity of symptoms makes it more difficult for epidemiologists to identify the existence of a problem and make it almost impossible for toxicologists to identify the causes accurately. Financial constraints often make it difficult for even the strongest class-action suits to be appealed if the initial verdict is unfavorable, since some of the claimants are too poor to wait. As one of the farmers in the Michigan PBB case said after the out-of-court settlement: "We all got a raw deal, all the way through. The lawyers were pushed into it, the same as we was. They were good lawyers but they were whipped. Our health is gone. Our credit is no good. As soon as you mention PBB, people back right off from you. Even the doctors here want nothing to do with it."[11]

CASE HISTORIES: METALS

Metals, being elements, can be neither degraded nor metabolized: they are an example of ultimate persistence. Most elemental metals do not easily enter living organisms, but metals can form both inorganic and organic compounds, which may differ markedly in their access to, and effects on, living organisms. Nonetheless, the essential toxicity of metal compounds is characteristic of the metal, modified by the degree to which a specific compound reaches specific tissues.

All metals are toxic if present in an organism in excess, but some metals are also essential trace elements, taken up actively by the organism. Of these, calcium, iron, and zinc are needed in the largest amounts. Selenium and copper are also essential elements but are needed in much smaller quantities than iron, zinc, and calcium. Absorption from the gut, as well as the percent absorption by target tissues, changes with the organism's need for a specific element at a given time. Under deficit conditions, a given metal will be absorbed more avidly. This increased absorption will frequently extend to other metals occupying the same relative position in the periodic table: we say that these metals "follow" essential trace elements to their target organ. We will be dealing primarily with the toxic effects of metals, but the chemical similarities in a series of metals (e.g., between

[11] From J. Egginton, *The Poisoning of Michigan*, W. W. Norton, New York, 1980.

zinc and cadmium, or between calcium and lead) are highly significant in determining uptake of the toxic metal.

The metals that have severely affected ecological or human health in the last quarter century include lead, mercury, cadmium, selenium, and tin. Of these, tin, selenium, and cadmium are new problems, in the sense that their presence in the environment had not previously been considered to be a hazard. The toxicity of mercury and lead, in contrast, were already known to Pliny the Elder in the first century of the current era.[12] The only surprise is that we continue to be surprised when these metals prove poisonous.

Cadmium

Cadmium is a silvery white brittle metal, with a melting point of 320°C, that readily forms a series of moderately water-soluble salts. Within organisms, cadmium follows zinc, both because of its position in the periodic table and because cadmium is almost invariably found as an impurity in zinc ores, constituting up to 1% of the metallic content of such ores.

Cadmium has no known role in living systems but is used increasingly in numerous industries. In 1988, cadmium production in the United States was 1900 metric tons, with a worldwide production of 19,800 metric tons. In 1968, U.S. production was 6664 metric tons, but worldwide production was 16,100 metric tons in that year. Major 1968 uses of cadmium were electroplating (6,000,000 lb), vinyl stabilizers (2,000,000 lb), pigments (2,800,000 lb, especially CdS), alloys (1,000,000 lb), nickel–cadmium batteries (400,000 lb), and fungicides (25,000 lb).

Release to the Environment.   As a result of the widespread use of cadmium-containing or cadmium-contaminated products, cadmium can be detected in many or most environments. Release of cadmium into the environment occurs in the waste streams created during its incorporation into products and in the degradation of those products in the environment with use. It is released into water or soil from the manufacture of pigments and vinyl stabilizers and into soil in sewage sludge from water treatment facilities. Fumes from electroplating processes release cadmium into the air. In addition, however, cadmium is released to air when coal is burned, because it is an impurity in coal, and as rubber tires wear against pavement, because the zinc oxide used in vulcanization contains 20 to 90 ppm cadmium as an impurity. In rural air in the United States cadmium levels are 0.003 $\mu g/m^3$, while urban air contains 0.02 to 0.05 $\mu g/m^3$. Zinc smelting is associated with particularly high levels of cadmium, with up to 0.5 $\mu g/m^3$ being recorded near a smelter in Japan. A Swedish factory using cadmium alloys produced air levels of 5.4 $\mu g/m^3$.

---

[12] Pliny the Elder was an inveterate collector of information, with a high tolerance for tall tales. He kept several scribes busy writing down the information he collected from travellers and from his own reading. Interspersed among stories of two-headed tribes and other wonders are well-organized compendia of then current knowledge of mineralogy, metalworking, agronomy, art, and all other human endeavors. An annotated edition of Pliny (one that provides modern words as well as the archaic terms for common items) is extremely enjoyable reading.

Cadmium enters water from sewage sludge, especially if industrial wastes are incorporated into the sludge, and also from fertilizers, in which it is an impurity. U.S. waters contain 1 to 130 ppb of cadmium. Unlike lead, cadmium is translocated into crops from soil and water, and the precise uptake depends not only on the crop but also on the pH of the soil. In Japan, rice grown in or with contaminated irrigation water averaged 0.37 ppm of cadmium, with a range of 0.05 to 1.02 ppm. Cadmium also enters the human environment in cigarette smoke. Some toxicologists consider the cadmium in cigarette smoke to be the heaviest exposure that American smokers have.

Biological Effects.   Cadmium is extremely toxic to plants as well as to animals. As little as 1 to 10 ppm has been shown to affect plant growth, while 5 ppm in drinking water shortened rats' lives by 15%. Inhalation of 40 mg, with retention of 5 mg, is fatal to humans. Contamination of the environment by cadmium has caused severe human disease. In Japan, cadmium from a smelter contaminated irrigation water; the rice crops irrigated by the contaminated water contained up to 1.02 ppm of Cd. Consumption of this rice caused a disease named *itai-itai,* or "ouch-ouch," because of the extreme pain. The disease affects primarily middle-aged and elderly women and is characterized by extreme fragility of the bones, with spontaneous fractures, bone deformity, and severe pain of bones and joints. Hundreds were affected. It is conjectured that multiple pregnancies and a diet marginal or frankly deficient in calcium caused the symptoms to be particularly severe in multiparous women; however, unlike most "environmental" diseases, no occupational exposure was involved—only consumption of rice grown in contaminated water.

At lower levels of intake, cadmium is bound to metallothionein and sequestered in the kidney. It has recently been postulated that when the kidney is "saturated," at about 50 years of age, cadmium is released into the bloodstream and may trigger malignant hypertension. Cadmium causes cancer in rats and has been linked epidemiologically to prostate cancer in humans. It has been noted that life expectancy is significantly decreased in women with itai-itai, but it is not yet clear whether this decrease is due to complications associated with repeated bone fractures, or whether some of the suspected cadmium-mediated effects are involved.

Lead

Lead (Pb, atomic weight 207; melting point, 327°C; specific gravity, 11.35) is a soft gray metal that readily combines with anions to form divalent salts. It is the most immutable of substances, the ultimate product of radioactive decay. Lead has been described as the most important toxic hazard in the development of civilization, a dubious distinction it owes to its ubiquity and versatility and to its consequent heavy use.

The availability and the ease of working lead, with its low melting point, make it one of the earliest metals to be worked and led to the use of the metal itself from Roman times on. Today, economically important lead compounds include lead chloride ($PbCl_2$), lead acetate or sugar of lead [$Pb[OC(O)CH^3]_2$], the insecticide lead acid arsenate ($PbHAsO_4$), pigments such as $PbCrO_4$, Pb_3O_4, and white lead ($PbCO_3$), as well as the organometallic tetraethyllead [$Pb(C_2H_5)_4$], long used as an antiknock additive in gasoline.

Production of lead in 1966 was 2.7×10^9 lb in the United States and worldwide production that year was 6×10^9 lb. That means that U.S. production—and use—accounted for almost half the worldwide consumption of lead. The automotive industry was the dominant consumer: in 1970, 1186 million pounds was used in storage batteries, 556 million pounds as gasoline additives, 144 million pounds in ammunition, 138 million pounds in solder, 196 million pounds in pigments, 48 million pounds in lead type, and 436 million pounds in miscellaneous metal products. Since then, production figures have changed considerably, because some uses (notably use of lead-based pigments for house paints) have been banned. Other uses, especially addition of tetraethyllead to gasoline, are being phased out, and less labor-intensive alternatives to lead type are used increasingly in printing. Even use of lead shot is being restricted, because birds try to use spent shot instead of pebbles in their gizzards and die of lead poisoning. In other countries, however, use of lead in gasoline continues to increase with the increasing use of automobiles. Moreover, despite the attempts to decrease lead use in the Untied States, this country still uses considerable quantities. In 1988, the United States *produced* 394,000 metric tons of lead, out of total world production of 3,381,300 metric tons (10% of production) but *consumed* (used) 1,201,000 metric tons, compared to worldwide consumption of 5,665,400 metric tons (over 20% of world use). Obviously, the consumption pattern have shifted, but world use of lead continues to increase.

Release to the Environment. Lead is released to the environment primarily during smelting and through the combustion of tetraalkyllead compounds in automotive engines, with the latter use accounting for 94.8% (300,000,000 lb) of all atmospheric lead in 1970, when U.S. gasoline contained an average of 2.6 g of tetraethyllead per gallon (Table 6.1). Despite increasing evidence that the airborne lead resulting from tetraethyllead was hazardous, U.S. use of lead as a gasoline additive was increasing 5% per year until catalytic converters were introduced as part of the air pollution cleanup in the mid-1970s. It is ironic that this most insidious form of lead as an environmental poison was eliminated almost by accident. The catalytic converter was introduced to decrease the emission of smog-producing NO_x by automobiles. But because lead poisons the platinum that is the catalyst of these devices, cars equipped with catalytic converters cannot burn leaded gasoline.

In gasoline augmented with tetraethyllead, lead is released as an aerosol of $PbCl \cdot Br$, with a median diameter of 0.2 μm. In Los Angeles, the fate of lead emitted from automobile engines was distributed as follows:

· 20% of the lead stays in the oil and motor of the car. Pollution will result when and if the dirty oil is disposed of into the environment. It could, however, be captured if the oil is recycled properly.
· 43% of the remainder of the lead enters the atmosphere as "near fallout," consisting of particles greater than 9 μm in diameter. Such particles precipitate from the atmosphere within 0.5 mile of the point of emission.
· 16% of the lead is emitted as "far fallout,"in particles greater than 0.3 μm but less than 9.0 μm in diameter.

Table 6.1 Atmospheric Lead Pollution and Estimated Absorption of Lead by Adults

Lead in Air ($\mu g/m^3$)	Absorption (μg Lead per Day)			Expected Blood Lead Level ($\mu g/g$)
	Air	Diet	Total	
0.0005[a]				
2.0	14	30	44	0.21
2.5[b]				
20.0	69	30	99	0.40[c]
24–38[d]				
50	345	30	375	0.72[e]

Source: Adapted from U.S. Environmental Protection Agency, *Air Quality Criteria for Lead,* EPA-600/8-77-017, 1977.

[a] Clean air.
[b] Low end of range in urban air.
[c] Significant ALA excretion begins here.
[d] Measured in downtown Los Angeles and near freeways during rush hour.
[e] Frank clinical toxicity starts in this range.

· 13% is more or less permanently airborne, being less than 0.3 μm in diameter. These particles may circumnavigate the globe or may be precipitated with snow, rain, or fog (see Chapter 4).

Over a 30,000-mile driving distance, 70 to 80% of the lead in the gasoline is released into the atmosphere; because of their small diameter, the aerosols remain airborne for long periods of time. Even the ice in the Greenland icecap is contaminated. Lead levels in the ice pack rose from 0.06 ppb in 1900 to about 0.22 ppb in 1970. Soil along heavily traveled roads contains >1000 ppm lead (see Figure 6.5), street dust contains up to 6900 ppm, and urban air contains from 0.5 to 40 $\mu g/m^3$. Lead levels in smaller communities can be illustrated by data from central Illinois in the early 1970s (Figure 6.5). Because lead with a particle size of <0.5 μm is more efficiently absorbed through the alveoli than through the healthy gastrointestinal tract, airborne lead is considered a major contributor to childhood lead poisoning.

Lead used as pigment in paint is a major source of local environmental pollution and has been considered an important cause of childhood lead poisoning, especially in old houses with peeling paint. Children with pica[13] eat paint chips. As paint ages, lead-containing dust in the air can be a source of inhaled lead; lead-containing dust on surfaces can be picked up on toddlers' hands and ingested. Despite the banning of lead in paint pigments, the hazard remains, since much of the 3.4

[13] *Pica* is the name given to compulsive consumption of nonfood items. It is seen more often in children and adults with nutritional deficiencies. In children, it may be an extension of the normal urge of toddlers to put anything into their mouths; in adults, the best example is the craving of pregnant women from certain rural cultures for clay. When urbanized, these women may consume cornstarch to the point of interference with absorption of nutrients.

Soil Transect of High and Low Traffic Volume

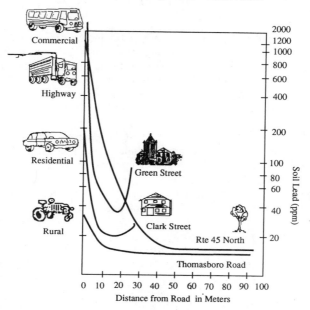

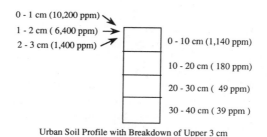

Urban Soil Profile with Breakdown of Upper 3 cm

Figure 6.5 Patterns of lead deposition in the top 10 cm of soil in Champaign–Urbana, Illinois, in 1973. Traffic on Green Street was below 1400 vehicles/day. (Drawn by G. K. Francis based on data from *The Lead Project, 1972–1975*. Institute for Environmental Studies, University of Illinois.)

million metric tons of lead-containing paint remains in place. Even if covered by several layers of unleaded paint, this lead can in the future crumble off the wall, yielding lead-containing dusts for inhalation and ingestion.

Lead also enters the human organism in food (0.4 ppm in cereals, 0.18 ppm in beverages, 0.5 ppm in fruit, 0.3 ppm in meats) and in water (0.02 to 0.92 ppm). Because lead was used in making pipes well into the twentieth century (the word *plumber* derives from the Latin word for lead, *plumbum*) and lead-containing solder was used in joining pipes until very recently, drinking water can be a source of lead poisoning. Fortunately, gastrointestinal absorption of lead is not efficient in healthy adults; the major danger in adults is from occupational exposure. In children and in adults with certain illnesses, however, absorption from the stomach is significant.

Biological Effects.   Lead is inherently toxic and has no useful function in the mammalian organism. Its toxicity has long been known. Pliny the Elder complained of merchants who adulterated wine with sugar of lead (lead acetate) to make it sweeter; he also commented on the high mortality associated with certain silver mines, probably due to the lead ore that is always found with silver. Lead's reproductive and neurological effects are credited by some with assisting the decline of the Roman empire.

Typical symptoms of acute lead poisoning or *plumbism* are intestinal cramps or "painters' colic," renal failure, sterility, and irreversible brain damage (cerebral palsy and mental retardation). Anemia results from impaired heme synthesis and hemolysis. In milder adult cases, mental effects present as irritability; in children, lead is suspected of contributing to hyperactivity, behavior disorders, and learning disabilities even in the absence of overt poisoning.

Evidence of chronic poisoning can be obtained from x-rays, since lead deposits in bones and teeth as lead phosphate, producing "lead lines" on the x-ray. Lead also binds to sulfhydryl groups and dithiols in proteins, which may account for its toxicity in tissues that do not contain specialized heme-proteins. The underlying biochemistry of lead-induced anemia probably begins with the inhibition of the enzyme delta-aminolevulinic acid dehydratase (ALA-D), which is required for heme synthesis. In the late 1970s, decreases in peripheral red blood cell ALA-D seemed to be the most sensitive indicator of lead exposure, detectable at blood lead levels as low as 10 μg per deciliter (dL), a level of lead exposure at which no other or adverse effects were detected. A few scientists argued forcefully that perturbations of the hematopoietic system by lead were "biochemical dysfunctions which probably cause harm" rather than "biochemical adaptations that may be rather harmless."[14] The National Academy of Sciences Report "Lead in the Human Environment" concluded, however, that although "the most readily measured biological change that can presently be associated with lead in the body is inhibition of enzymes involved in the biosynthesis of heme . . . (the) significance to health of biochemical changes without associated evidence of illness is uncertain, and the question of what degree of inhibition of heme synthesis should be considered a detrimental effect is controversial, both scientifically and in policy debates."[15]

The same report of the National Academy of Sciences listed the lowest blood lead level associated with cognitive deficits in children to be 50 to 60 μg/dL, again echoing the 1970s consensus that clear damage to health in children occurred at blood levels above 400 ng/mL (40 μg per dL). Not that the Committee took the risk of lead poisoning lightly: they were fully aware that childhood blood levels

[14] C. Patterson, a member of the National Academy Committee on Lead in the Human Environment, wrote a powerful minority report arguing that "the mining and smelting of lead and dispersal of manufactured leaded products is actually a monumental crime committed by humanity against itself."

[15] National Academy of Sciences, "Lead in the Human Environment," a Report Prepared by the Committee on Lead in the Human Environment, Environmental Studies Board, Commission on Natural Resources, National Research Council, 1980.

of 40 μg/dL were not rare. The report notes that, in a 1970 study, 10% of Chicago children had blood lead levels greater than 60 μg/dL. The potential human costs in terms of mental retardation and shortened life span were clearly enormous.

Unfortunately, the 1970s consensus was too optimistic. Since 1980, numerous epidemiological studies have shown that neurological damage occurs in children with as little as 10 μg of lead per deciliter of blood. In 1991 the Center for Disease Control (CDC) redefined the level at which a child is considered to need intervention from 25 μg/dL to 10 μg/dL. At this blood lead level, inhibition of ALA-D is not statistically significant, making childhood cognitive dysfunction the most sensitive indicator of lead toxicity. Acceptance of the lower threshold for childhood neurological damage means stricter federal regulations of lead in drinking water, and extensive efforts in many cities to remove lead-containing paint from old dwellings. Some scientists now argue that there is no threshold for the developmental toxicology of lead, suggesting that the blood lead levels at which intervention is recommended may decrease yet further.

Between the optimistic 1970s and the 1990s, it has also become clear that childhood lead poisoning is not restricted to large cities, or to the poor. Lead-containing paints and putties were used in houses of all size well into the 20th century. The ubiquity of lead-containing paints, battery cases, and lead-contaminated soil makes lead poisoning a common source of poisoning in domestic and farm animals, even in rural areas. Low temperature glazes on pottery, hobbies involving the use of lead solder or lead-containing pigments, and (for adults) storing brandy in crystal decanters, have also been implicated as sources of actual or potential lead intoxication. Moreover, the threshold levels for lead toxicity keep decreasing as investigators become more conscious of the risks and as both behavioral assays and exposure assessment become more precise.

Mercury

Mercury is a silver metal, liquid at room temperature, with a density of 13.6, a melting point of $-38.9°C$, and a boiling point of $357.3°C$. It is a relatively rare element, making up approximately 0.08% of the earth's crust by weight. It is not often found as the element, but most commonly as its sulfide, in the ore called cinnabar. The free metal, which is extracted by burning the ore, has a high vapor pressure, leading to considerable air levels when mercury is in an open container even at room temperature.

Use and Release to the Environment.   Mercury has been and continues to be used in numerous products. The metal itself is the most used form of mercury. Its major uses are in electrical goods (mercury vapor lamps, mercury switches, batteries) and in the chloralkali industry, where it functions as the cathode in the electrolysis of NaCl to produce chlorine gas and NaOH. Metallic mercury is also used in dental amalgams, in thermometers, and as a catalyst in the plastics industry. Inorganic salts are relatively little used. Use of organic mercury compounds, while significant, has perhaps decreased due to the many episodes of pollution involving mercury.

Because it is an element, mercury is indestructible. It has long been known to be toxic. The "mad hatter" of Lewis Carroll's *Alice in Wonderland* is thought to have been "mad as a hatter" because he was a victim of the mercury then used in felt production. Nonetheless, mercury is so versatile and so valuable that it continues to be used. Moreover, although many of its uses have been eliminated, the wastes that have accumulated from past production cannot easily be cleaned up. A major source of mercury waste is the paper industry: much of the mercury was dumped into rivers or holding ponds. Once in the environment, mercury can be methylated by microorganisms, producing the biologically much more available *methylmercury*.

Toxicity.   Organic mercury compounds are highly toxic to most organisms and have been used as fungicides in agriculture, as slimicides in paper production, and as mildew retardants in latex paints. The latter use is still permitted, but agricultural uses of mercury were restricted during the 1950s after several episodes of human consumption of treated seed. In Iraq in 1973, 459 people died and 6530 illnesses were reported, following consumption of methylmercury treated grain during a famine. It is considered probable that the actual incidence of illness was much higher, but because the government attempted to end the poisoning by making it a crime to have the treated seed, considerable underreporting of illness occurred.[16] This tragedy and several similar, although smaller episodes in other countries led to the outlawing of mercury fungicides on edible seeds. The major uses of mercury at present are in the chloralkali industry, where mercury is the cathode in the electrolysis of NaCl to produce chlorine gas (and NaOH).

As described in the case history in Chapter 2, methylmercury has caused documented illness and death in purely environmental exposures because it accumulates in the food chain. It remains a major hazard for fish-eating populations along contaminated rivers. In particular, the indigenous peoples of Canada's Northwest, with traditional reliance on fish as a major source of protein, continue to suffer from mercury poisoning.[17]

---

[16] The somewhat disturbing question of why people would assume that their government wished to starve them has been answered by the events that followed the Persian Gulf War. Given the hostility between Kurds and the central government in Baghdad, it is not surprising that the people thought the presumably benevolent warning against eating the contaminated grain was an attempt at genocide.

[17] The ravages of methylmercury poisoning are described in "The Huckleby hogs," pp. 253–272 in Berton Roueché's *The Medical Detectives*, Washington Square Press, Time Books, New York, 1982.

7

HEALTH EFFECTS OF ENVIRONMENTAL CHEMICALS

> If you want to explain each poison correctly, what is there that is not poison, all things are poison and nothing [is] without poison. Solely the dose determines that a thing is not poison.
>
> —Paracelsus[1]

DESCRIPTIVE TOXICOLOGY

Theophrastus Bombastus von Hohenheim (1493–1541), usually known by his eponym of Paracelsus, was a physician who aroused great controversy in his own time, and has fascinated historians, doctors, and even poets ever since.[2] Today he is variously credited with being first to recognize the chemical basis of medicine and with emphasizing the importance of observation in medicine. Modern toxicologists are wont to call him the father of toxicology, because he observed that toxicity is a relative rather than an absolute quality. If there is no essential difference between toxic and nontoxic substances except dose, all substances are potential poisons. If one accepts this axiom—as toxicologists do—it becomes necessary to evaluate the *relative* toxicity of substances rather than arbitrarily describing them as "safe" or "poison." Toxicology then becomes not the science of identifying poisons, but the science of quantifying the levels at which substances are toxic, and then of describing various types of toxic effects.

Endpoints

The first decision to be made in quantifying toxicity is the definition of the "toxic effect": the choice of an endpoint to be measured. The second decision is how to measure the chosen effect. These decisions are related, because the choice of an endpoint determines how one measures it.

[1] Quotation from W. B. Deichmann, D. Henschler, B. Holmstedt, and G. Keil, *Archives of Toxicology* 58:207–213, 1986.
[2] A play titled *Paracelsus* is included among the poetical works of Robert Browning.

Historically, the first and most reliable measure of toxicity was the dose that caused death, and death remains the least equivocal and least controversial evidence of toxicity. It is not the only endpoint toxicologists use, but it is usually the first, and it always serves as a benchmark. Death is irreversible; it is readily recognized; it is undeniably a toxic effect. If eating a plant, being stung by an animal, or swallowing a chemical can kill, then the plant, animal, or chemical is surely toxic: poison. Moreover, chemicals, toxins, and poisonous plants can be compared by how much it takes to kill a person or an animal.

As soon as comparisons of toxicity are systematically made, it becomes clear that several factors influence the lethal dose. A major factor is the *route of administration*. Some chemicals are as deadly when applied to skin as when given by mouth, while others are relatively harmless if applied to the skin. Most often, however, an agent is less toxic if given by mouth than if injected. Routes of administration that place an agent directly and quickly into the organism's tissues—inhalation, intravenous injection—are often (but not invariably) the most toxic. Slower delivery of a chemical to tissues—by intramuscular injection, or by percutaneous exposure—tend to be less toxic. Enough variability exists, however, that each agent must be tested by each route of administration.

Even within routes of administration, differences in toxicity may be seen. Generally, a poison administered by mouth is more effective, and acts more quickly, if given on an empty stomach than if given after a meal. Injured skin allows the passage of more chemicals than does intact skin. And of course, the lethal dose depends on the size of the victim. Not only does the same dose that kills one rat kill several mice, but large rats require larger doses than small rats. Because of this ability of larger animals to withstand larger absolute doses of a poison, all dosing is done on the basis of the animal's weight: usually, milligram of toxicant per kilogram of body weight (mg/kg).

Toxicologists use animals to determine the toxicity of chemicals and have devised protocols that give reasonably reliable results. These protocols take into account the different sensitivities of individuals to a poison as well as the possibility of weakening an animal by repeated doses. Because species differ in many ways, including their metabolic capabilities, extrapolation between species is always less certain than extrapolation within species. Such differences between species are more commonly seen with some endpoints (e.g., developmental deficits) than others (neurotoxicity), and usually increase with increasing phylogenetic distance between species. Thus rats are a better model for humans than are insects.[3]

---

[3] It is said that Lucretia Borgia tested her potions on servants and peasants before poisoning her relatives. This is probably more reliable than using rodents, but the Borgias are not considered good role models for toxicologists. Human testing of chemicals is allowed only after the major outlines of toxicity have been identified in nonhuman species, and only with the *informed consent* of the volunteers and the approval of a designated committee.

Median Lethal Dose

How does one determine the lethal dose of a chemical? If one gave increasing doses to separate animals (i.e., the first animal gets 1 mg/kg, the next gets 5 mg/kg, and so on, until a killing dose is found), it will often happen that the animal given a certain dose x dies, while the animal given the next higher dose, $x + n$, survives. Although it is easy to say that the latter animal is more resistant to the chemical, it is not possible to decide which dose is the "true" lethal dose.

As an alternative, one might consider giving increasing doses of a chemical to a single animal until a dose is found at which the animal dies. At doses that cause illness without actually killing an animal, it is obvious that enough time must be allowed between doses to permit the animal's full recovery. Unfortunately, in testing newly discovered or recently synthesized chemicals, it may be difficult to define "enough time." If insufficient time is allowed for recovery between consecutive doses, the effect of several doses may be cumulative. A new animal given the supposedly lethal dose might remain quite alive.

To avoid such vexing arguments about the actually lethal dose, toxicologists treat groups of animals at each of several doses and measure the *percent* or *proportion* of animals reaching the endpoint. To determine lethality, the range should extend from the invariably nonlethal to the invariably lethal. But the invariably lethal dose and the invariably nonlethal dose are very difficult to define.[4] In contrast, the *average* (either the median or the mean)[5] of a statistical distribution can be reliably identified by finding values above and below it, and interpolating the 50% level. Therefore, the lethality of a compound is usually defined in terms of the *median lethal dose*, LD_{50}. Note that determining LD_{50} involves the use of statistics, and statistics is itself a science. We cannot deal adequately here with the problems of using statistics correctly, but be aware that the correct interpretation of statistics often makes the difference between true and false conclusions.

Nonlethal Measures of Toxicity

Although death is in many ways the easiest endpoint to define, it is not necessarily the most appropriate. We may well be interested in the dose

---

[4] If an agent causes 1% mortality at a given dose, it can be *expected* to kill 1 animal of every 100 dosed. If the usual sample of 20 rats per dose is tested, one death is already 5% mortality. More plausibly, no deaths would occur, even though the *true* incidence is greater than zero.

[5] The *mean* is the sum of all the values in a set, divided by the number of values. If N is the number of animals and mg/kg_i is the lethal dose for each animal i, the mean is the sum of all the lethal doses divided by N, or: $\Sigma(mg/kg_i)/N$. In contrast, the *median* is obtained by ranking all values in ascending or descending order and finding the middle value: that is, it is the middle dose. The difference is that the mean is sensitive to unusually large or unusually small values. A familiar example is shown by looking at the mean and median for salaries in consecutive years. If you earn $2000, $3000, $4500, $5500, and $15,000 in five successive years, your *mean* annual income is $6000; your *median* annual income is $4500.

needed to kill a pest organism when we are determining the toxicity of an insecticide or a rat poison. But when considering other chemicals, such as drugs, household cleaners, or food additives, we would worry about far less serious effects. Other measures of toxicity would then be appropriate. Most of these other measures are based on the degree of illness that is seen at a given dose. The protocol is nonetheless similar to that used for determining LD_{50}: a series of doses are administered to groups of animals, and the *median effective dose*, ED_{50}, is calculated. The range should include one or more doses that almost never cause the endpoint as well as at least one dose that almost always causes it.[6]

Death is a discontinuous variable: an all-or-nothing effect. An animal is dead or alive; it cannot be moderately dead. Illness, on the other hand, is a continuous effect, ranging from "slightly ill" to "very ill." In quantifying toxic effects, continuous variables are difficult to score. Therefore, one commonly measures the degree of illness as a discontinuous variable by dividing illness into discrete stages. Unless there are well-defined markers for such stages, the distinctions can get subjective, but the gradations are remarkably robust when numerous animals are evaluated. The various gradations of illness are referred to as clinical toxicity. It will still be difficult to discriminate between a chemical that causes 25% of treated animals to be "slightly" ill, 50% to be "moderately" ill, and 25% to be "very" ill and another chemical that causes 50% to be "slightly" ill and 50% to be "very" ill; one may have to adjust the criteria in terms of the purpose of the test compound. If, for example, the chemical is for use in cancer therapy, it would be preferable to have 75% of patients become no more than moderately ill, with only 25% becoming "very" ill. If the chemical is to be used as a rat poison, one might select the chemical that causes 50% of the rats to be very ill and see whether a somewhat higher dose can push the rest of the population into the seriously affected category.

As techniques for identifying effects of chemicals have become more sophisticated, toxicologists have often identified biochemical effects that occur well before obvious illness is seen, or that occur at levels well below the doses at which a particular toxicant causes death or even mild illness. Such biochemical markers may be either qualitative, such as the presence of a cell constituent that is not normally present; or quantitative, such as an increase in the level of an enzyme that is always present to some extent. These biochemical effects constitute a category called *subclinical toxicity*. There is no overt illness; often the effect can be measured only with instruments. But there is an effect.

Alcohol (ethanol; C_2H_5OH) provides an excellent illustration of the several endpoints of toxicity. The median lethal dose (LD_{50}) of orally administered ethanol is above 5000 mg/kg in humans, putting this chemical into the "mini-

---

[6] Absolutes are statistical impossibilities. One can determine the probability that an event occurs, but when animal testing is involved, the probability is *never* zero. . . . Or hardly ever?

Table 7.1 Classification by Acute Toxicity

Toxicity Class	Probable Lethal Dose (mg/kg, human)[a]	Example	Actual LD_{50}
6. Supertoxic	<5	Botulinum toxin	10^{-9}
		TCDD	10^{-3}
5. Extremely toxic	5–50	Parathion	1–10
		Dieldrin	10
4. Very toxic	50–500	DDT	100
3. Moderately toxic	500–5,000	Ethanol	>5000
2. Slightly toxic	5,000–15,000	PCBs	
1. Practically nontoxic	>15,000		

Source: Adapted from G. Zbinden, *Progress in Toxicology*, Vol. 1, Springer-Verlag, New York, 1973.

[a] If human data are not available, it is assumed that the LD_{50} value from animal experiments can be substituted for the mean lethal dose in humans.

mally toxic'' category (see Table 7.1). Well below the lethal level, however, alcohol induces several toxic effects, including nausea, vomiting, and stupor. These are examples of clinical effects. At even lower levels, it is still possible to discern behavioral effects and changes in reaction time to stimuli. The more pronounced behavioral effects, such as badly slurred speech and the inability to walk a straight line, would be considered clinical effects. Milder behavioral effects, such as a slight decrease in critical faculties or a slightly greater difficulty in speaking clearly, would probably rank as subclinically toxic effects because of their subjective nature.

Alcohol also shows the fairly common phenomenon of *tolerance*. Tolerance means that repeated doses result in diminished exhibition of toxic effects (or, conversely, progressively higher doses are needed to cause the same level of toxicity). Tolerance is caused by the induction (increased production) of liver enzymes, which are active in the degradation (detoxification) of alcohol. The presence of such elevated enzyme levels can also be considered a toxic effect of alcohol, even in the absence of the subtlest behavioral effects and even in the absence of any measurable dose of alcohol in the body. What, then, is the ``toxic dose'' of alcohol? Is it the dose that causes stupor? Or the dose that interferes with a drinker's ability to walk a straight line and enunciate clearly? Or is the induction of alcohol dehydrogenase a toxic effect, regardless of the presence of other symptoms?[7]

---

[7] Note that in testing urine for illegal drug use, it is almost invariably *metabolites* of the drugs of interest that are present. In some cases these metabolites are detectable days or weeks after the last dose of the parent compound. Thus an employer who tests employees' urine for marijuana in the interests of a ``drug-free workplace'' also identifies weekend use—even if the use was earlier than the last weekend.

The questions concerning the toxicologic significance of biochemical markers become especially important in considering the effects of chronic exposure, or otherwise delayed effects of chemicals. Few people would claim that alcohol is toxic at all levels of consumption, and some studies suggest benefits from moderate drinking (up to two ounces of alcohol per day). However, the potential significance of subclinical markers of exposure is extremely important in determining public policy toward specific toxicants, and large amounts of money are at stake in controversies over definitions of toxicity. One example of a controversy over the significance of enzyme inhibition by a toxicant is seen in the case history of lead (Chapter 6).

A second example is found in the irreversible inhibition of acetylcholinesterase (AChE) by organophosphorus ester (OP) insecticides (see Chapter 8). Over 50% of AChE must be inhibited before symptoms of intoxication occur; death can occur once 80% of AChE is inhibited. Many OPs are absorbed through the skin, and death can occur within minutes of a heavy exposure to one of the more toxic OPs. Resynthesis of the inhibited AChE is a slow process, taking up to a month. People who are exposed to an OP with less than 50% inhibition of AChE may not realize that exposure has occurred or may assume that the absence of symptoms means there has been no damage. But their lower levels of AChE will persist for some weeks, making them more vulnerable to subsequent exposures. Clearly, therefore, subclinical inhibition of AChE carries a risk, and monitoring of such subclinical inhibition in situations where exposure is probable may prevent deaths.

OPs also inhibit a number of serum esterases whose functions are not known and whose inhibition is not associated with clinical symptoms. In occupational settings with a high risk of OP exposure, inhibition of a significant fraction of these serum esterases is sometimes used as an indicator of ongoing or potential AChE inhibition and as a reason for removing workers from the possibility of exposure. Thus, even the inhibition of clinically insignificant enzymes can be used to prevent life-threatening toxicity.

Even when there is incontrovertible, undisputed toxicity, however, there may still be difficulties in measuring it. Consider the least controversial measurement of toxicity,[8] the median lethal dose, LD_{50}. LD_{50} is defined as the dose that has a 50% probability of killing a given individual and/or kills 50% of a population. First, LD_{50} is specific to the route of administration. If this is not specified, one can usually assume that it is oral (*per os,* or *po*). Second, LD_{50} is species specific. Species vary in their response to specific chemicals, a fact that is the basis of a thriving pesticide industry. For the purposes of this book, as in most of the literature pertaining to health effects, the rat is the reference species. If the species is not specified, assume that

---

[8] There are, in fact, controversies over the *use* of the LD_{50} to measure toxicity. But the controversies over the LD_{50} deal less with its significance or reliability than with the large numbers of animals used to determine it. Careful use of statistical techniques can minimize the numbers of animals that must be killed to quantify the median lethal dose. Additional reductions can result if regulatory agencies do not insist on extreme precision in estimating the LD_{50}.

it is the rat. This is the standard regulatory assumption. Therefore, the unqualified statement that "the LD_{50} of compound x is y mg/kg" ordinarily refers to the oral LD_{50} of compound x in rats.[9] In principle, the species used should be that species most similar to the human with respect to the target or mechanism of action of the agent. In practice, that information is rarely available. Species are selected on the basis of size, cost, and availability. Besides rats, species commonly used in toxicity testing include the mouse, guinea pig, and hamster. Rabbits are favored for evaluating irritation to the eye, and dogs are used when a larger species is required. Primates are used in rare instances, especially if there is a controversy about interspecies variation in the disposition of a chemical.

Toxicologic assessment for ecological purposes is even more complex and is often tailored to the anticipated environmental presence of the chemical. Only recently have federal laws required that industrial chemicals be tested for environmental fate and persistence. For some ubiquitous contaminants (e.g., lead, cadmium, and PCBs), data on a large number of species are available. For many less notorious chemicals, most of the data remain unpublished, available only in the documentation presented to regulatory agencies. Even if published in a government document, such data are often difficult to find.

Pesticides, because of their inherent toxicity and because of the very high probability that nontarget organisms are exposed, have been extensively examined for their toxicity to a large number of species, ranging from soil microorganisms to wild birds. It is nonetheless difficult to make direct comparisons between compounds, because the tested species vary so greatly. This is inevitable, given the range of organisms (from bacteria to birds) that may be exposed, the varied uses to which pesticides are put, and the different ecosystems to which they are applied. Table 7.2 illustrates the toxicity of some pesticides to different animals.

Since most pesticides eventually reach soil and water, some testing of their effects on prokaryotes is common. This varies from determining whether rather general soil processes such as ammonification or nitrification are inhibited or enhanced to identifying toxic effects on specific microorganisms in culture. Comparisons among different tests are often difficult, because it is not clear to what extent differences in the media (ranging from different soil types to agar or nutrient broths) affect results. Pesticides behave differently under different soil conditions (e.g., cation exchange capacity, oxygen content) as well as in different soils (sand, clay, loam, and combinations), and at different temperatures. Moreover, differences in the particular species of microorganisms present will affect overall effects on soil processes.

---

[9] This vague description is not, however, a notation to be recommended. It is important to give the reader as much information as possible about the experimental situation. Rather than "the LD_{50} of compound x is y mg/kg," one should say that "the *oral* LD_{50} of compound x in *10-week-old male Swiss-Webster rats* is y mg/kg." For many compounds, strains within species, sex within strain, and/or age affect the LD_{50}.

Table 7.2 Toxicity of Pesticides to Different Phyla

Pesticide	Invertebrates 96 hr LC$_{50}$ (ppb water) Daphnia[c]	Freshwater Fish[a] 96 hr LC$_{50}$ (mg/liter water)		Birds: Acute p.o. LD$_{50}$ (mg/kg body wt.)		Birds: Dietary[b] (ppm diet)		Mammals: Acute p.o. LD$_{50}$ (mg/kg body wt.)
		Bluegill	Rainbow Trout	Bobwhite[d]	Mallard[e]	Bobwhite[d]	Mallard[e]	Rat[f]
Aldrin[g]	32	0.013	0.0026	6.6	52		155	20–70
Atrazine[h]	3600	17	8.8			5760	19650	3000
Azinphos-methyl[g]		0.0046	0.02	90	136	488	1940	
Captan[i]	9960	0.111	0.08					>8000
Carbaryl[g]	6.4	6.76	1.95		2179	>5000	>5000	40–540
Chlordane[g]	590	0.0748	0.090	83		331	858	280–500
DDT[g]	0.36[j]	0.008	0.002[k]					113
Diazinon[g]	0.522	0.079	0.635	10	3.5	245	191	75–120
Fenthion[g]	4.0[j]			5.9				250
Heptachlor[g]	42	0.0026	0.0074		>2000	200	480	100
Linuron[h]	4000	16	16					4000
Malathion[g]			0.17		1485			≥1000
Paraquat[h]	4000		38.7	176		981	4048	100–400

Parathion-methyl[g]	2.0	5.72	4.74	10			♀24; ♂14
Parathion[g]	15	0.047[l]	2.65	2.0			♀13; ♂3.6
Toxaphene[g]	≥1000	0.018	0.05[l]	70.7	834	536	90
2,4-D (acid)[g,m]	≥1000		250[m]				400

Source: Data are taken from D. Pimentel, *Ecological Effects of Pesticides on Non-target Species*, U.S. Office of Science and Technology, Executive Office of the President, 1971, and from several sources reviewed in J. R. Sanborn, B. M. Francis and R. L. Metcalf, *The Degradation of Selected Pesticides in Soil, EPA 600/9-77-02*, 1977.

[a] The bluegill (*Lepomis macrochirus*) represents warm-water fish; the rainbow trout (*Salmo gairdneri*) represents cold-water fish.
[b] Pesticide was administered in the diet for 5 days, followed by clean feed for 3 days, at which time mortality was assessed.
[c] Species was *Daphnia magna* except as indicated.
[d] *Colinus virginianus*.
[e] *Anas platyrhynchos*.
[f] *Rattus rattus*.
[g] Insecticide.
[h] Herbicide.
[i] Fungicide.
[j] The test species was *Daphnia pulex*, not *Daphnia magna*.
[k] Species tested was brown trout (*Salmo trutti*).
[l] 48 hour LC$_{50}$.
[m] The salt, or ester, and the formulation are critical in determining the toxicity of 2,4-D.

It is expected that herbicides will be toxic to plants, and the range of that toxicity would be part of the target organism testing carried out to determine their potential market. Thus the phytotoxicity of herbicides is well-documented for both crop and weed species, but not usually for wild plants without commercial importance (e.g., wild orchids). The toxicity of herbicides to animals is less predictable, and the types of testing to be carried out would depend on the uses to which a particular herbicide is put. A herbicide such as paraquat, which is often used for aquatic weed control, would be tested for toxicity to model aquatic invertebrates such as water fleas (*Daphnia magna*) or planaria (*Dugesia dorotocephala*). Additional species tested might include oysters (*Crassostrea viginica*) in coastal areas and the immature aquatic stages of caddis flies (*Trichoptera* species), mayflies (*Ephemoptera*), or dragonflies (*Libellula* species) in forested areas. Conversely, an insecticide would be routinely tested for toxicity to useful insects such as honey bees (*Apis mellifera*), which are often needed to pollinate the crop, or to predatory mites. The toxicity to aquatic invertebrates other than insects will be determined for any insecticide applied to or near water, but the test organisms will differ. Toxicologists on the Atlantic coast might examine effects on oysters and clams (*Venus mercenaria*). In forested freshwater ecosystems, mayflies and caddis flies would be tested because of their importance as food for fish.

The vertebrate species that are tested will also differ. Among birds, mallards (*Anas platyrhinchos*), bobwhite quail (*Colinus virginianus*), and pheasant (*Phasianus colchicus*) are often tested because of their importance as game birds; conversely, the Japanese quail (*Coturnix coturnix japonica*) is favored because of its small size, even though it is not a North American species. Among fish, the species used include those that are conveniently kept in the laboratory, such as mosquitofish (*Gambusia affinis*) or guppies (*Lebistes reticulatis*); and those that are commercially important, such as trout (*Salmo trutta*, *S. gairdnerii*), bluegill (*Lepomis macrochirus*), and cat fish (*Ictalurus punctatus*). The species tested, especially if it is commercially important, may also depend on the area of use: that is, Canadian scientists would test different species than Floridians would, and Californians different species than New Yorkers. Testing of insecticides for phytotoxicity would probably be limited to assuring that crops will not suffer.

In addition to the oral route, other routes of administration include *intraperitoneal* (ip), *subcutaneous* (sc), *intravenous* (iv), *intramuscular* (im), *inhaled* (IHL), *dermal* or *percutaneous,* and *gavage* (given directly into the stomach). In principle, the route of administration should mimic the expected route of exposure. In practice, gavage is the most common route of exposure in research.[10] Other techniques are considered to be too imprecise (inhalation,

---

[10] Oral dosing too often allows the compound to dribble out of the animal's mouth, increasing the variability of the administered dose and of the observed effects. Gavage provides more certain delivery but also has drawbacks. Gavage is more stressful to the animals and requires greater skill on the part of the handler.

subcutaneous, dermal) or too cumbersome (inhalation, dermal, intravenous) for most applications. For the same reason, gavage is the most common route of administration when only incidental human exposure to the test compound is expected: for example, pesticides or gasoline additives. Pharmaceuticals, however, are tested using the expected route of administration, even if this requires using inhalation chambers or percutaneous dosing. Similarly, a pesticide that is suspected of having caused human illness will be administered to animals by the presumed route of human exposure in order to confirm (or disprove) the activity.

One measures the LD_{50} by giving *predetermined groups*[11] of animals carefully calibrated doses of the test compound, recording the number at each dose that die within a predetermined length of time (usually 24 hours). If graphed, the results usually look like Figure 7.1, which is really the cumulative frequency of all the animals that respond at each dose. The S-shape of the curve when dose is graphed against mortality without transformations of scale is difficult to use in calculations; therefore, the data are typically transformed into response (mortality) against log dose. Notice the importance of covering the full range. If only the bottom of the response curve is analyzed, the slope looks far different, and extrapolation to 100% (or 50%) mortality is quite deceptive. The meaning of the curve is simply that we are

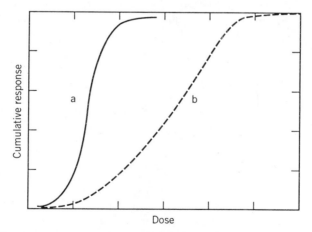

Figure 7.1 A dose–response curve shows the number of organisms responding at increasing levels of treatment. A homogeneous population displays exhibits a steep slope in the cumulative response curve (a), while a heterogeneous population exhibits a shallow slope to the dose–response (b).

[11] Allocating animals to experimental groups can be done "randomly". Statistically, this means that animals are assigned to groups using a table of random numbers. Assigning animals haphazardly, for example in the order in which they are grabbed out of the cage, is *not* a statistically random assignment. In principle, random assignments are ideal. In practice, it is usually preferable to set up treatment groups such that the major variables (e.g., mean weight and/or range of weights), are similar between groups. Otherwise the results may be difficult to interpret.

assuming a normal distribution: that is, one in which the mean ± the standard deviation accounts for 68% of the population. *This is not always the true underlying distribution,* but it is remarkably often a good approximation even for small samples, and *almost* invariably for large samples.

Figure 7.2 shows the distribution of the incremental response for each increase in dose. At low doses, a very few sensitive individuals in the population respond—think of the people who fall asleep after one drink. A correspondingly small number at the other extreme are remarkably resistant to the compound and do not respond even at very high doses. The extremes of a distribution, those values lying more than 3 standard deviations from the mean, called its "tails," are difficult to measure precisely. In mammalian toxicity tests, one usually misses the tails entirely because the sample sizes are too small. In many cases, however, it is precisely the tails that are critical. If the compound being tested is a headache remedy, it is not enough to know that 95% of people who take it do not die. With a potential market of 250 million in the United States alone, it is important to know that not even one person in a million will die. On the other hand, when applying pesticides to a corn field, the other tail—survivors—becomes interesting. The population of an insect pest species may reach 1 million per acre. If the insecticide kills 99.99% of the pest, 1 insect in 10,000, or 100 per million, survive. Because their generation time is measured in days rather than years, the 100 survivors can repopulate the field and destroy the crop. Moreover, because resistance to pesticides is often genetic, the offspring will probably inherit resistance to further applications of the same insecticide. Thus these survivors in the tail of the distribution are critically important in pest control strategies.

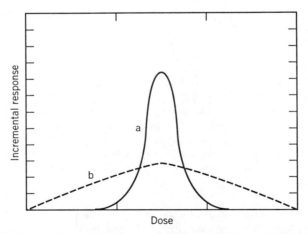

Figure 7.2 Distribution of the incremental number of individuals responding at each dose. A homogeneous population displays little scatter about the median effective dose (has a small standard error), whereas a heterogenous population has a large standard error.

The shape of the response curve is determined by the variability of the population (Figures 7.3 and 7.4). A heterogeneous population will show much more scatter around the mean (larger standard deviation). A homogeneous population produces a tight curve, with most of the population going from no response to response over a very small dose range. Sources of variation include environmental factors (weight, age, health) and genetic factors. On occasion, one sees a bimodal distribution, which signifies two underlying populations. More often, the real distribution is bimodal, but the appearance in toxicity testing is that of a very widespread curve. If sensitivity or resistance is determined genetically, single-factor inheritance would produce a bimodal curve[12] while multifactorial inheritance would produce the apparently monomodal curve with a large variance. A "tail" at one end or the other may also indicate genetic variation within the population.

The shape of the dose–response curve depends on the species in which it is being tested, and is also specific to the chemical being tested. For some chemicals, there are only very small differences between a dose that does not kill and a dose that kills essentially 100% of the population. For other chemicals, there may be a very large difference between these doses. For pesticides, a small difference (i.e., a very steep dose–response curve) is desirable, since it makes it possible to calculate a dose that will kill >99% of the insects. For pharmaceuticals, a large difference between the therapeutically effective dose and the lowest dose at which adverse effects occur is desirable: that is, a shallow dose–response curve is preferable.

The great advantage of statistics in analyzing data is that it becomes possible to quantify various relationships. Unfortunately, statistics can also be used to "add artistic detail and so give verisimilitude to an otherwise bald and unconvincing tale," as Koko notes in Gilbert and Sullivan's operetta *The Mikado*. Users of statistics must understand the assumptions underlying the various tests they use. It is equally important that statistical consumers understand the underlying assumptions of the statistical methods used in the papers that they read—and check whether those assumptions are being violated. It is easy to lie with statistics, intentionally or unintentionally.

Measuring the Toxicity of Chemicals: Lethality

Assume, for the purpose of simplifying arithmetic, that the weight of adult humans ranges from 50 kg (112 lb: a small woman) to 100 kg (225 lb, a moderately large man). One teaspoon = 5 g, or 5000 mg. The LD_{50} of salt is 5000 mg/kg. Therefore, an LD_{50} value of 5000 mg/kg means that 50 tsp of salt, eaten all at once, would kill half the 50-kg women eating it, and

---

[12] An example of a bimodal distribution in humans is acetylation of the antituberculosis drug isoniazid. Fast acetylators degrade the drug rapidly, whereas slow acetylators do not. The latter are more easily cured of TB than the former, because they have higher blood and tissue levels of isoniazid for longer periods.

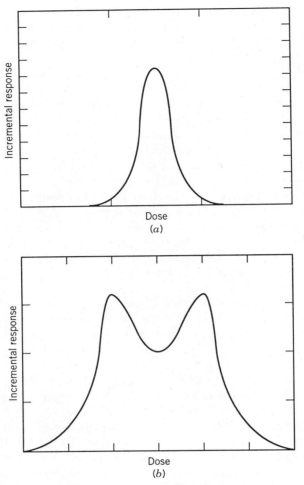

Figure 7.3 Response curves of different types of populations to increasing doses of a toxicant. (*a*) In a homogeneous population, most individuals respond within a small dose range; (*b*) a population that includes two subgrups with very different responses may produce a bimodal distribution; (*c*) an asymmetric or skewed distribution is also due to two subpopulations with differing responses; (*d*) a broad monomodal curve indicates a heterogeneous population, with two or more subpopulations differing in response.

100 tsp would be needed to have a 50% probability of killing our moderately large man.[13] But a single teaspoonful of salt would kill a 1-kg rabbit 50% of the time. For the insecticide aldicarb, on the other hand, with an LD_{50}

---

[13] In this instance it is the weight, not the sex, of the victim that determines the lethal dose. There are, however, chemicals that are twice as toxic to one sex as to the other, at least in rats.

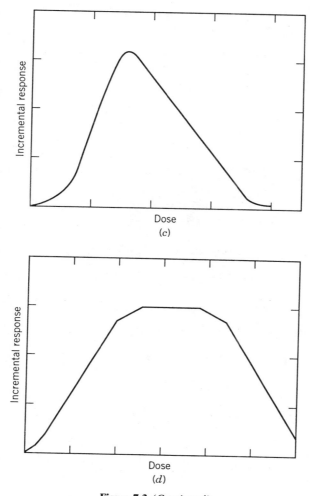

Figure 7.3 (*Continued*)

of 0.5 mg/kg, a dose of 25 mg would be the median lethal dose for the 112-lb woman; 50 mg would kill half the 225-lb men. That corresponds to 1 teaspoon as the amount of aldicarb needed to kill 50% of 100 to 200 people (1 tsp = 5 g = 5000 mg).

The endpoint determines the type of data needed. For insecticides, one might wish to know how much will kill 100% (nominally) of the insects exposed. Therefore, the LD_{95}, the dose needed to kill 95% of the population, is the minimum level of interest (Table 7.1). For humans, on the other hand, one is worried by far smaller death rates and really wants to know (even for pharmaceuticals) the LD_0: that is, the maximum dose that does not cause *any* deaths. For other compounds, such as insecticide residues in food, or

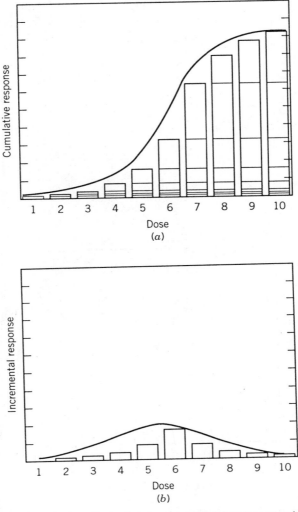

Figure 7.4 Discrete doses and small samples are common in laboratory experiments, producing a discontinuous response. (*a*) Cumulative number of individuals responding; (*b*) the incremental increase in response for each increase in dose. The sketched curve is the underlying distribution in the populations from which the samples are drawn.

food additives, one wants to be sure that the expected exposure levels have no (recognizable) toxic effects (Figure 7.5). This is a matter of prudence as well as of decency: the possibility of acute nausea and/or very low probabilities of cancer led to lawsuits over the malathion spraying of California against Mediterranean fruit flies ("Medflies"). If even 1% of the population had become ill, there would have been serious political and social consequences.

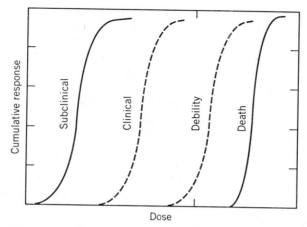

Figure 7.5 A family of response curves exists for each toxicant. The shape of the curves at the extremes is not well defined. The mean is the most robust measure of an effect and coincides with the median if the distribution is normal. Note also that the response curves need not be parallel or the same shape.

Therefore, one often calculates (or estimates) the *no observable effect level,* or NOEL: that dose which does not cause any of the particular toxicity one is measuring.

In practice, a toxicology study should include at least one dose at which no adverse effects are seen. The highest such dose is called the "no observed adverse effect level" or NOAEL. The lowest dose at which an adverse effect is seen is the "lowest observed adverse effect level" or LOAEL. The LOAEL and NOAEL depend on the spacing of doses in the study: the more closely spaced the doses are, the closer the NOAEL and the LOAEL will be. Thus, although intuitively one assumes that the "true" (population) threshold for adverse effects lies between the NOAEL and the LOAEL, the NOAEL must be calculated to take into account the sample size, the amount of variability in the sample, and the number and spacing of doses. The calculated lowest adverse effect level is often below the NOAEL. Note that in all of these measurements, it is necessary to define the endpoint.

Measuring the Toxicity of Chemicals: Other Endpoints

- TD_{50} is the *median toxic dose,* or dose that is toxic to 50% of the population. The nature of the toxic effect must be specified. A minimal toxic effect in pregnant animals, for example, is decreased maternal weight gain in the absence of any obvious illness.
- ED_{50} is the *median effective dose,* usually used as an index of therapeutic efficacy. The ED_{50} of aspirin is the dose that relieves headaches in 50% of the sample.

· LC_{50} is the *median lethal concentration,* which is particularly pertinent in aquatic studies, since acute exposure of fish to chemicals is almost impossible and certainly is not very realistic. Fish are generally exposed to chemicals by immersion in water contaminated with the given substance; the LC_{50} mimics this exposure. To avoid the problem of indeterminate or differing lengths of exposure, one typically uses a definite time span for the exposure, and this is included in the acronym: LC_{96} means the median lethal dose after 96 hours of exposure.[14] The LC_{50} is also important in inhalation studies in mammals.

The difference between endpoints is really the question of what risks we are willing to take or to have assigned to us. This concept of *permissible risk* comes up again and again in toxicology. It means that under different conditions, different hazards are tolerated (see Chapter 13). The risk that one is willing to take in prescribing a drug against serious illness is far different from the risk that one is willing to take in one's food additives. It is not uncommon, nor excessive, to have 1 death per 10,000 prescriptions of an anticancer drug. But 1 death per 100,000 for a painkiller caused an immense outcry and the drug's disappearance from pharmacies. Even 1 death per million for a food coloring would obviously amount to criminal negligence. The bisulfites, used to keep certain vegetables from browning after they have been peeled or chopped, were banned after fewer than 10 documented deaths (out of millions of salads served); their presence in wine is labeled because of *one* documented death.

Finally, there are other forms of toxicity than death. Most of modern toxicology focuses on these "other" toxicities: not because there are not enough lethal substances around, but because death is a measurable, identifiable, and in most cases avoidable, side effect. Other side effects of chemical exposure are less easily defined (e.g., nerve damage) or less easily identified (e.g., genetic damage, cancer) and therefore less easily avoided and more easily ignored.

Having defined endpoints, it becomes interesting to ask why the endpoint occurs: What is the mechanism of toxicity? In toxicology as in most sciences, the descriptive phase preceded the analytic phase, but emphasis is now on mechanisms of toxicity rather than on mere occurrence. A mechanism is, of course, many things to many people. A biochemist may be satisfied to know that compound X inhibits enzyme Y. A pathologist will be interested in the selective mortality that the chemical causes to cell type Z in organ W and whether Y mediates this toxicity. A cell biologist will consider the effects of the chemical on (for example) intracellular calcium levels in the cell and on the role that Y plays in this effect. A geneticist will want to know whether

---

[14] The potential confusion between LC_{50} as median and as 50 hours' exposure is solved by using multiples of 24 hours for exposures: 48, 72, 96.

inhibition of Y is due to direct interaction with the DNA or whether inhibition of Y leads to alterations in DNA. The complete mechanism could be described as the sum of the effects of the chemical on the cell, from binding to a receptor that allows it to enter the cell, through interaction with cytoplasmic and/or nuclear cell constituents, through the cascade of interactions among cell constituents that results from the initial effect of the chemical, to the ultimate fate of the cell, tissue, and organism.

At the simplest level, death occurs because an essential function fails for a period of time sufficient to cause the collapse of other essential organs. The most critical functions are the circulatory system and the respiratory system. Failure of these for more than a few minutes is usually lethal, because cells cannot obtain oxygen and die. The central nervous system is also critical, because it controls autonomic functions, including respiration. Moreover, because brain cells are extremely sensitive to oxygen deprivation, the brain is the first system to die if respiration or circulation are interrupted. Failure of the gastrointestinal (GI) tract, the kidneys, and the liver can be survived for days. Fat, bone, and skin are even less critical in the short term, while sensory organs are not strictly essential to life. From the point of acute toxicity, much of us is dispensable.

Events in Toxicosis

A typical sequence of events in acute toxicosis is:

1. *Access of Toxicant to Organism.* Unless an agent reaches an organism, there can be no toxicity. The most common routes of exposure to environmental chemicals are percutaneous, oral, and inhalation. This is equivalent to a knock at the door: a chemical *on* the skin is not *in* the organism. Even if an agent is swallowed, it is not, strictly speaking, *in* the body, far less *in* the cells.

2. *Absorption of Toxicant into Bloodstream.* Unless an agent reacts physically or chemically with cells on the skin, in the lungs, or in the gastrointestinal tract, it must be transported into the bloodstream, which carries it to the *target organ*: the site where it acts. Each route of exposure has a characteristic speed of absorption. Transport into blood is fastest via the lungs, slowest through skin.

3. *Transport to Organs, Including Target Organ(s).* Blood is in contact with all cells of the body, but the first organs an agent reaches depend on the route of entry. From the lungs, blood goes directly to the brain; from the stomach and intestines, the liver is the "first pass" organ. Percutaneously absorbed chemicals will first reach tissues underlying the skin they deposit on. Since blood, including both cells and plasma, has metabolic capabilities, it is possible for an agent to be inactivated (or activated) during the time of transport.

4. *Transport from Blood to Cytoplasm of Target.* Like transport into the bloodstream, this requires the agent to move across cell membranes.

5. *Metabolism to Toxic Form (Activation).* Activation can occur before, during, or after transport to a target organ. Some chemicals need not be activated, some can be activated by almost any cell; others can be activated only in a few cell types. The liver is the single most important metabolic organ, but the metabolic activity of lungs, nerves, and the bloodstream should not be discounted.

6. *Binding to Target and/or Inactivation of Target (= Toxicity) Organelle or Enzyme.* If the toxicant binds to an enzyme, binding typically causes inhibition, which may be reversible or irreversible. If the toxicant binds to a receptor, additional processes may be triggered, including transport of the toxicant to DNA or the synthesis of a new protein. Binding to a structural component of the cell may cause cell death. On the other hand, binding to an enzyme may also inactivate the toxicant before it reaches the "true" target enzyme. This appears to be the role of butyryl (serum) cholinesterase in organophosphorus ester poisoning (Chapter 8).

7. *Metabolism of Toxicant to Nontoxic Metabolite (= Inactivation).* Like activation, this process can occur at any time before, during, or after entry of the toxicant into the bloodstream. Inactivation can compete with activation. In certain cases, metabolic processes that should function to make a chemical more water soluble, so that it can be excreted more easily, also activate it.

8. *Excretion of the Toxicant, Its Toxic and/or Inactive Metabolites.* Removal of a chemical from the body is enhanced by increasing its water solubility, since urine and sweat are two major excretory routes. Gases are readily lost through exhalation. Fat-soluble compounds may be excreted in the bile, but a significant fraction of material excreted in bile may be reabsorbed from the intestines. Note that feces are largely composed of substances that never entered the body because they never crossed the gastrointestinal walls.

9. *Repair of Toxic Effects.* Most cells and organs have a tremendous capacity for repair. Subcellular mechanisms of repair include increased rates of synthesis of enzymes or other cell constituents that have been irreversibly inhibited until levels return to the normal, or pretoxicosis, state (see *Adaptation,* below). Damage to a single strand of DNA is readily repaired by "reading" the complementary strand, but repair of double-stranded damage may induce error in the genetic code (Chapter 11). Cell death is compensated by replacement of cells, with an efficiency specific to the damaged tissue. The organ least capable of replacing cells is the nervous system, since fully differentiated nerve cells have lost the capacity to divide. Large scale damage to tissues capable of replacing cells may result in replacement cells that lack

some of the function of the original tissue (e.g., scar tissue in skin), leading to permanent decrements in function. Such functional impairment can eventually cause death (liver failure, for example) or, if the organ is not essential, "merely" a chronic health problem: diabetes, blindness, or organ transplant.

Combinations of these steps and variations on the sequence of events can occur. Activation may occur either before the toxicant gets to the target, or at the target. The toxicant may bind to many targets other than the one producing the toxic effect, resulting in smaller quantities reaching the target. Competing with these processes are processes that inactivate, remove, or sequester the toxicant; to a considerable extent, detoxifying and activating reactions occur simultaneously and may compete. Inactivation of the target (organ or enzyme) frequently leads to compensatory action by the cell; if sufficiently rapid, this may mitigate the toxicity.

It is important to realize that any organ of the body can be a target of toxicity. In Chapter 4 we considered the relationship between inhalation and pulmonary toxicity, noting that a chemical may be toxic to lung cells because it is inhaled and therefore reaches the lungs at high concentration, or because of an intrinsic toxicity to a particular cell type, regardless of the route of administration. The nervous system is the target for organophosphorus esters, regardless of the route of administration (see Chapter 8). It must be emphasized that, for every tissues and organ, there are chemicals with a specific toxicity for that cell or tissue type. Examples include methanol, which is particularly toxic to the human optic nerve; paraquat, to the lungs; dimethylnitrosamine, to the liver; mixtures of aspirin and phenacetin, to the kidney; TCDD to the skin; pyriminil, to the β-cells of the pancreas. There are also chemicals that attack each of these organs only because the route of administration sends the agent to the tissue at high levels, and most of the organ-specific agents listed here also attack other tissues if somewhat higher doses are administered. Pharmacologic toxicologists investigate the mechanisms of *target organ toxicity,* the mechanisms by which a particular agent attacks a particular cell or tissue.

CHRONIC TOXICITY

The term *chronic toxicity* is used to refer to two separate entities. First, it refers to the occurrence of chronic symptoms, whether or not due to a single exposure. Many of the health effects of concern to toxicologists are in this category: kidney or liver damage, nerve damage, diabetes. The rodenticide pyriminil induces diabetes after a single exposure because it destroys the insulin-producing cells of the pancreas. A single high dose of methanol can destroy the human optic nerve, producing permanent blindness. And as described in Chapter 8, some OP insecticides can cause paralysis after single exposures. Such "chronic toxicity," which actually consists of various non-lethal irreversible effects, is a very real danger, but there is little difficulty in assessing such effects when they result from a single and/or recent, high-dose exposure. More difficult to identify is the long-delayed effect. The most obscure and—from an evolutionary point of view the most serious—of long-delayed effects is the occurrence of a mutation in the germ cells. Such a mutation can be transmitted to the next generation and may lead to wide-spread genetic disease many generations later (see Chapter 10). The most feared long-term effect is cancer, which may result from a single high-level exposure to a carcinogen, but is more often associated with the long-term low-level exposures discussed below. Because of its significance in our society, cancer and carcinogenesis are discussed separately in Chapter 11. Other delayed effects include some forms of neurotoxicity (Chapter 8) and birth defects (Chapter 9).

Chronic toxicity also refers, however, to the consequences of *chronic exposure* to a toxicant. Such consequences, whether reversible or irreversible, are often difficult to associate with their cause, because the exposures are typically long-term, often to very low levels of a chemical. Many of the exposures causing illness fall into this category: occupational exposure to pesticides, formaldehyde or vinyl chloride; lung damage due to smoking or other air pollutants. Because of the uncertainties associated with long-term exposures, there is often disagreement over the relationship between types of exposure and the effects observed. It is therefore worth examining the kinetics of uptake and elimination of toxicants from the body.

KINETICS OF EXPOSURE

Uptake of a compound may be acute, as in swallowing a pill. Blood levels, the usual measure of absorption, rise abruptly after acute exposure to a chemical, and fall gradually as metabolism and/or excretion proceeds (Figure 7.6). If a second dose follows the first, a second rise in blood levels occurs. If the second dose is given before the first dose has been fully cleared from the system, blood levels will probably rise to a higher level than the initial peak. Subsequent doses eventually result in a maximal or peak level that occurs after each dose, with troughs between doses as the compound is metabolized or eliminated (Figure 7.7).

When exposure is continuous—for example, the inhalation of carbon monoxide by politicians in a smoke-filled room—uptake is directly proportional to the concentration of the chemical in the external environment and is said to be *first order* with respect to the chemical. In general, the rate of uptake of the toxicant P in the blood can be expressed as

$$\frac{dp}{dt} = kP_{\text{ext}}$$

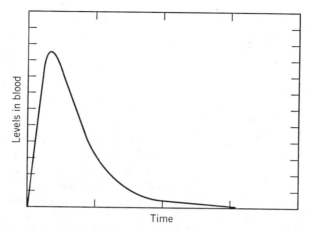

Figure 7.6 Kinetics of a single exposure. Blood levels, the usual measure of absorption, rise abruptly after acute exposure to a chemical and fall gradually as metabolism and/or excretion proceed.

where dP/dT is the rate of increase of P in the blood, and P_{ext} is the concentration of P in the external environment. P is specific for each agent.

In the ordinary way, one soon leaves the smoke-filled room, and P_{ext} goes essentially to zero. Concentrations of P in the blood then decrease in the same way as from the peak after the last dose of a drug. However, many toxicants are so widespread in the human environment that we are constantly

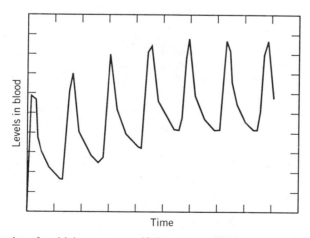

Figure 7.7 Kinetics of multiple exposures. If there are multiple exposures to a chemical and the intervals are not great enough to clear the previous dose completely, there is a rise in blood levels with each exposure, probably to a higher level than the initial peak. Subsequent doses eventually result in a maximal or peak level that occurs after each dose, with troughs between doses as the compound is metabolized or eliminated.

exposed to them. They are part of the food we eat, the water we drink, the air we breathe (cf. lead, cyclodienes, PCBs). We can consider them to be present in the environment at a steady level of P_{ext}. The rate of their uptake is

$$\frac{dP}{dt} = k_{intake} \times P_{ext} = K$$

Obviously, the minute-to-minute environmental concentration of P varies, and therefore the rate of uptake does, too. However, over a period of weeks the environmental concentration of P and its uptake can be treated as constant, and this constant is designated K. At the same time, excretion of the chemical from the body (by exhalation, by secretion through the skin, by excretion through the kidney and bile) is competing with the continuing uptake of toxicant into the blood. The rate of elimination by these routes is directly proportional to the concentration of P *in the blood*:

$$\frac{dP}{dt} = -k_{out} \times P_{int}$$

where P_{int} is the concentration of P in the blood and k_{out} is the proportionality constant. The minus sign denotes decreasing levels in the blood. Since the two processes, uptake and elimination, are each working on levels of P in the blood, the *net* level of P in the blood is their sum:

$$\frac{dP}{dt} = K - k_{out}P$$

Since we have assumed that uptake of P is constant, it seems logical to conclude that there will eventually be an equilibrium level. That is, the level of P in the blood will rise until the product $k_{out}P$ equals K. At this point, the concentration of P in the blood will no longer change, because P is removed from the blood as fast as it is taken up into the blood. The concentration of P in the blood at equilibrium is a constant that can be determined from K and k_{out}:

$$\frac{dP}{dt} = 0 = K - K_{out}P$$

Therefore, the level of a chronic toxicant in the blood is determined by the relative efficiency of uptake and elimination. If uptake is inefficient or if elimination is very efficient, blood levels will be low. If *elimination* is very inefficient (k_{out} is very small), blood levels can become quite high.

K_{out} can be determined experimentally by giving a single dose of P so that the blood level rises in a short time, and then allowing P to be eliminated while no new P is taken up. Intravenous (iv) dosing is best for such studies, since it delivers the agent directly to the blood, eliminating the lag caused by absorption from stomach. During elimination, levels of P in the blood are measured repeatedly. Then

$$P = \frac{K}{k_{out}}$$

Integrating yields

$$\ln \frac{P_t}{P_0} = -k_{out}t$$

or

$$P_t = P_0 e^{-k_{out}t}$$

Another useful value is $t_{1/2}$, the time required to clear 50% of P from the blood:

$$t_{1/2} = \frac{0.693}{k_{out}}$$

Accumulation

Obviously, there are numerous variations on the theme of uptake and elimination. That blood levels of xenobiotics have been used to study pharmacokinetics is due as much to the ease of sampling blood as to its role in transporting endogenous and exogenous chemicals to target organs. In terms of lipid-soluble compounds such as DDT and PCBs, the level of P found in fat is more significant than the levels seen in blood, because blood levels of lipid-soluble compounds do not rise very high, no matter what the exposure level. In the case of chemicals that are toxic to a specific organ or tissue, the levels in that tissue are more important than blood levels. Thus the toxicity of paraquat, which is specifically toxic to pulmonary tissue, depends on the level at which it reaches the lung, while the toxicity of parathion depends on its concentration at the synapse.

This closer focusing on levels of toxicant at the site of action has been given the name *physiologically based pharmacokinetics* (PBPK). PBPK is a refinement of the often simplistic estimate that blood levels provide for the distribution of a chemical in an organism. In some cases, differences in toxicity between two closely related chemicals depend on a structural difference that enables one congener to reach the target tissue while the other is excluded. This is true, for example, of paraquat analogs that are not actively taken up into lung tissue (e.g., the related herbicide diquat). Often, however, seemingly minor structural changes that cannot be associated with large differences in access to the target do make great differences in toxicity. In the case of the diphenyl ethers, there is no evidence that nitrofen (2,4-dichlorophenyl 4'-nitrophenyl ether) reaches the fetus in smaller quantities than the far more fetotoxic 2,4,5-trichlorophenyl 4'-nitrophenyl ether (Chapter 9). Similarly, both paraoxon and fenitro-oxon reach the synapse at similar levels. But the *meta*-CH$_3$ of fenitro-oxon decreases its affinity for mammalian acetylcholinesterase, rendering it far less toxic than paraoxon (Chapter 8). In sum, the presence of a chemical at a target is necessary for activity. Demonstrating that a chemical is not present at a target is sufficient explanation of no effect, but showing that a chemical reaches a target does not explain *why* it causes its effect. The fact that two closely related chemicals may reach a target at the same concentrations, but only one of them produces the desired (or undesired) effect, is the mother lode of both the pesticide and the pharmaceutical industries.

In the case of *chronic exposure* to a toxicant, a constant level will eventually be seen in blood (Figure 7.8). If some organ has a high affinity for the compound (i.e., if, for some organ, uptake significantly exceeds elimination), that organ will accumulate the toxicant. This is a sadly frequent phenomenon. DDT accumulates in fat because it is fat soluble; strontium accumulates in bone because it is taken up like calcium; mercury accumulates in the brain. Over long periods of time, appreciable levels of the toxicant can build up in such a depot tissue. Although differential uptake is often the result of the general structure of a chemical (i.e., its lipophilicity or water solubility, which

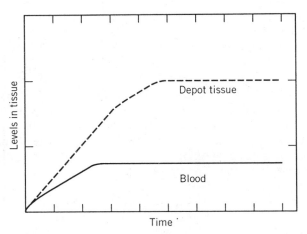

Figure 7.8 Level of toxicant in the body: chronic exposure. Blood levels reach steady-state levels when excretion equals intake. Each tissue or organ will have its own equilibrium, differing with each chemical. Tissues with high equilibrium levels are called *depot tissue,* since the chemical accumulates there. When exposure ends, blood levels are maintained by chemical released from the depot tissue.

determine the ease of transport into various tissues), occasional instances of active uptake by one organ are seen, as in the case of paraquat, which is actively transported into the human lung.

Adaptation

When a toxicant interacts with a target receptor in the body, it alters the receptor's function, often by inhibiting the receptor. Living systems are dynamic and will ordinarily compensate for such loss of function by increased synthesis of the functional component. For example, if a mammal loses blood, the body compensates for the decreased oxygen-transport capacity by synthesis of more hemoglobin and red blood cells. If OPs inhibit AChE, more is synthesized. In chronic exposures, there may be continuing destruction or inhibition of a receptor molecule, coupled with compensatory resynthesis. The end result, in the case of chronic exposure to a toxicant, is an increased turnover of the receptor. The increased turnover of receptor enables the organism to deal with sudden short-term increases (bursts) of toxicant entering the body. Synthesis of enzymes that metabolize (deactivate) the toxicant can also be stimulated. The sum of these feedback responses is called *adaptation*. A net increase *or* a net decrease in total receptor function can occur. It is the phenomenon of adaptation that accounts for enhanced tolerance to poison by people who deliberately take small doses of the poison (snake handlers, medieval monarchs, poisoners in murder mysteries). Also, all else being equal, the habitual drinker has a greater capacity for alcohol than does an otherwise identical teetotaler.[15] Some evidence of tolerance is also seen for OP exposure. It is possible, by giving repeated small doses of the insecticide EPN, to work up to a daily dose that is lethal in hens that were not pretreated. On the other hand, repeated doses of OP insecticides will gradually decrease the available levels of AChE, leading to symptoms after repeated asymptomatic exposures, because new synthesis is relatively slow.

A side effect of adaptation is that the organism's response to other substances may also be affected. The system of enzymes known as *microsomal oxidases, mixed function oxidases,* or *cytochromes P-450* is not specific to single compounds, but acts on classes of similar agents. These enzymes are the major system for metabolizing xenobiotics.[16] Exposure to chemicals metabolized by cytochrome P-450 usually results in increased synthesis of specific subsets of these enzymes. The P-450 gene superfamily contains more

---

[15] The greater drinking ability of the alcoholic undoubtedly includes several factors. Metabolic capacity may be minor relative to learned behaviors that minimize the appearance of drunkenness, adaptations of brain receptors and of neurotransmitters to the presence of alcohol, and eventually, liver damage that decreases tolerance sharply.

[16] This is not their only function, however. They also metabolize many endogenous agents, including steroid hormones.

than 150 genes; is found in plants, animals, and prokaryotes; and metabolizes both endogenous and foreign chemicals. Although metabolism of endogenous substrates is very specific, there is considerable overlap between endogenous and xenobiotic substrates. Thus, housing rats in cages with bedding of pine shavings results in activation of cytochrome P-450 in the liver and shortens the duration of response (sleep time) after a given dose of phenobarbital. In humans, exposure to chemicals such as solvents can enhance metabolism of endogenous hormones and also of steroid contraceptives.

The complexity of interactions between cytochrome P-450 and xenobiotics is beyond the scope of this book. One example must suffice. When the herbicide nitrofen (2,4-dichlorophenyl 4'-nitrophenyl ether) was administered to mice, it induced different isozymes of cytochrome P-450 isozymes in pregnant females than in nonpregnant females. Moreover, the quantities of these isozymes already differed between pregnant and nonpregnant females in the absence of treatment. Thus the initial metabolic capacity of pregnant mice appears to differ from their nonpregnant counterparts, and the introduction of a xenobiotic alters these basal states in different ways.

IRREVERSIBLE TOXICITY

The most serious form of irreversible toxicity due to chemical exposure is undoubtedly death, but the term is usually reserved for nonlethal but debilitating consequences of toxic insults. In those cases in which death (or other forms of toxicity) follow exposure by minutes or hours, it is—at least in theory—relatively easy to control exposure and so prevent the untoward consequences. On the other hand, when toxic effects are delayed for weeks, months, or years, considerable exposure can occur—both on an individual and on a population level—before the toxicity becomes apparent. Even when the toxic effects are eventually recognized, the delay between exposure and effects may make it difficult to identify the cause. Three mechanisms for irreversible damage occur, differing in their consequences as well as in their origins:

1. *Cell Death.* In adult organisms, irreversible toxicity occurs when an exposure—acute or chronic—damages an organ or tissue too seriously for recovery of full function, yet does not cause death. Given a high level of medical sophistication, almost any organ is susceptible to irreversible but nonlethal damage, although the relative sensitivity of organs will differ in response to different agents. Certain organs are more susceptible to damage than others, and damage to certain organs is more likely to result in death in a short period of time than is damage to other organs. (It is not yet possible to live without a functioning

liver, but bodily functions can be maintained in the absence of most brain functions by using feeding tubes and respirators.) Familiar ailments associated with nonlethal damage to different organs are: diabetes, caused by nonfunctioning of the beta cells of the pancreas; blindness; and kidney failure. Although not producing a disease state in the affected person, reproductive failure can be due to damage to ovaries, testes, or other reproductive organs. Because the human nervous system is highly developed and because we place great demands on it, the nervous system (and especially the brain) is particularly vulnerable to toxic insult. Because the incapacitation of an organ or tissue depends on the number of cells killed, even irreversible damage due to cell death follows the canons of classical toxicity: *the severity of the damage in the individual is reasonably proportional to the exposure.*

2. *Developmental Toxicity.* In developing organisms, the interactions between cells are so delicately balanced that even minor perturbations of cellular activity, if occurring at a critical time, result in cascading disruptions of normal patterns and lead to permanent abnormalities of development. Because of the extreme complexity and rapidly changing sensitivity of different organs and tissues with the time of development, relatively low exposures may have severe effects. Moreover, the effects of exposure to a single agent may differ sharply, depending on the relationship between exposure and stage of development. For a single agent and a single time of administration, however, *either the severity of the effect and/or the probability of its occurrence is correlated with the intensity of the exposure.*

3. *Genotoxicity.* Mutations, or alterations in the genetic code, can have serious consequences for the entire organism if the mutation is not lethal to the cell in which it occurs. Such alterations will be propagated to all the progeny of the original cell. If the mutation occurs in a germ cell, the alteration will be transmitted to the next generation. If the alteration occurs in a somatic cell, the consequences are limited to the individual in whom the alteration occurred. Because the initial toxic effect is dichotomous—either a mutation occurs or it does not—*the severity of the consequences* in an individual *is not correlated with the intensity of the exposure, and dose–effect relationships are only seen at the population level.* The *probability* of a mutation occurring is directly correlated to the dose in the individual, however.

Identifying the chemicals that cause irreversible toxicity, working out the mechanisms by which they cause it, and learning to prevent or repair the damage, are the major challenges to toxicology today. Because the consequences of incurable toxicity are so burdensome to individuals and to society, every aspect of this challenge is steeped in controversy. In the remaining chapters of this book we examine the biological mechanisms underlying several common forms of irreversible toxicity and the means for identifying which chemicals cause them. We will also look at the ways in which our society determines the risk posed by such chemicals and the methods by which it controls those risks.

8

NEUROTOXICITY

So perhaps our task . . . is to make clear that there is no technical solution, no monetary solution. The only solution is not to repeat the pollution anywhere.
—Miyamoto, 1975[1]

INTRODUCTION

A functioning nervous system is absolutely essential to animals because the nervous system controls involuntary functions such as heartbeat and breathing, without which cells quickly die. A functional nervous system depends not only on the integrity of the cells that make up the system, but also on unimpeded electrical transmission along the surface of the cells and on the presence of chemical neurotransmitters that communicate between nerve cells and between nerve cells and effector organs (sensory organs, muscles). Thus it is possible to disrupt transmission of nerve impulses by killing enough neurons in a given tract, *or* by changing the electrical potential of the cell surface, *or* by changing the concentrations of neurotransmitters at the synapse or at the interface between a nerve cell and its neighbor.

Most of us do not use our nervous system to its limits, so we do not immediately notice when the limits are reduced, little by little. An analogy can be drawn between the way most people use their voices and the way a singer does. Few of us coddle our throats; we still speak euphoniously. A professional singer, using her voice to its limits, notices the effects of dry air, of marginally too little sleep, of traces of smoke in the room, as a decrease in her singing ability. Similarly, when a retired academic announced that he did not have the "energy" to complete his book, his family readily accepted the explanation. Only months later did they realize that his memory was failing. The scientist, however, had realized that he could no longer keep the overall plan of his work in mind while writing, even though his brain still managed ordinary, less demanding, activities.

That there can be neurological consequences of toxic exposure has been known for centuries, both because plants contain hallucinogens and because

---

[1] P. Kruus and I. M. Valeriote, eds., quoting Dr. Miyamoto, in *Controversial Chemicals*, Multiscience Publications, Montreal, Quebec, Canada, 1979.

many metals are potent neurotoxicants (Table 8.1). The effects of lead on motor nerves lead to palsy; the central nervous system effects of mercury are thought to have inspired the phrase "mad as a hatter." On the whole, however, recognition of the vulnerability of the nervous system to irreversible toxic insult is a twentieth-century phenomenon. The development of organometallic pesticides provided new routes of exposure to old toxicants such as mercury, as well as revealing the potential toxicity of tin. The synthetic insecticides not only attack insect nervous systems, but owe their mammalian toxicity to their effects on the same neurotransmitters in the mammalian nervous system. The organophosphorus ester insecticides exhibit not only the acute neurotoxic action resulting from their inhibition of acetylcholinesterase (below), but also induce a delayed neurotoxicity that causes irreversible damage to the long nerves of the legs. In addition, several solvents (hexane, methyl butyl ketone), monomeric precursors to plastics (acrylamide), pharmaceuticals (Enterovioform), and a fragrance enhancer (AETT) proved to be neurotoxic at high doses. Plant-derived neurotoxins also continue to be identified (buckthorn, lathyrogens).

Although some of these neurotoxicants were identified before human illness occurred, human damage has occurred from prolonged occupational exposures to high levels of hexane, methyl butyl ketone, and acrylamide monomer, and from long-term high-level therapeutic exposure to Enterovioform. Lathyrogens, atypical amino acids found in certain legumes, have caused nerve damage when poor people, despite awareness of their toxicity, ate these legumes during prolonged periods of famine.[2] In the cases of both occupational exposure and therapeutic exposure, long delays have occurred before the cause of the outbreaks have been identified. In one case, authorities initially considered the outbreak of weakness, fatigue, and muscle aches to be due to "mass hysteria" among the mostly female workers. The women were subsequently found to be suffering from solvent-induced neuropathy. In other episodes, doctors initially considered the victims to be suffering from idiopathic diseases such as multiple sclerosis, or from life-style toxicoses such as alcoholism. In all of the cases of occupational toxicity, resolution of the problem has been straightforward once the source of toxicity was identified, although in each episode some workers remained permanently damaged.

The nervous system is extremely vulnerable to injury as a result of cell death because nerve cells do not divide once they have differentiated fully. Therefore, when a nerve cell dies, it cannot be replaced. If enough cells in a given tract of the system die, function is impaired and cannot be recovered.

---

[2] The lathyrogens act on glutamate, which functions as a transmitter in the central nervous system. There is some evidence that the same excess of glutamate occurs in idiopathic amyotrophic lateral sclerosis ("Lou Gehrig's disease" or ALS), which causes progressive paralysis and eventual death when the breathing muscles fail. Familial ALS has been traced to a mutation in the gene for superoxide dismutase, suggesting oxygen damage as a mechanism in ALS.

Table 8.1 Sampler of Neurotoxic Agents[a]

Physical Agents	Pharmaceuticals
Anoxia	Glutamate
Pesticides	Hexachlorophene
	Isoniazid
Organochlorines	Malonitrile
DDT	Pyrithiamine
Dieldrin and aldrin	Vinca alkaloids
Chlordecone (Kepone)	Nicotine
Pyrethroids[b]	Lathyrogens
Fenvalerate	
Pyrethrum	"Life-Style" Chemicals
Carbamates[c]	Opioids
Carbaryl	Solvents
Aldicarb	Cocaine
Organophosphates	Marijuana
Reversible only	Alcohol
Parathion	
Malathion	Metals
Irreversible and reversible	Lead
Leptophos	Thallium
EPN	Mercury
Chlorpyrifos	Manganese
Nerve Gases	Organometals
Organophosphorus esters	Gold thioglucose
Solvents and Industrial Intermediates	Triethyltin
	Methylmercury
Acrylamide	Tetraethyllead
Hexane	
Methanol	Miscellaneous
Ethanol	Carbon monoxide
Tri-*ortho*-cresyl phosphate (TOCP)	Acetylpyridine
	Azide
	Carbon disulfide
	Cyanide

[a] Although far from exhaustive, this list gives some indication of the variety of substances that can damage the nervous system.
[b] Although pyrethroids are insecticidal by virtue of their neurotoxicity, they rarely cause neurological effects in mammals because they rarely reach the mammalian nervous system. Only reversible neurotoxic effects have been observed.
[c] Neurotoxic effects usually due to inhibition of acetylcholinesterase, and therefore reversible. Carbaryl may cause ataxia, but that syndrome is poorly characterized and seems to be reversible as well.

Even in the absence of toxic insult, nerves die as one ages. Fortunately, considerable redundancy is built into the nervous system, as into most essential organs. Ordinarily, there is no loss of function until a significant fraction of cells in a given tract is destroyed. For example, normal aging processes do not result in obvious loss of neuronal function until late in life, and not even then for many people. Numerous chemicals will accelerate destruction of neurons, however. Many of these chemicals are extremely specific in the types of nerve cells that they damage, leading to very specific types of functional deficits. It cannot be emphasized too strongly that killing fully differentiated nerve cells results in permanent damage to the nervous system. Even if no symptoms result at the time of the injury, the possibility of cumulative effects must be taken into account.

In addition to cytotoxic chemicals, numerous agents exist that temporarily disrupt the transmission of signals in the nervous system. Foremost among these are the natural toxins used by animals as defensive weapons. Almost all snake and spider venoms are neurotoxic,[3] as are the poison of the puffer fish (tetrodotoxin or fugu poison) and the toxins produced by microscopic organisms and accumulated by some shellfish (saxitoxin or dinoflagellate poison) or fish (cigueratoxin). A few plants have also developed defenses based on neurotoxicity.[4] Among these plant-derived chemicals is physostigmine, which facilitates transmission of impulses across the synapse.

Human ingenuity has added to the catalog of neurotoxic agents: almost all insecticides function by disrupting neurotransmission. Moreover, in contrast to the natural neurotoxins, many of which act by *inhibiting* the transmission of nerve impulses, most insecticides act by *facilitating* transmission, causing death as a result of hyperstimulation and hyperactivity. In many cases the chemicals affect the mammalian nervous system as well as the insect's, as illustrated by parathion (see below). It is intriguing that if one survives acute exposure to chemicals that act on neurotransmitter levels, there is usually no permanent damage. The danger from snakebite, from cigueratoxin, and from physostigmine is quick death, but survivors recover fully within weeks. The effects of most commercially available synthetic neurotoxic insecticides are similarly reversible. In some cases, however, irreversible damage occurs.

---

[3] F. E. Russell, in the chapter on animal toxins in *Casarett and Doull's Toxicology: The Basic Science of Poisons* (3rd ed., 1986, Macmillan, New York, edited by C. D. Klaassen, M. O. Amdur, and J. Doull), warns of the danger of designating a natural toxin as "neurotoxic," "cardiotoxic," and so on, when it may have significant components of several types of toxicity, any of which may be lethal under given circumstances.

[4] This is not to suggest that plants have few chemical defenses, only that a minority are clearly neurotoxic. Terrestrial plants, unable to move, protect themselves from predators by elaborating toxic or unpalatable chemicals. The use of such defensive chemicals is so well developed among terrestrial plants it is suggested that the xenobiotic-metabolizing capacity of the cytochrome P-450 enzymes is a response to the transition of animals from aquatic to terrestrial life.

In the past decade, two neglected aspects of the neurotoxic effects of chemical exposure have emerged as major areas of research: behavioral effects resulting from damage to the nervous system during development, and premature aging of the nervous system. The former is best illustrated by the effects of lead but may occur as a result of exposure to PCBs as well. The latter is exemplified by the so-called senile dementias such as Alzheimer's disease, and also by Parkinson's disease and Huntington's chorea. It has recently been demonstrated that the street drug MPTP caused sudden, severe Parkinson's disease in young adults by killing large numbers of neurons in the region of the brain known as the *substantia nigra.* Cell loss in the same region is correlated with Parkinson's disease at the normal age of onset, late in life. The rapid development of Parkinsonism, combined with destruction of the substantia nigra in the young adults who used MPTP, supports Spencer's hypothesis that the neurological diseases of aging originate in accelerations of normal cell death in the nervous system.[5] It is quite reasonable, then, to ask whether even earlier chemical exposures—between conception and puberty—might not affect the functioning of the nervous system, and especially of the brain, in adulthood or old age.

While such questions are at the forefront of neurotoxicology, old problems continue to haunt us. The Romans knew of the toxicity of lead, but in the United States we are still grappling with lead damage to children. Japan banned certain organophosphorus ester (OP) insecticides in the early 1960s because their acute toxicity made them too dangerous: OPs still rank as the leading cause of acute poisoning in California fields. The delayed paralysis caused by some OPs, first described in 1930, is still not understood. Considerable evidence exists that OPs may produce other, still not fully defined damage to the nervous system.[6] Thus the OPs, together with lead (Chapter 6), serve to illustrate the role of chemical exposures in neurotoxicity.

CASE HISTORY: ORGANOPHOSPHORUS ESTERS

Parathion and malathion (Figure 8.1) provide excellent illustrations of the mechanisms involved in acute toxicity. Both chemicals are insecticides of the class of organophosphorus esters. *Exposure* is typically percutaneous or by inhalation; oral intake also occurs. *Absorption* into the bloodstream is rapid by any route,

---

[5] P. S. Spencer, P. B. Nunn, J. Hugon, A. C. Ludolph, S. M. Ross, D. N. Roy, and R. C. Robertson, "Guam amyotrophic lateral sclerosis-Parkinsonism-dementia linked to a plant excitant neurotoxin," *Science* 237:517–522, 1987. A narrative account of Guam disease and the efforts to find its cause is found in Berton Roueché's "This obscure malady," published in *The New Yorker,* October 29, 1990, pp. 85–113.

[6] Evidence for such effects are scattered in the scientific literature and may be buried in the files of numerous lawsuits settled out of court. A case history is presented by B. Roueché in "The fumigation chamber," *The New Yorker,* January 4, 1988, pp. 60–65.

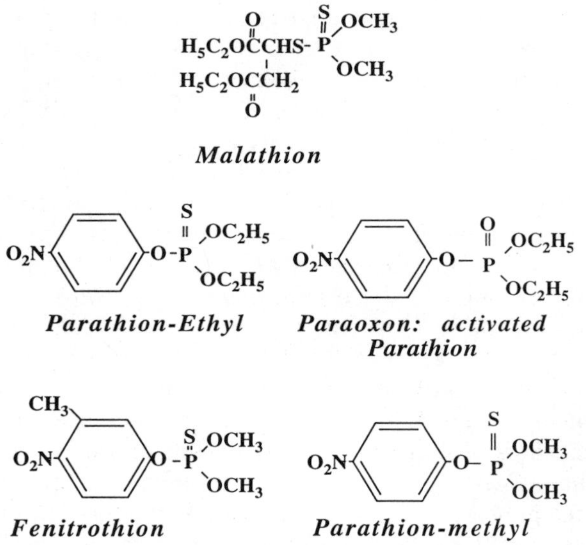

Figure 8.1 Structures of malathion and parathion, and of two organophosphorus ester insecticides related to parathion, parathion methyl and fenitrothion.

but percutaneous uptake is slower than other routes.[7] In the blood, OPs bind to assorted esterases (known variously as aliesterases or pseudocholinesterases) in serum and to acetylcholinesterase in the red blood cell, thus mitigating toxicity because relatively little of the agent reaches the nervous system. *Activation* occurs primarily in the liver, where cytochrome P-450 converts the P=S to P=O, the oxon or toxic form. Activation can also occur in the nervous system, however. *Transport* to the true target, the nervous system, must occur before any symptoms are seen. *Inactivation* occurs in the liver, either before or after the P=S to P=O conversion occurs, and to some extent by sequestration in binding to esterases other than acetylcholinesterase, primarily outside the nervous system.

Mechanism of OP Toxicity.   Transmission in the nervous system of all animals that have nervous systems combines electrical transmission along the surface of

---

[7] "Slow" and "fast" are relative. The story is told of a scientist who wanted to describe symptoms of organophosphorus ester toxicity. He put a syringe with the antidote by his hand, took an oral dose of parathion, and began to record his symptoms. He was found dead: the effects were too rapid for him to inject the antidote before he lost consciousness. In another case, a pesticide applicator reached into a jar of parathion, realized his error, washed immediately, and was dead in 30 minutes. There are, however, organophosphorus esters that produce symptoms only hours after oral exposure. In our experience, one organophosphorus ester administered orally to hens caused death in less than 5 minutes; others did not produce symptoms for several hours. Similarly, the time between onset of symptoms and recovery varied greatly. When onset of toxicity was rapid, it usually abated in a few hours. On the other hand, cholinergic shock began only several hours after EPN was administered together with the antidote atropine, but lasted up to 4 days.

the nerve membrane with chemical transmission of the impulse between nerves. A major chemical transmitter in all animals from annelids to humans (and including insects) is acetylcholine (ACh). To modulate transmission of impulses, the space between nerve cells contains large amounts of an enzyme, acetylcholinesterase (AChE), that breaks acetylcholine into its constituent parts, acetate$^-$ and choline$^+$. The result is that when an electrical impulse reaches the junction between two nerves (synapse), a packet of ACh is released, moves across the synapse, and binds to the ACh receptor in the postsynaptic membrane (Figure 8.2). This initiates the electrical impulse in the next nerve. Degradation of ACh by AChE limits the length and/or intensity of the impulse.

OPs bind irreversibly to AChE in the synapse, with the result that ACh is not degraded at the postsynaptic membrane, and synaptic transmission of impulses is not terminated. The result is continuous stimulation of nerves and their target

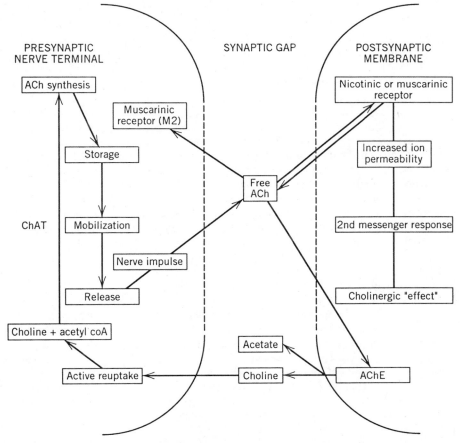

Figure 8.2 Schematic outline of cholinergic transmission at the skeletal neuromuscular junction. (Adapted from J. K. Marquis, *Contemporary Issues in Pesticide Toxicology and Pharmacology*, S. Karger, Basel, 1986.)

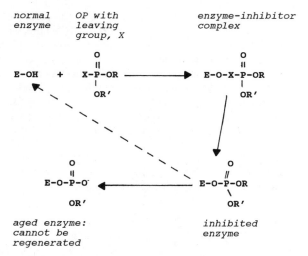

Figure 8.3 Inhibition of acetylcholinesterase by organophosphorus esters (OPs). The OP forms a complex with acetylcholinesterase (E), displacing the leaving group (X). The inhibited enzyme may regenerate (dashed line) if the OP is displaced. Alternatively, the complex loses another moiety (R), resulting in an aged enzyme that cannot be regenerated.

muscles. Cholinergic nerves include both voluntary motor nerves and the involuntary nerves controlling the gastrointestinal tract and bronchial secretions. Inhibiting AChE results in continuous peristalsis, leading to vomiting and defecation, massive respiratory and gastric secretion, and continuous voluntary muscle contraction ranging from fine tremors to convulsions. Death results from exhaustion or from choking on the copious quantities of mucus that are secreted. For diagnostic purposes, the mildest (or earliest) symptoms typically consist of pinpoint pupils and/or nausea. Subclinical symptoms in people synthesizing OPs have been described as a "pricking" of the eyelids, perhaps due to exposure to fumes[8] (Figure 8.3).

Dose–Response Relationships.   There is considerable redundancy in the cholinergic system, as is usual for critical functions of an organism. A tremendous excess of AChE is present in the synaptic cleft, both to ensure that ACh at the postsynaptic membrane is broken down rapidly (which enhances the calibration of signal strength in nerve impulse transmission) and to protect the system from collapse in the event that minor amounts of AChE are lost. A 50% reduction in AChE is required before noticeable symptoms appear. A 70% reduction in AChE results in clinical toxicity (illness), and 80 to 90% reduction is lethal in the absence of treatment. If death does not occur, AChE is resynthesized by the organism. Some regeneration of inhibited enzyme apparently also takes place for up to a

[8] The symptoms of AChE inhibition can be extremely subtle. Scientists with decade-long experience say that they have mistakenly ascribed mild symptoms to other causes, such as flu or mild food poisoning. Such errors can be lethal if exposure continues.

week. However, AChE levels remain depressed for up to 30 days, even if symptoms have abated long before.

Comparative Toxicology of Organophosphorus Esters.   Michaelis initially synthesized several OPs between 1903 and 1915, but the first uses of this class of chemical were as nerve gases, with civilian uses being given a subsidiary role (Figure 8.4). Thus the toxicology of OPs is inextricably bound up with the history of twentieth-century warfare. Their extreme toxicity attracted the attention of both Germany and England in the years before World War II, but the German chemist Gerhard Schrader is usually given credit for their development, since he recognized the insecticidal as well as the chemical warfare potential of the class. Schrader, at the huge German chemical company of I.G. Farben, investigated the insecticidal properties of OPs beginning in 1934; he patented the first OP, tetraethylpyrophosphate (TEPP), in 1937, followed by dimefox in 1940, schradan[9] in 1942, and parathion in 1944. During the same period, Saunders, in England, investigated the application of OPs to chemical warfare and examined the nerve gas potential of DFP (diisopropylfluorophosphate) as a 1% mist.

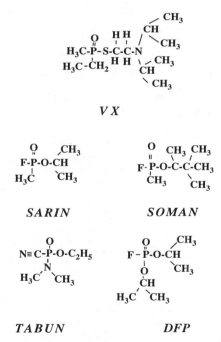

Figure 8.4 Structures of the major OP nerve gases: sarin, soman, tabun, and VX, and the related DFP.

_____

[9] Schradan is of particular interest because it is actively taken up by plants: such systemic insecticides are much sought after.

In Germany, chemical warfare research during World War II focused on three compounds, each with LD_{50} to humans below 1 mg/kg: tabun, sarin, and soman. By war's end, Germany was producing 100 tons/month of tabun, and two sarin plants were under construction. The German data on OPs provided the postwar impetus for British and American development of both nerve gases and insecticides.[10] The major improvement in the former category has been the development of binary gases such as VX, which are assembled in shells containing inactive precursors that mix when the shell is fired, releasing VX on impact. These binary gases, together with sarin and soman, are apparently the major components of the U.S. nerve gas arsenal of the 1990s.

The original American stocks of nerve gases were visibly decomposing at the beginning of the 1980s, leading to considerable controversy over the safety of the military's plans for disposing of these stocks. Meanwhile, new stocks were being generated, purportedly in response to the USSR's efforts in the same direction. As the Cold War began to thaw in 1989–1990, the Soviets suggested eliminating chemical warfare agents from the superpowers' arsenals. After much hesitation, the U.S. president, George Bush, offered to destroy *existing* stocks of nerve gases. This was not a very dramatic concession, since he did not commit the United States to stop production of new stocks, and it was known that existing stocks had in any case to be destroyed for safety reasons. In the late 1980s, Germany, in the face of large demonstrations by its own population and the obvious thawing of the Cold War under Gorbachev, requested removal of chemical warfare agents from its territory. These were burned. The subsequent outbreak of the Gulf War in 1991, and the continued emphasis on Saddam Hussein's willingness to use chemical warfare agents, further complicated the issue.[11] The new generation of OP chemical warfare agents in the United States are often binary gases: relatively innocuous precursors mix to combine the lethal agent when the shell hits the target.

The commercial exploitation of OPs began with the marketing of tetraethylpyrophosphate (TEPP, Bladan) in 1944. TEPP hydrolyzes within hours after application, however, and was not terribly useful in the field. Structure–activity studies

---

[10] Notwithstanding this adoption of his inventions, Gerhard Schrader was for years denied a visa to visit the United States. He was convicted of war crimes for using slave labor in the nerve gas factories.

[11] Despite the comfortable myth that chemical warfare agents have not really been used since World War I, and despite the denial of almost all the users, they were used in most decades of this century. It is generally accepted that chemical warfare was used (successfully) by Italy in Ethiopia in 1936, and unsuccessfully by Egypt in the Sudan in the 1950s. The United States used herbicides in Viet Nam in the 1960s and is unique in admitting the use of chemical warfare agents. Iraq used toxic gases against Iran early in the 8-year-long Iraq–Iran war and, later in the 1980s, against Kurdish villagers. The consensus is that chemical warfare agents are not effective against troops that expect their use but are psychologically useful in demoralizing unprepared troops, although the only example of effective use seems to be the Italian campaign in Ethiopia. The use of protective gear decreases the efficiency of an army, but this is an advantage to the users only if they can be sure that their own troops need not be protected. The use of chemical warfare agents against civilian populations seems to be viciously effective in terms of mortality, but like saturation bombing, fails to achieve the more desirable goal of destroying resistance.

led to the development of paraoxon, which was an improvement but still too easily hydrolyzed under alkaline conditions. Alkaline conditions were quite common in field use, since older pesticides such as Bordeaux mixture contained lime. Further study identified the stability of the thiophosphates [P=S], and of parathion in particular. An additional advantage of the thiophosphates was the slightly slower onset of toxicity: given the extreme toxicity of so many early OPs, a 20-minute delay in the onset of symptoms could be lifesaving.

The extreme toxicity of parathion (LD_{50} = 1 to 10 mg/kg = $< \frac{1}{5}$ teaspoon for 200-lb man) led to a search for less toxic insecticides. Much of the work was done by trial and error, using structural analogs of known insecticides or nerve gases and testing their toxicity to mammals and insects. In retrospect, it is known that two routes to selective toxicity are available: selection of compounds that inhibit insect, but not mammalian, AChE; or selection of compounds that are degraded by mammalian, but not insect, metabolism before reaching the nervous system.

The selective fit of OPs to AChE is illustrated by comparison of the toxicities of parathion and its analog fenitrothion: the latter, with the 3-methyl addition, does not fit mammalian AChE well, but apparently fits the slightly different insect AChE nicely. In contrast, the selective toxicity of malathion, which is almost 100 times as toxic to insects as to mammals, is due to differences in metabolic pathways between insects and mammals (Table 8.2). Malathion is metabolized by mammals more rapidly than it is activated to the P=O form. The necessary esterase is present in mammals but not in insects, rendering it relatively safe. Under conditions in which the mammalian enzymes are inhibited (e.g., concurrent use of monoamine oxidase inhibitors), the mammalian toxicity of malathion is enormously enhanced (Figure 8.5).

Antidotes to OPs consist of two types: those that block the effects of acetylcholine at its receptor and those that regenerate the acetylcholinesterase by removing the OP from its active site. The best known of the AChE regenerators is pyridine-2-aldoxime methiodide (2-PAM), although newer analogs exist. These regenerating antidotes must be given quickly, since OPs "age"—undergo further metabo-

Table 8.2 Risks of Poisoning by Parathion and Malathion During the 1960s[a]

Database	Poisonings	Deaths
For Parathion		
Worldwide (from WHO)	500,000	20,640
Japan, 1960–1969		
Application related	2,059	110
Accidents	113	56
Suicides	3,243	3,040
For Malathion		
Japan, 1960–1969		
Application related	100	31
Suicides	1,624	844

Source: World Health Organization data.

[a] Annual production of these pesticides in the United States at this time was 100,000,000 lb.

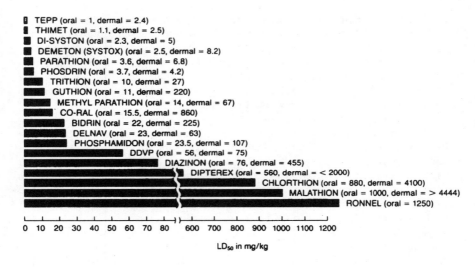

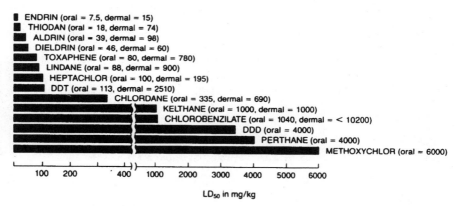

Figure 8.5 Range of acute toxicity of organophosphorus ester and organochlorine insecticides. Although many organochlorine insecticides are only "moderately toxic" and many OP insecticides are in the "extremely toxic" category, there is considerable overlap between the two categories. (From the Bureau of Occupational Health, State of California Department of Public Health.)

lism—in the OP–AChE complex, and cannot then be regenerated. The rapidity with which aging occurs is highly variable between OPs, but can to a considerable extent be predicted from structural considerations.

The second (and older) antidote for OP intoxication is atropine, which acts by blocking the acetylcholine receptor at the postsynaptic membrane. As a result, the excess acetylcholine present cannot stimulate muscle contraction/nerve transmission. By itself, atropine is a potent neurotoxicant. In the presence of OPs, however, two poisons add up to survival.[12] Although in most cases atropine need not be administered for more than a day, some of the more persistent OPS require atropinization for longer periods. In one case, treatment was continued for 30 days before symptoms abated.

Even though the OP insecticides were already being developed before DDT was commercially available, use of OPs lagged until the combination of insect resistance and growing concern over the environmental consequences of the organochlorine insecticides led to the gradual phasing out of these compounds. If one compares the *human* costs of parathion use to those of DDT use, it will seem that the use of DDT should have been continued. Certainly, parathion has been responsible for more deaths than any other single insecticide. During the 1960s, half the children who died of poison in Dade County, Florida, died of insecticide poisoning. This cause of childhood mortality dropped precipitously in the 1970s, when parathion was restricted. Much can be made of this toxicity and of the human cost of banning DDT. It should be remembered, however, that parathion was not the only possible replacement for DDT. The far less toxic fenitrothion has a range of insecticidal activity that is very similar to that of parathion and is similar enough in structure to be produced in the same factories. The difference in precursor chemicals might make it somewhat more expensive, but it is heavily used in other countries, including Japan, where parathion has been banned, and Canada. Had fenitrothion been introduced as a DDT replacement, the human costs would be much less serious than those for parathion, and the benefits of banning DDT quite obvious.[13]

Organophosphorus Ester–Induced Delayed Neurotoxicity

The first recognized episode of organophosphorus ester–induced delayed neurotoxicity (OPIDN) was the "ginger jakes," an epidemic of poisoning that occurred in the United States in 1930. It is estimated that over 10,000 people were paralyzed by a solvent, tri-*ortho*-cresyl phosphate (TOCP; Figure 8.6), that was used to adulterate a ginger extract. The ginger extract was apparently widely used as an alcoholic beverage during Prohibition. The episode takes its name from the poisoned *Jamaican Ginger,* and from the "jakes" (shakes), or flaccid paralysis, suffered by the victims. It is now known that the active neurotoxicant in TOCP poisoning is not TOCP itself, but a metabolite: saligenin cyclic phosphate.

---

[12] During the early stages of the 1991 Gulf War, Israelis who feared that Saddam Hussein had mounted nerve gas on the Scud missiles were made ill by injecting themselves with the antidotes. In case of genuine OP poisoning, however, it is essential to administer atropine in *massive* doses: ordinary medicinal doses are almost useless. The advice is to "titrate the victim until the pupils of the eye dilate" (i.e., until symptoms of atropine toxicity appear). It is also essential to realize that even when symptoms abate, AChE levels do not return to normal until new AChE has been synthesized: up to one month. A second exposure during this period—even if much less severe than the initial exposure—can be lethal.

[13] Many uses of parathion were banned by the U.S. EPA in 1991. The remaining uses were relevant mostly in California, and millions of gallons of parathion were shipped to that state in the wake of parathion's sharply restricted label. California, in direct consequence, has begun proceedings to ban all use of parathion within its borders.

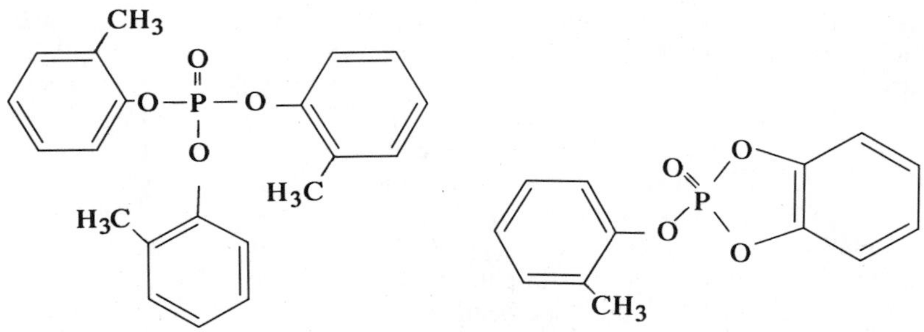

(a)   (b)

Figure 8.6 Structures of TOCP and saligenin cyclic phosphate.

The syndrome was characterized by an ascending ataxia/paralysis, generally restricted to the legs, that had its onset 8 to 30 days after exposure. The symptoms were to a large part irreversible; the characteristic time course of the syndrome includes (at the least) an 8-day latency, followed by the onset of ataxia, followed by slow recovery for up to one year (in humans). Ataxia remaining at the end of 1 year is probably permanent; neither further deterioration nor significant improvement usually occurs thereafter.

Rodents are almost totally resistant to OPIDN, although they are excellent models for the acute toxicity of OPs. Adult hens were rapidly identified as the best experimental species; this is still true, despite numerous efforts to find a mammalian species—other than the human—of equal sensitivity. Even in hens, young birds remain unaffected by the ataxia. Because of the source of the TOCP, few if any children were exposed in the ginger jakes episode. In a much later outbreak, occurring in Meknes in Morocco in 1959, the source of the TOCP was food oil that had been adulterated with TOCP. In this episode, which again affected over 10,000 people, it became tragically apparent that human children are also sensitive to the paralysis, although they seemed to be less severely affected than adults. It is certain that human beings are at least as sensitive as hens, and we may well be the most sensitive species.

Except for the identification of saligenin cyclic phosphate as the active neurotoxicant and the recognition of the sensitivity of children, this outline of the syndrome was complete within 5 years of the ginger jakes outbreak. Earlier awareness that saligenin cyclic phosphate, and not TOCP, was the proximate neurotoxicant might have altered the history of the investigation, since TOCP does *not* inhibit AChE, whereas saligenin cyclic phosphate does. In the event, the connection between AChE inhibitors and the delayed paralysis was not made for over 20 years.

Recurring episodes of OPIDN in the 1930s and 1940s were invariably due to TOCP exposure. The settings ranged from industrial exposure to TOCP fumes when TOCP was used as a lubricant or solvent, to environmental exposure to TOCP in contaminated drums that were being used to hold water, to TOCP-adulterated foods. In one bizarre case, the source of exposure was a parsley-derived abortifacient, Apiol. Actually a purgative that is used in the hope of inducing abortion, Apiol itself is still used in Africa. The TOCP may have been used as a solvent for the other ingredients or to extract the parsley derivative.[14]

In 1959, the last major episode of TOCP occurred in Meknes, Algeria. The source was a cheap cooking oil, adulterated by TOCP; 10,000 people were affected. Characteristic of OPIDN is the incident reported from the Meknes tragedy: in one family the oil was recognized as being of poor quality and was first fed to the dog. When the dog remained healthy the next day, the entire family used the oil. All, including the dog, became paralyzed a week later. It is also from Meknes that we know that children become ataxic, unlike the young of other species, although there is a suggestion in the data that very young children are at least partially protected.

During the 30 years between ginger jakes and Meknes, precautions on the use of tricresyl phosphates were developed: primarily, that industrial mixtures contain little or no tri-*ortho*-cresyl phosphate. Meanwhile, World War II saw the development of OPs as nerve gases and the introduction and proliferation of organophosphate pesticides (cf. parathion case history). The extreme toxicity of the early OP insecticides led to conscious attempts to introduce safer OP insecticides. Unfortunately, because TOCP does not inhibit AChE, no connection had been drawn between ginger jakes and cholinesterase-inhibiting organophosphorus esters; attention focused on the tri-*ortho*-cresyl configuration.

Among the safer insecticidal OP compounds being developed in the early 1950s was mipafox (Figure 8.7), a selective insecticide so safe to humans that the developers kneaded the formulation with their hands. Three became paralyzed, and mipafox was immediately withdrawn from commercial development. Despite its small scale, this tragedy led to OPIDN testing of all OP insecticides in Britain, and to an unwritten but scrupulously honored ban of OPIDN-causing pesticides by European chemical companies. The World Health Organization (WHO) formally banned OPIDN-causing pesticides from its malaria-eradication program, leading to a further reduction in the incentive to market such compounds, since WHO provided considerable testing of candidate insecticides as well as a sizable market for insecticides actually used. In the United States, which had not been directly affected by mipafox, OPIDN-testing of OP insecticides was included in the protocols of the Food and Drug Administration (FDA) and the U.S. Department of Agriculture (USDA).

A second wave of research on OPIDN followed the mipafox incident and established that:

1. Most or all phospho*fluoridates* are delayed neurotoxicants, and other OPs can also cause paralysis. Among the OPs identified as causing paralysis were the nerve gas sarin and several candidate insecticides, such as CELA-K43, with an LD_{50} of >1 g/kg in rats.

2. *All* paralytic OPs inhibit AChE, but the correlation between the two forms of toxicity is terrible: parathion (or, to be precise, its oxon) is an excellent AChE inhibitor but does *not* cause paralysis. The nerve gas sarin does, as does the laboratory chemical diisopropylfluorophosphate (DFP). The nerve gas soman, on the other hand, has not been shown to cause OPIDN.

[14] See also R. L. Metcalf, "Historical perspective of delayed neurotoxicity," in *Delayed Neurotoxicity*, J. M. Cranmer and E. J. Hixson, eds, Intox Press, Little Rock, AR, 1984.

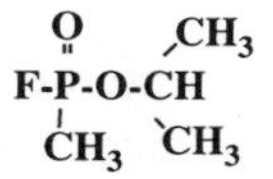

$$\begin{array}{c} O \\ \| \\ F\text{-}P\text{-}O\text{-}CH \\ | \\ CH_3 \end{array} \begin{array}{c} CH_3 \\ \diagup \\ \diagdown CH_3 \end{array}$$

$$\begin{array}{c} O \quad CH_3 \\ \| \quad \diagup \\ F\text{-}P\text{-}O\text{-}CH \\ | \qquad \diagdown CH_3 \\ O \\ | \\ CH \\ \diagup \quad \diagdown \\ H_3C \quad CH_3 \end{array}$$

SARIN 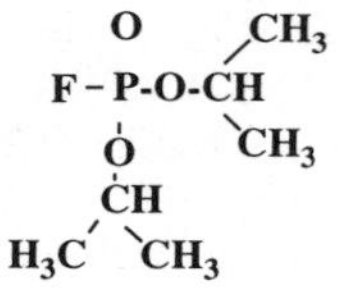   *DFP*

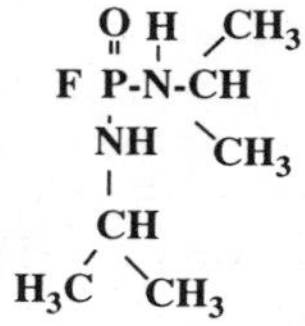

$$\begin{array}{c} O \; H \quad CH_3 \\ \| \; | \quad \diagup \\ F \; P\text{-}N\text{-}CH \\ | \qquad \diagdown CH_3 \\ NH \\ | \\ CH \\ \diagup \quad \diagdown \\ H_3C \quad CH_3 \end{array}$$

MIPAFOX 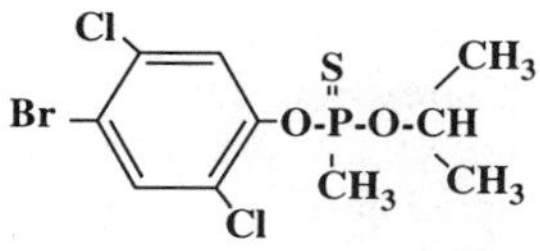   *CELA K-43*

Figure 8.7 Structures of sarin, mipafox, DFP, and CELA-K43. Sarin has an LD_{50} well below 1 mg/kg. Toxicity is decreased several orders of magnitude by substituting *either* the $NCH(CH_3)_2$ of mipafox for the $OCH(CH_3)$ of sarin, *or* the halogenated-phenyl group of CELA-K43 for the F of sarin. Unfortunately, the substitutions do not affect delayed neurotoxic potential. All of these compounds induce OPIDN.

3. The paralysis is not, as was assumed until then, due to primary demyelination, but to axonal degeneration followed by demyelination.

In 1970, the U.S. Environmental Protection Agency (EPA) was formed, and as part of its function, took over pesticide safety evaluation from the FDA and USDA. The initial rules for pesticide testing were essentially the same as they had been under USDA. Safety studies were required before pesticides could be registered, but the choice of what studies to carry out, *and which of the data to submit to EPA,* was still left up to companies to a considerable extent. Somewhere in the transfer to EPA, the USDA/FDA requirement of testing OPs for OPIDN was lost. Against this background, the Velsicol Chemical Company developed leptophos.

CASE HISTORY: LEPTOPHOS[15]

In 1965, chemists at Velsicol Chemical Company synthesized a promising OP insecticide, which was field-tested on a small scale in 1966, followed by large-scale testing on deciduous fruits, potatoes, tomatoes, rice, soybeans, sugar beets,

[15] Information about the efforts to register (and to prevent the registration) of leptophos is taken mostly from the appendix of the hearings held by the U.S. Senate Subcommittee on the Judiciary concerning the functioning of the U.S. EPA. The appendix of this document consists of all memos, letters, and documents gathered in the investigation of the registration of the OP insecticide leptophos (*O*-2,5-dichloro-4-bromophenyl *O*-methyl phenylphosphonothionate). Additional information was obtained from R. L. Metcalf of the University of Illinois.

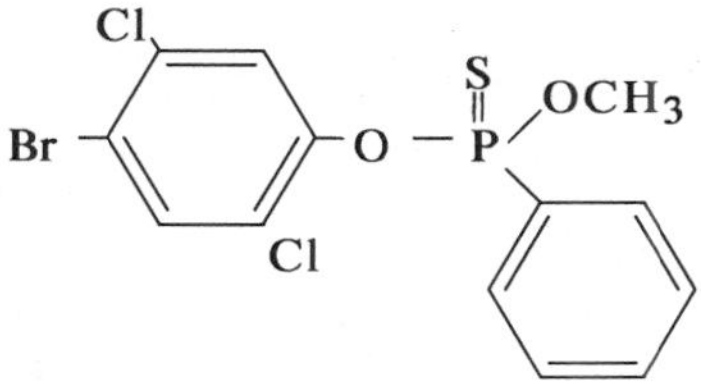

Figure 8.8 Leptophos structure.

and cole (cabbage family) crops. The compound proved to act best against larvae of the Lepidoptera (caterpillars) and was therefore given the generic name *leptophos* (Figure 8.8). The brand name was *Phosvel*. Because it was not then required for a company to submit all testing data to EPA, there is no way of knowing all the tests that were carried out on leptophos. It is known, however, that safety testing included at least one two-generation reproduction study (see Chapter 9), a 28-day feeding study in steer, a 2-year chronic dosing study in dogs and in rats, and a demyelination (OPIDN) study in hens. Only OPIDN study data are available. This demyelination study was carried out by Industrial Biotest, then the largest independent testing laboratory in the United States.[16]

The OPIDN study was carefully designed, and included two treatments, spaced 21 days apart, for each hen. There were six hens at each dose level and three doses. Despite the small number of animals at each dose, the study was thorough, including daily observation, repeated weighing, and measurement of the food consumption of each hen throughout the study. Tissues from the hens were also examined histologically. Shorn of verbiage, the results were:

· After two doses, 21 days apart, of 100 mg/kg, all hens remained normal.
· After two doses of 200 mg/kg, one hen became permanently ataxic, one became mildly ataxic, with apparent recovery, and three hens exhibited only the acute effects of acetylcholinesterase inhibition.
· After a single dose of 400 mg/kg, four hens of six became ataxic, and a fifth became ataxic after the second dose of 400 mg/kg. One hen never showed any but acute effects. Four of the hens died before the end of study.[17] Histology showed no demyelination, but the histological techniques used were not very sensitive.

Despite the data in the body of report, the abstract that formed the first page of the IBT report to Velsicol stated that *there was no evidence of delayed neurotoxicity*. Velsicol forwarded the complete report, with the abstract as its first page, to EPA as part of the evidence for the registration of leptophos.

Field testing of leptophos proceeded abroad while registration efforts proceeded in the United States. Cotton, because it is a cash crop, grown intensively, and

---

[16] That Industrial Biotest is no longer in existence is a direct consequence of the hearings on leptophos and the further investigations prompted by the story that follows.
[17] Ataxic hens cannot reach food and water in the usual chicken battery, and so die. In our laboratory, ataxic hens often, although not invariably, survive indefinitely if feed and water are kept within reach.

subject to numerous pests, provides a considerable challenge to insecticides. This is especially true in tropical climates, where several crops a year are grown and there is no cold period to decrease insect numbers. Resistance to insecticides occurs rapidly among pests of cotton, and there is always a market for new chemicals.[18] In 1971, during leptophos trials in Egyptian cotton fields, an estimated 1200 water buffalo were paralyzed and died after exposure to leptophos, although the cause of the outbreak of ataxia was not immediately identified.

In October 1971, Velsicol petitioned for registration of leptophos in the United States. At the same time, a Velsicol factory in Bayport, Texas, was converted to full-scale leptophos production. Actual synthesis of the insecticide from its precursors occurred in a closed system, with minimal exposure of the workers. Once synthesized, however, the leptophos was poured into flat tins or large drums and left to cool and solidify (the melting point is 71 to 72°C). The cooled "cakes" were chopped into chunks with hatchets, and the chunks were dumped into an open grinder to produce the powder that could be formulated with inert ingredients to produce the commercial insecticide. Permanent workers used—but also reused—gloves, hats, and safety glasses. Temporary help used no protective clothing. Workers frequently felt nauseated, which they attributed to the odor of the leptophos: actually, nausea is a typical symptom of OP poisoning.

In 1971, at the time of the registration petition, a memo was sent from the World Health Organization (WHO) to EPA, notifying EPA that leptophos was a delayed neurotoxin that should not be registered. Robert L. Metcalf of the University of Illinois, a member of the WHO pesticide-evaluating team, was alerted to the delayed neurotoxicity of leptophos by Wendell Kilgore of California, who knew about the water buffalo. EPA took no official action. However, the WHO memo may well have been the reason that Ronald Baron of EPA undertook OPIDN testing of leptophos. In 1973, Baron notified the Toxic Chemicals Branch of EPA that leptophos was a TOCP-like neurotoxin, that it caused demyelination in hens, and that he was planning further studies. As far as is known, those further studies were never done, or at least never reported.

In the same year, Velsicol requested crop tolerances for leptophos of 10 ppm on lettuce and of 2 ppm on tomatoes. Such tolerances are the legal maximum residues of a pesticide that can be present in a food crop for the crop to be sold. At their best, tolerances represent a balance between the persistence of a chemical and its hazard: more persistent pesticides need higher tolerance of residue levels, but these can be granted only if the residues do not present a hazard. Leptophos is a relatively persistent OP, as Velsicol obviously knew. We now know that the residue levels they proposed *might* have led to observable nerve damage in consumers; Velsicol probably did not know that. They did know, however, that more water buffalo died in Egypt during 1973. Certainly, the information was spreading among pesticide toxicologists, and also between companies.

In 1974 the first published data on the delayed neurotoxicity of leptophos appeared in *Experientia*.[19] The study also included negative results for two other

---

[18] The resistance problem is so bad that cotton growers in several areas have abandoned the crop because the cost of insecticides make cotton unprofitable. In other areas, cotton growers have turned to *integrated pest management* (IPM), which minimizes pesticide use by drawing on knowledge of the pests' life cycles to devise control strategies.

[19] M. B. Abou-Donia, M. A. Othman, G. Tantawy, A. Z. Khalil, and M. F. Shawer, "Neurotoxic effect of leptophos," *Experientia* 30:63–64, 1974.

insecticides that had been suggested as culprits in the water buffalo deaths. Kilgore reported his information about the paralyzed water buffalo to Gunter Zweig of EPA, including a report on controlled feeding studies in which water buffalos were fed leptophos-treated forage, proving that leptophos was actually responsible for the paralysis. A conference was then held between Zweig (EPA), Kilgore (University of California, Davis), and Clara Williams (head of the EPA toxicology branch). Williams concluded after the conference that EPA should approve Velsicol's requested tolerances. She wrote: "If sometime in the future we get data which . . . lead us to feel that these tolerances should be revoked, we will take action at that time." She considered the feeding studies in water buffalo to be "inadequate . . . no convincing evidence . . . ."

Several memos were sent in 1974 and remained in the files, including one from Metcalf at the University of Illinois to Donna Kuroda of EPA, warning that leptophos would pose a hazard if exported to Indonesia, and a second from J. Barnes at WHO to EPA, warning that leptophos had proven to be a delayed neurotoxin. As a result of the pressure from these world-renowned pesticide toxicologists, a memo from the EPA Coordination Branch concluded that Zweig and Kuroda were worried because Metcalf and Kilgore were worried, and that Metcalf and Kilgore were worried because Barnes was worried. The memo's authors concluded, however, that, although there was evidence for ataxia in both hens and water buffalo, there was no evidence of "permanent" nerve damage because there was no *histological* confirmation of axonal degeneration.[20] A recommendation of possibly restricted use was made, and close review of data was suggested.

In another memo, Metcalf reported to EPA "rumors" of human nerve damage.[21] EPA's Williams dismissed the new information with a reiteration of the "anecdotal" nature of the information and referred to the possibility that another insecticide, not leptophos, was responsible for the water buffalo deaths. As a result of the slow but definite progress of leptophos through the regulatory maze, European chemical companies became aware of the plausibility of marketing OPIDN-inducing insecticides. CIBA management asked their toxicologist why CIBA refused to market its OPIDN-inducing chemicals when the Americans obviously did not care. Calo, chief toxicologist for Velsicol, went to Egypt and sent back reports claiming that he could not confirm the water buffalo rumors. Velsicol even suggested that the incident was the result of "industrial espionage." At EPA, Baron sent a four-page internal memo to Kuroda, saying (among much else)

---

[20] This emphasis on sophisticated techniques is a form of pseudoscience. Rather than accepting the simple observation that water buffalo were paralyzed and that the results could be reproduced in two species by exposure to leptophos, a complicated technique was demanded as evidence that the cellular cause of the observed paralysis is the same as that observed when other OPs cause OPIDN. In effect the regulators were saying that the occurrence of paralysis was less important than the name of the syndrome. Such exertions should be reserved for demonstrations of lack of effect: that is, had the hens *not* become paralyzed, one should have checked for subclinical damage at the cellular level. Industry might object that this is double jeopardy, but when there is obvious risk of irreversible harm, strict safety standards should be enforced.

[21] These rumors have never been confirmed in published data but have been confirmed by Egyptians involved with the leptophos incident. Unfortunately, the Egyptian government has refused, even to date, to allow studies of affected workers and their families. It is said that considerable human exposure occurred because farm workers are accustomed to taking the insecticides from the field to their home for use against mosquitos.

that "although the test did show toxicity at high level after second dosing, the predominant 'tunnel vision' . . . suggests that demyelination was not a problem. . . ." In this memo he also claimed that leptophos caused no demyelination in hens, despite his own earlier data showing demyelination.

In 1975, new tolerances, lower than those originally requested by Velsicol, were published in the *Federal Register*. Although not synonymous with registration, the publication of such tolerance levels usually precedes actual registration by a very short period. Metcalf says of that period that whenever he or other concerned toxicologists attempted to delay leptophos registration, they only seemed to speed it up. Nonetheless, time was actually running out for leptophos.

It was in 1975 that an internal Velsicol memo about the Bayport plant that produced leptophos described a "series of unusual CNS illnesses . . . among employees at this plant . . . ." The symptoms included weakness, ataxia, bizarre behavior, central nervous system symptoms, impotence, nervousness, wheezing, and choking. A number of workers had gone to private doctors, and the diagnoses included asthma, multiple sclerosis, encephalitis, psychosis, and alcoholism. A more accurate diagnosis was provided by the other workers in the Bayport factory, who referred to the affected men as "Phosvel zombies." Unlike the private doctors, the workers saw more than one case at a time and recognized the occupational link.[22]

Despite the internal memo, and in the same month, Velsicol requested a National Academy of Sciences panel to adjudicate leptophos registration, complaining that the EPA was too slow. Almost simultaneously, a prestigious consulting firm hired by Velsicol to look at the functioning of the Bayport plant warned of problems in the factory and recommended changes in production with the goal of decreasing worker exposure to chemicals. In November 1975, Velsicol informed EPA that there were illnesses in the Bayport plant.

In 1976 the Occupational Safety and Health Administration (OSHA) became involved because of a "tip" from an informant, whose identity was not recorded in the available memos. Also in 1976, Velsicol's doctor at the Bayport factory wrote a memo to Velsicol's management in Illinois, advising against informing workers of the hazards of leptophos exposure, because "where permanent neurological damage has occurred . . . there can probably be no additional benefit in the treatment by definitive diagnosis. That is to say, toxic destruction of a nerve ends up being the same no matter what the toxic material. . . . I see no benefit to the people . . . in your employ by your making an active attempt to identify the toxicity of your products to them. . . ." In August 1976, Velsicol withdrew its petition to register leptophos.

Postscripts.   Industrial Biotest, the company that claimed negative results on an abstract when data showed positive for OPIDN, was forced out of business by disclosures of numerous badly done—or simply undone—studies, as well as of mislabeled and falsified data, of which the leptophos study was only one example. Industrial Biotest was the largest "independent testing laboratory" in the country. Several executives were prosecuted for criminal actions. IBT's misdeeds forced

---

[22] Eventually, 53 of 163 workers were found to be affected.

reform of reporting requirements: today a company must report all safety data to EPA, whether self-generated or acquired from others. Moreover, approximately 20% of pesticide registration data should be recertified, since it rests on now-suspect IBT studies. Since EPA did not by any stretch of the imagination have the necessary resources to carry out this process expeditiously, reregistration proceedings continue into the 1990s.

After ending production of leptophos, Velsicol converted the Bayport plant to making EPN, which was first identified as a delayed neurotoxicant in 1956. EPN is still on the market in 1991, on the basis of essentially the same arguments that Clara Williams used for leptophos: we do not know of any human damage, so let's wait until we get data saying that we have to change our minds.[23] In 1983, after a fire broke out in a Chicago plant making EPN (not a Velsicol plant), OSHA investigated the neurological status of nine workers and found that seven of the nine had nerve impairment, although none were overtly paralyzed. At the scene of the fire, a Stauffer spokesman asserted that there was no risk to firefighters and rescue workers from the chemicals in the plant; 24 hours later, many of the workers and bystanders were recalled to have their cholinesterase levels tested. When the spokesman was asked about the risk of OPIDN, he responded that there was no risk, since no chickens worked in the plant.

A major consequence of the leptophos episode was a resurgence of research on OPIDN. Data that had long been neglected were dusted off. It was realized that there were other neurotoxic compounds in use as insecticides: EPN, cyano-fenphos, trichlornate (Agritox). Additional suspected OPIDN agents include tri-chlorfon; dichlorvos, the insecticide in "No Pest" strips; the cockroach insecti-cide, chlorpyrifos (sold as Dursban for household use and as Lorsban when used in agriculture as a soil insecticide); and the canine flea treatment fenthion. Other commercially used OPIDN compounds include the defoliants DEF and merphos, the herbicide DMPA, the laboratory chemical DFP, and the nerve gas sarin. A number of advances were made in our understanding of OPIDN. Not the least of these was the realization that the irreversible neurological effects of OPs are not restricted to paralysis of the legs. Some OPs, like leptophos, also affect the central nervous system in a chronic, perhaps irreversible, manner. This effect led to the appellation of "Phosvel zombies"; it, rather than OPIDN, may account for the bizarre symptoms observed in veterinarians exposed to fenthion.[24]

---

[23] To my own knowledge, there have been two reviews of EPN by the U.S. EPA since the leptophos fiasco. Neither has resulted in banning the chemical, despite studies showing that repeated subclinical doses of EPN are more effective than single doses at causing ataxia, despite evidence that percutaneous exposure is more likely to cause ataxia, and despite data showing that workers in the Stauffer EPN factory had suffered nerve damage.

[24] Some of the data exist only in case history form, often anecdotal. Several women worked in a building in North Carolina that was repeatedly treated with chlorpyrifos to exterminate cockroaches. One woman developed weakness of her arms, numbness in her hands, and some breathing problems. Several other women developed parasthesias, muscle weakness, or back-aches. Diagnoses included multiple sclerosis, but symptoms in all the women abated when chlorpyrifos treatment was stopped. Because of the atypical pattern of ataxia, toxicologists did not measure acetylcholinesterase in the victims, so there is no proof that OP intoxication occurred.

9

DEVELOPMENTAL TOXICOLOGY

> If we accept the statement that the ability to produce developmental abnormalities is a property of all drugs, our purpose is then to decide the laboratory circumstances under which a drug, in exhibiting significant teratogenic activity, might be expected to have similar effects on the human fetus.
>
> —D. A. Karnofsky[1]

HISTORICAL ASPECTS

From earliest times, the birth of malformed children has profoundly affected those who saw them. In many primitive (and not so primitive) societies, the birth of an obviously malformed infant was considered to be the result of witchcraft, and mother and child were summarily killed. In other societies, such infants were considered omens or portents: warnings of great changes. From this comes the use of *teratology* (from the Greek word *teraton,* meaning "wonder" or, by derivation, "monster"), for the study of malformations. It is conjectured that many of the monsters that populate ancient myths are transformed from congenital malformations. Today, many of the malformations bear the name of the monsters they resemble: cyclopia, sirenomelia, and so on. In common speech, resemblance to animals leads to names like "harelip" for a cleft lip or the German *Wolfsrachen* (wolf-throat) for cleft palate.

If one excludes the possibility of witchcraft or cohabitation with demons, the common Western theory of the origin of congenital malformations is that of *maternal impressions:* that the fetus is influenced by activities of the

---

[1] From D. A. Karnofsky, "Mechanisms of action of certain growth-inhibiting drugs," in J. G. Wilson and J. Warkany, eds., *Teratology: Principles and Techniques,* University of Chicago Press, Chicago 1965. This is the continuation of the quote often referred to as *Karnofsky's law,* which states: "It is proposed as a law, which cannot be disproved, that any drug administered at the proper dosage to embryos of the proper species—and these include both vertebrates and invertebrates—will be effective in causing disturbances in embryonic development." Karnofsky's law is often cited as evidence of the frustration of teratology.

mother or by what she sees during pregnancy. In its most benign form, this theory led women to spend their pregnancies listening to good music or looking at uplifting art. It also led to the general avoidance of strawberries (which were thought to cause birthmarks), of rabbits because they have a cleft upper lip, and of all deformed adults.

With the advent of Mendelian genetics (Chapter 10), the professional view of the sources of congenital malformations shifted drastically. Judicious use of the newly discovered laws of Mendel, plus the later concepts of incomplete penetrance, epistasis, and mutation, made it possible to attribute all malformations to genetics. Thus although the onus of devil worship was removed from the woman, the parents were now held to be responsible, through their genes, for the malformation. This viewpoint never really caught on with the population as a whole. The theory of maternal impressions was more comfortable and far less threatening to most people. Even for the professional, however, the totally genetic origin of congenital malformations was hardly formulated before it was shattered.

In the 1930s, Gregg, an Australian ophthalmologist, noticed that an epidemic of congenital cataracts in his practice followed the occurrence of an epidemic of rubella (German measles), a mild disease characterized by a rash and a 2 to 4-day low-grade fever. The children with cataracts were born 4 to 9 months after the rubella epidemic. Rubella was thus identified as the first human teratogen. It is remarkably typical of subsequent human teratogens in its characteristics. The disease is not serious in adults or in children; in fact, many women do not have any clinical illness. The effects in the fetus are not correlated with the severity of effects in the mother—women with clinically detectable disease did not have more severely affected children than those with clinically unapparent disease. Affected offspring are, however, more frequent among women who have had clinically apparent disease than among women who were exposed without developing rubella themselves. Women in the latter category might or might not have an affected offspring, since this group would include women who were already immune, and women who, although in contact with an active case of rubella, were not actually "infected," as well as women with infections so mild they never noticed an illness. Only the last subgroup is at risk of having an affected child.

There was a strong correlation between the frequency of malformations in the infants and the time of exposure during pregnancy: whereas women who were infected with rubella during the first month of gestation had a 50% chance of having an affected child, those infected in the second month had only a 30% risk. For the third and fourth months the risks to the fetus had decreased to 10% and 5%, respectively. We now know that congenital rubella is an ongoing infection of the fetus and newborn: it is the probability of the virus crossing the placenta that decreases as pregnancy progresses. Like several other infectious agents that cross the placenta (e.g., toxoplasmosis), rubella continues to act on the baby after birth, so that the status of the infant can deteriorate. Other infections act perinatally: the transmission of maternal disease during birth can have disastrous consequences of the baby. The most famous example is congenital syphilis, which causes blindness and is the reason that babies' eyes are still treated with silver nitrate after birth. Genital *Herpes simplex* infection can cause encephalitis, with ensuing brain damage, if there is a flare-up of the mother's disease at delivery.[2]

At the same time that Gregg was demonstrating the effects of rubella on human fetuses, another Australian, Hale, was demonstrating the effects of avitaminosis A on pigs. The feeding of vitamin A–deficient diets to sows led to a high incidence of eyelessness (anophthalmia) in piglets. Hale's work was followed, in the next decades, by elegant demonstrations of the terato-genic effects of vitamin deficiencies and vitamin excesses on the fetuses of numerous laboratory species. Further work in the 1940s and 1950s also showed the effects of metabolic inhibitors, hormones, antineoplastic chemi-cals and a host of other chemicals on the fetuses of laboratory animals.

In humans, however, the only major teratogens identified before 1950 were rubella and radiation. It was first suspected in the 1920s, and generally accepted by 1940, that exposure of the human fetus to high doses of radiation led to microcephaly (small head) and mental retardation. Thus two major teratogens, one physical and one viral, had been identified before 1950. Nonetheless, throughout the 1950s, teratologists more or less assumed that the human fetus was protected from chemical insults. The value of teratology was deemed to be in its application to mechanisms of development. It was hoped that the elucidation of mechanisms of activity of animal teratogens could explain the numerous spontaneous malformations seen in human in-fants.[3]

Unfortunately, there was no general recognition of the potential for mass epidemics of malformations due to the proliferation of chemicals in Western society and the increasing use of pharmaceuticals in Western medicine. There was, if not among teratologists, at least among the general medical profession and the public, a comfortable certainty that the uterus provided an essentially impermeable barrier to toxins (other than rubella and radiation, of course). The attitude persisted even though several chemotherapeutic agents were identified as animal teratogens in the 1950s, and even though aminopterin,

---

[2] This disease is most easily transmitted during the baby's passage through the vagina. Since the woman may not realize that the viral infection is active, it is probably advisable for women who have genital herpes to deliver by cesarean section.

[3] The major advances of classical embryology culminated in the identification of embryonic organizers as chemicals between 1932 and 1935. The frustrating finding that almost anything could act as an organizer in embryos of the right stage essentially ended the golden age of classical embryology. Thereafter, although many details were elucidated, no obvious progress was made on the major questions of vertebrate development until homeobox-containing genes were identified in vertebrates in 1984. Thus the idea of using chemicals to probe development was an exciting proposition, and attracted geneticists and embryologists as well as pediatricians, who deal with the consequences of abnormal development.

used as an abortifacient, was identified as a human teratogen in the early 1950s. Complacency was shattered once and for all by the thalidomide tragedy.[4]

Stated briefly, thalidomide was an inadequately tested sedative and analgesic that was heavily advertised as being totally safe and nontoxic. Based on results in adult laboratory animals, the manufacturer (Chemie Gruenenthal, Germany) advertised thalidomide as being without an identifiable LD_{50} and without side effects. Its purported harmlessness made it particularly attractive for the relief of nausea, tension, or sleeplessness during pregnancy, and it was heavily advertised for this use. For many months it was marketed "over the counter": that is, no prescription was required. The advertisement of a lack of side effects continued even after the company knew that the drug often caused irreversible nerve damage in adults, and despite the total lack of data on the ability of thalidomide to cross the placenta or on its effects on embryos.[5]

Approximately 8000 infants were born between 1959 and 1962 with severe thalidomide-induced malformations, which included no arms, no legs, deformed ears, paralyzed faces, and internal malformations. About one-third of the children died in early infancy, primarily of internal malformations. Most of the thalidomide children were born in Germany, Britain, Australia, and Japan, where thalidomide was most heavily used. In the United States, Frances Kelsey of the FDA insisted on tests of thalidomide in pregnant animals before she would approve the marketing of thalidomide. Before the American licensee was able to complete the additional testing, the drug's teratogenicity had been identified by a German pediatrician, Widukind Lenz,[6] and it never reached the American market. Between 50 and 100 thalidomide children were nonetheless born in the United States. Most of these children were the result of drug tests carried out in advance of licensing, but some women obtained thalidomide in Canada or Europe.

The immediate result of the thalidomide disaster was a heightened awareness of the vulnerability of the human fetus to environmental insult and a new

[4] A detailed history of the thalidomide tragedy is provided in the book *Suffer the Children: The Story of Thalidomide*, edited by The Insight Team of the *London Times* from information gathered by their investigative reporters, Viking Press, New York, 1979. The account offered by the *London Times* differs sharply from the conventional story, which excuses Chemie Gruenenthal's failure to test for teratogenicity as acceptable by the standards of the times. The *Times* also details efforts made by Chemie Gruenenthal to silence doctors who reported adverse effects of thalidomide in adults.

[5] It has in fact been suggested that pregnant women became a major marketing target after (and because) the company knew that long-term use in adults could lead to irreversible neuropathy. Since the nausea of pregnancy rarely persists for more than 3 months, the manufacturer thought the neurological side effects would not be relevant in this population.

[6] Lenz identified thalidomide, under its German trade name Contergan, at a pediatric meeting. Almost simultaneously, the British journal *Lancet* published a note by W. McBride of Australia, also identifying thalidomide as the cause of malformations in Australian infants.

emphasis on reproductive testing of drugs, pesticides, and other chemicals to which pregnant women might be exposed. Since thalidomide, numerous chemicals have been demonstrated to cause malformations in humans. A partial list includes antiepileptic drugs, anticoagulants, alcohol, smoking, and the synthetic estrogen DES (Table 9.1). The recognition that growth retardation, fetal death, and functional deficits such as mental retardation must be included among teratogenic phenomena has widened the field dramatically. The addition of DES-induced transplacental carcinogenesis as a delayed effect of prenatal exposure (Chapter 12) emphasizes the principle that development does not stop at birth, but that prenatal exposures may continue to affect us until senescence.

Despite increasing sophistication in the regimens used for testing chemicals, despite voluminous information on nearly a thousand chemicals tested in one or more species, there remains the basic frustration expressed by *Karnofsky's law:* Every substance, if given at the right dose to the right species at the right time of development, is teratogenic. It remains impossible to predict human teratogenicity from the results of animal experiments. Moreover, the species specificity of teratogens almost certainly reflects a corresponding specificity in vertebrate development.

Table 9.1 Current Status of Chemicals Suspected of Causing Human Malformations

Known to Cause Human Birth Defects		Strongly Suspected of Causing Human Birth Defects	Probably Do Not Cause Human Birth Defects
Rubella	1930s	Quinine	LSD
Radiation	1930s	Amphetamines	Sulfonamides
Toxoplasmosis	1950s	Hypoglycemics	Adrenocortical steroids
Aminopterin	1952	Insulin	Antihistamines
Androgens	1959	Tranquilizers	(meclizine)
Thalidomide	1961	Cocaine[a]	Bendectin
Methylmercury	1960s	Aspirin	
Warfarin	1960s	Marijuana[a]	
PCBs	1968	Cadmium	
Smoking	1970s	Dioxin (TCDD)	
Alcohol	1973	Barbiturates[a]	
DES	1974	Narcotics[a]	
Diphenylhydantoin (Dilantin)	1970s		
Methadone	1980s		
Valproic acid	1981		
Acutane	1983		
Vitamin A derivatives	Ongoing		
Lead	Ongoing		

[a] Almost all psychoactive drugs cause behavioral deficits in animals and are suspected of causing these in humans.

Recent advances in developmental biology (the synthesis of embryology and molecular genetics) illustrate the unity and diversity of developmental processes across species and even across phyla. The homeobox, a 180-base pair sequence that is strongly conserved in species from yeasts to humans, encodes a DNA binding motif. Although many types of genes contain the homeobox sequence, a subset of such genes, related to the *antennapaedia* gene in the fruit fly, are fundamental to normal pattern formation in all species examined, including the fruit fly, *Drosophila melanogaster,* in which they were first identified and studied, and in mice, amphibians, birds, humans, and plants. It is now clear that the conserved DNA sequences are utilized in laying down the anterior–posterior axis in all animal species studied. Thus there is a fundamental unity to development, a unity that nonetheless gives rise to an incredible diversity of shapes in different species. The complexity of the developmental sequences that cascade from the initial patterning can be seen in limb formation. The *antennapaedia* gene in *Drosophila* is a homeobox-containing gene that affects fly limb development. Very similar sequences affect limb development in vertebrate species. Even setting aside the large evolutionary distance between insects and vertebrates, *antennapaedia* and its homologs produce vertebrate limb patterns that range from the legless snakes to birds, with wings as forelimbs; to seals, with shortened flipper limbs; to the specialized human hand, with its opposable thumbs. Clearly, there are idiosyncracies of development within species that determine these variations, and many teratogens will act at the level of these specializations. Thus *most* interspecies extrapolation of teratogenic activity will always be dubious.[7] Nevertheless, there are very definite principles in teratology, whether experimental, clinical, or epidemiological.

BASIC PRINCIPLES OF TERATOLOGY

Access of Teratogens to the Fetus

Access of the agent to the developing tissues is essential to the action of any teratogen. For physical agents, this access essentially requires penetration of maternal tissues and/or circumvention of maternal buffering systems. As a

---

[7] There are some chemicals that act on processes so fundamental that they have similar developmental effects on all species studied. These agents are called *universal teratogens.* Potent mutagens, which act at the level of DNA and its replication, are the obvious examples. Less obvious, but very exciting, is retinoic acid, which appears to be one of the primary signals used by the embryo in pattern formation. An excellent overview of retinoic acid in its role as a possible *morphogen* is found in "Retinoids, homeoboxes, and growth factors: toward molecular models for limb development," *Cell* 66:199–217, 1991, by C. J. Tabin. A good review of the classical view of retinoic acid teratogenesis is provided by the 4th edition of *Casarett and Doull's Toxicology: The Basic Science of Poisons* (Macmillan, New York, 1991) in the article on teratology by J. Manson and H. D. Wise.

result, *most* physical agents affect the fetus only at maternally toxic levels. Radiation, anoxia, and hypothermia are the major physical agents that affect the fetus. Smoking, a definite teratogen, may act as a physical agent that causes anoxia in the fetus, or as a chemical teratogen because of the numerous chemicals found in smoke and tars (e.g., nicotine, carbon monoxide).

For chemical agents, the ability to cross the placenta is critical. Very few teratogens have been shown *not* to cross the placenta: trypan blue is the best-studied example. For agents that cross the placenta, teratogenicity may well be a function of the interaction between the rate at which they reach the fetus and the rate at which the maternal organism or the fetus can metabolize the agent. These degradative rates need not be the same. In general, fetal liver is less capable of metabolizing xenobiotics than is the adult (maternal) liver. In consequence, those chemicals that require activation before being teratogenic may be metabolized by the dam before teratogenic concentrations build up in the fetus. Alternatively, a substance that is not noxious to the adult organism because it is rapidly metabolized may well affect the fetus, which does not yet have the degradative ability, if the removal from the fetus across the placenta is slow. Similarly, substances that are sequestered in fat (PCBs, DDT) in the adult cannot be so disposed of in the fetus, with its essentially nonexistent depot fat.

Interaction of Genes and Environment

Environmental and genetic factors interact in causing developmental defects. That is, the susceptibility of an organism to a teratogenic agent depends on the fetal genotype and on the manner in which this interacts with the adverse environmental factors. Sensitivities differ between species, between strains within species, and even between littermates (or siblings).

The degree of genetic influence varies with the agent. For thalidomide, there is large interspecies variation (see Table 9.2), but an extremely high percentage of the human embryos that were exposed during the sensitive period (see below) were malformed. In humans, therefore, thalidomide-induced malformations exhibit maximum environmental and minimum genetic influence. The genetic influence is seen only in *inter*species variation in sensitivity. For the malformation complex of anencephaly/spina bifida (ASB), on the other hand, there is a considerable genetic component, in that mothers of one ASB infant have 10 times the background risk of producing an ASB infant in subsequent pregnancies, and women who have borne two ASB infants have an even higher probability of having a third such child.[8] However, 95% of ASB infants are born to women with no previous history of ASB infants, and there are pronounced seasonal, geographical, and temporal variations in the incidence of ASB, which argue strongly for environmental

---

[8] The risk of recurrence is proportionally higher both in geographic regions with high background incidence and/or in high-risk ethnic groups, regardless of location.

Table 9.2 Relative Sensitivities of Species to Induction of Structural Malformations by Thalidomide

Species	Smallest Dose Inducing Malformations (mg/kg/day)	Largest Dose Not Producing Malformations (mg/kg/day)
Human	≤ 1.0	?
Cynomolgous monkey	5	—
Rabbit	30	50
Mouse	31	4000
Rat	50	4000
Armadillo	100	—
Dog	100	200
Hamster	350	8000
Cat	—	500

Source: H. Kalter, *Teratology of the Central Nervous System,* University of Chicago Press, Chicago, 1968, p. 11.

Table 9.3 Incidence of Spina Bifida and Anencephaly in Various Ethnic Groups and Geographic Locations

Population	Incidence per 100,000 births	
	Spina Bifida	Anencephaly
American Indians		
British Columbia	24	8
U.S.	74	30
Negroes		
Pretoria	110	50
U.S.	53	40
Caucasians		
Belfast	426	440
New York	117	80
Hawaii	100	60
Japanese		
Japan	21	60
Hawaii	64	50
Hawaiians		
Caucasians	100	60
Japanese	64	50

Source: Adapted from J. R. Miller, in *Proceedings of the 2nd International Conference on Congenital Malformations,* M. Fishbein, ed., International Medical Congress, New York, 1964, pp. 335, 337.

factors (Table 9.3 and Figure 9.1). Finally, even those congenital malformations that are influenced primarily or exclusively by single genes can be considered in the framework of teratology, since many genetic syndromes are also influenced by environmental factors. In Down's syndrome, for example, the developmental deficits are due to the presence of an extra chromosome 21: a purely genetic etiology. The pronounced effect of maternal age on the incidence of Down's syndrome is, on the other hand, an environmental influence, and therefore the occurrence of a child with Down's syndrome represents the interaction of genetic and environmental influences.

The interaction of genotype and environment in the occurrence of a congenital malformation can be illustrated (Figure 9.2) as the shifting of an unknown distribution across a threshold that consists of the point at which the outcome for a continuous variable shifts from "normal" to "abnormal." In a sensitive genotype, a certain portion of the population may already fall into the malformed range of the curve; these are the "spontaneous" malformations that are seen in 2% of human births. Addition of an environmental insult pushes the entire population toward the malformation threshold, leading to an incidence of "induced" malformations in addition to the spontaneous group. In a resistant genotype, there will be few or no spontaneous malformations of the specified type, and relatively few induced malformations even in the presence of the environmental insult.

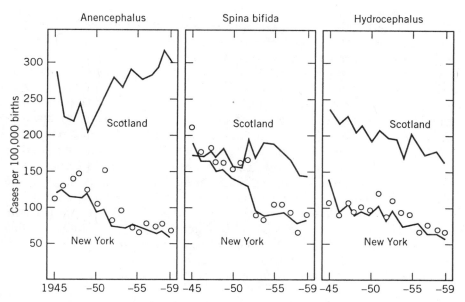

Figure 9.1 Incidence of central nervous system defects in Scotland and New York between 1945 and 1959. Note different temporal trends between defects in Scotland and in the incidence of anencephaly between New York and Scotland. (From L. Saxén and J. Rapola, *Congenital Defects,* Holt, Rinehart and Winston, New York, 1969, p. 11.)

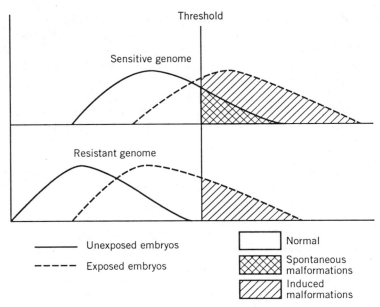

Figure 9.2 Threshold model of teratogenesis. (Adapted from F. C. Fraser, "Some genetic aspects of teratology," in *Teratology: Principles and Techniques*, J. G. Wilson and J. Warkany, eds., University of Chicago Press, Chicago, 1965, p. 34.)

The new molecular synthesis in developmental biology provides an elegant illustration of such interactions at the level of gene expression. On day 12 of gestation in the mouse, the mRNA for numerous *Hox* genes are expressed in the developing somites. There is an anteroposterior sequence of expression that can be correlated with the shape of the vertebrae and with the presence or absence of ribs on a specific vertebra, at birth. Kessel and Gruss[9] postulate that the shape of a given vertebra, and the presence or absence of an attached rib, depends on the *Hox* "code"—the combination of *Hox* genes present in each prevertebral segment (somite). Some variability may occur in the exact segment in which a certain *Hox* gene is expressed, leading to variability in the vertebral pattern in untreated mice. In mice treated early in gestation with retinoic acid, *Hox* gene expression is permanently altered, and vertebrae shift their identity anteriorly or posteriorly. The direction of the shift in identity corresponds to the shift in *Hox* gene expression. Moreover, the greatest frequency of retinoic acid–induced changes of *Hox*-gene expression, *and* of changes in vertebral identity, are seen at the cervical–thoracic border. In untreated, normally developing mice, there is a low incidence of the same

[9] M. Kessel and P. Gruss, "Homeotic transformations of murine vertebrae and concomitant alteration of *Hox* codes induced by retinoic acid," *Cell* 67:89–104, 1991. Another view is presented by E. M. De Robertis, E. A. Morita, and K. W. Y. Cho, "Gradient fields and homeobox genes," *Development* 112:669–678, 1991.

shift, demonstrated both by the shape of the last cervical vertebra and by the occasional existence of a "cervical rib." Spontaneously occurring "lumbar ribs"—extra ribs at the posterior end of the rib field—are quite rare in this strain of mice. Retinoic acid only rarely changed the *Hox*-gene expression and only occasionally induced extra lumbar ribs. Thus in the mice used by Kessel and Gruss, an underlying genetic variability in *Hox*-gene expression can be magnified by treatment with exogenous retinoic acid. Intuitively, one would predict that in mice that have a higher incidence of spontaneously occurring lumbar ribs but that rarely exhibit supernumerary cervical ribs, the *Hox* code would show instability at the thoracic–lumbar (rather than at the cervical–thoracic) border, and retinoic acid might well produce a very high incidence of lumbar ribs.

It must be emphasized that the identification of a genotype as "sensitive" or "resistant" is specific for the malformation *and* for the agent and cannot be generalized. For example, the C57BL/10 mouse strain, which is highly resistant to cortisone-induced cleft palate, is quite sensitive to aspirin-induced cleft palate.

Manifestations of Developmental Toxicity

The four manifestations of deviant development are death, malformation, growth retardation, and functional deficit. In sum, it is not only the structural malformation, the visible crippling, that is subsumed under developmental toxicology, but also the subtle defects of function, from mental retardation to dyslexia or mild motor disturbance manifesting as clumsiness. This broadening of the field from the original emphasis on structural malformations has emphasized the inadequacy of animal models. Although it is impossible to extrapolate between species with respect to the occurrence of structural malformations, it is excruciatingly difficult even to develop an animal model for some of the effects that are of concern in human children. Again, the ramifications of mental retardation come to mind, but similar limits of the animal model are seen for motor and sensory deficits. For example, at least one inbred line of mice was totally deaf for many generations before experimenters noticed the trait: in humans, even a 10% hearing loss is of considerable concern. Learning deficits too subtle to be defined in humans nevertheless cause children to fail in school: How shall one investigate dyslexia in rats?

There is an increase of malformations with increasing dose of a teratogen, unless or until increasing fetal mortality obscures it. It is not, however, axiomatic that embryonic death is simply a more extreme form of developmental toxicity than structural or functional abnormality. There are agents such as the fungicide captan, that cause no malformations in surviving hamster fetuses even at doses that cause well over 50% embryonic mortality. There are also agents such as retinoic acid that may cause no prenatal mortality even at doses at which over 50% of the offspring cannot survive postnatally.[10] Similar independence must be assumed for other pairs of endpoints in developmental toxicology. Although it is very often true that a chemical that causes malformations at high doses causes growth retardation at lower doses, exceptions exist. Some chemicals also cause embryonic growth retardation without ever inducing malformations until maternal deaths occur.

Thresholds in Developmental Toxicology

In contrast to mutagenesis (Chapter 10) and genotoxic carcinogenesis (Chapter 11), *structural teratogenesis is a threshold phenomenon*. As shown in Figure 9.2, there is an effective threshold for congenital malformations, a dose below which no defects occur. It is generally accepted that nonmutagenic structural teratogens must be present in significant concentrations to exert a deleterious effect.[11]

The teratogenic threshold, defined as the dose at which the agent induces malformations, may be above, below, or identical with, the *embryolethal threshold*, which is the dose at which the agent causes excessive mortality among fetuses. Embryolethality and/or teratogenicity, in turn, can occur above, below, or at the same dose that causes symptoms of toxicity in the dam. Really vicious teratogens are those that have a threshold for maternal toxicity well above the thresholds for fetal toxicity. In this situation, the cautious respect given to toxic substances is absent; the agent is assumed to be harmless, and few if any precautions are taken to minimize or prevent fetal exposure. Among teratogens with relatively low adult toxicity are thalidomide and diphenylhydantoin. The latter, which has the trade name Dilantin, has been touted as an all-purpose antidepressant and (in phrasing reminiscent of the thalidomide tragedy) as having "no side effects." More obviously toxic teratogens, such as the anticoagulant warfarin, antineoplastic agents, and so on, are prescribed relatively infrequently to women of childbearing age. They are always dispensed under close medical supervision and with considerable worry over their side effects, which are not limited to teratogenicity. Because structural malformations are induced in the human fetus

---

[10] The most cogent study is that of R. J. Kavlock, N. Chernoff, and E. H. Rogers in *Teratogenesis, Carcinogenesis and Mutagenesis* 5:3, 1985, on the relationship between maternal toxicity and teratogenicity, which demonstrates that all combinations of maternal toxicity, embryolethality, and malformations occur. The subject is reviewed by N. Chernoff, J. M. Rogers, and R. J. Kavlock in *Toxicology* 59:111, 1989.

[11] This statement includes the exception that mutagens can, in theory, alter DNA if even one molecule of the mutagen is present; they could similarly act as teratogens via their effect on fetal DNA. This theoretical action of one molecule presumes that the molecule reaches the DNA. The probability that exposing an organism to a single molecule would result in exposure of the DNA to that molecule is obviously very small, perhaps vanishingly so. Any one of numerous metabolic, transport, and excretory systems would probably eliminate the molecule before it reached its target. Thus, for practical purposes, there is a threshold even for mutagens.

during the time when the mother is hardly aware of the pregnancy, it is imperative that extreme caution be used in prescribing or administering chemicals to any woman of childbearing age who may be fertile.

Recent evidence that blood lead levels at or below 10 μg/dL are correlated with cognitive dysfunction (Chapter 6) has led some scientists to say that there is no threshold for the developmental toxicity of lead. The difficulty of identifying cognitive decrements in individuals (rather than populations) makes this hypothesis even harder to prove or disprove than the argument that there is (is not) a threshold for carcinogenesis.[12]

Susceptibility Varies with Age

Susceptibility to teratogenesis varies with the developmental stage. Reference has already been made to the "sensitive period" in thalidomide teratogenesis. It is generally accepted that every teratogen acts at specific times of development, and that these vary, within limits, between agents. The limits are determined by the developmental sequences within the embryo. It is not likely that structural heart malformations will be induced after the heart has formed, and the induction of anencephaly/spina bifida cannot occur later than the closing of the spinal column over the neural tube. For agents that cause multiple malformations, the spectrum of malformations will often vary with the day of exposure. For example, in rats, 100-rad radiation on day 8 of gestation causes 100% of the fetuses to be malformed; the same dose on day 10 causes 75% of the fetuses to be malformed. On day 11, no malformations occur. The nucleotide analog 5-azacytidine causes extra ribs if administered to pregnant rats on day 9 of gestation, but missing ribs if administered on day 11, while limb malformations are seen at a low level on day 11 and with high frequency on day 12.[13]

The stages of gestation can be divided, at first approximation, into three time periods: preimplantation, organogenesis, and the fetal period.

1. During the *preimplantation* period, the embryo consists of relatively few cells, most of which are still *totipotent*. This means that if they

---

[12] There are disagreements over the effects of prenatal exposure to lead. In pages 8–10 of "Symposium Overview: An update on exposure and effects of lead," D. B. Beck argues that "at this point, a threshold is not readily discernable for the association between development and either prenatal or postnatal exposure to lead" (*Fundamental and Applied Toxicology* 18:1–16, 1992). In "A critical review of low-level prenatal lead exposure in the human: 2. Effects on the developing child," C. B. Ernhart argues that "if there are effects of prenatal lead exposure at the levels described in these studies, that effect is small and, apparently, not persistent in the massive milieu of other conditions that influence children" (*Reproductive Toxicology* 6:9–19, 1992). Both papers review the same body of epidemiologic literature on the postnatal consequences of prenatal lead exposure.

[13] The data are taken from M. B. Rosen et al., "Teratogenicity of 5-azacytidine in the Sprague–Dawley rat," *Journal of Toxicology and Environmental Health* 29:201–210, 1990.

are moved to a different site on/in the embryo, they will become what their new position dictates. A teratogen that causes significant disruption of the embryo before implantation is *usually* lethal. Less drastic degrees of cell death can be compensated, especially in the mammalian embryo, which obtains nutrients from the maternal organism.

2. *Organogenesis and early differentiation* are characterized by rapid growth and the differentiation of most organ systems. Extremely rapid cell division leaves the embryo particularly susceptible to teratogens, since any decrease in nutrient levels, as well as cytotoxicity, can lead to too few cells being present. Moreover, the increasing number of cells permits survival of the embryo even after significant damage. The period of organogenesis is the time when most structural teratogens act (Figures 9.3, 9.4, and 9.5).

3. *The fetal period* is marked by growth, maturation, and the functional differentiation of the organs laid down in the preceding stage. Interference with development at this period leads to more subtle structural malformations and to *functional deficits,* of which the most dramatic is probably mental retardation. Functional deficits can be induced in most organs even in the last trimester. Because the brain continues to develop throughout prenatal life and for some years after birth, mental retardation can result from postnatal exposures, including in maternal

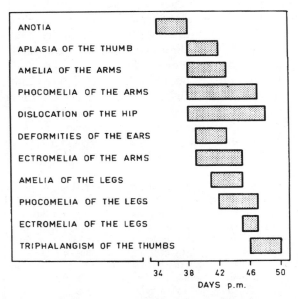

Figure 9.3 Relationship between time of exposure to thalidomide and the malformations observed. (From L. Saxén and J. Rapola, *Congenital Defects,* Holt, Rinehart and Winston, New York, 1969, p. 202.)

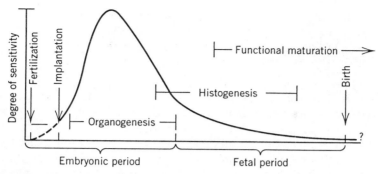

Figure 9.4 Developmental stages and their relative sensitivity to structural malformations. (From J. G. Wilson, "Environment and teratology," in *Pathophysiology of Gestation,* Vol. 2, N. S. Assali, ed., Academic Press, New York, 1972.)

milk (cf. methylmercury, Chapter 2). Growth retardation is also a consequence of disturbance of this phase of development. The duration of the human fetal period is from approximately the fourth through the ninth month of gestation. Agents that act during the human fetal period include methylmercury, alcohol, and cigarette smoking. Sulfa drugs act during the last trimester.

Mechanisms of Teratogenesis

Teratogens act in specific ways (mechanisms) on developing cells and tissues to initiate altered sequences of developmental events (pathogenesis). Many

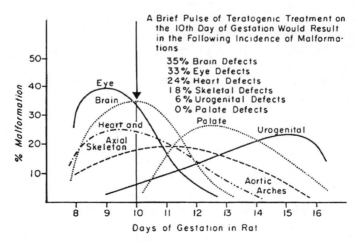

Figure 9.5 Changing incidence of malformations in rat embryos exposed to a hypothetical teratogen. (From J. G. Wilson, *Environment and Birth Defects,* Academic Press, New York, 1973, p. 19.)

teratogens act on processes common to all cells and are devastating to the specific structures within the embryo primarily because that structure is vulnerable at the time of exposure (e.g., because its cells are rapidly dividing). The general mechanisms of teratogenesis include:

1. Selective cell death. In rats there are 10 mitoses between days 8 and 10 of gestation, resulting in $N \times 2^{10}$ cells at the end of this period if N is the number of cells at the start of organogenesis. The average length of a cell cycle is thus approximately 5 hours; division is faster than for any adult tissue. Disruption of this rapid cellular multiplication is almost certain to disrupt the ratios of cells in the newborn drastically, since different areas of the embryo are multiplying rapidly at different times within organogenesis. Moreover, selective cell death may lead to the failure of inductive sequences. For example, when the ureter does not reach the metanephrogenic tissue, the kidney does not develop. If the ureter does reach the metanephrogenic tissue, at least a small kidney will develop, but its size depends on the degree of branching that the tip of the ureter undergoes, since only cells in proximity to a branch will differentiate into kidney. Similarly, the optic cup (which becomes the eye) is determined by the action of the underlying roof of the archenteron on the neural plate. Once the optic cup is determined, it will produce retinal and pigmented epithelial tissues, which then act as inducers of the lens from epidermal cells. This sequence can be prevented, with multiple malformations resulting, if the optic cup is not induced.

2. Altered biosynthesis can lead to structural or functional malformations, because alterations in DNA production or protein synthesis can lead to skewed production of structural or regulatory elements of the organism; differentiation of tissues can be prevented.

3. Energy inhibition will act primarily by slowing growth. But whereas growth retardation during late embryogenesis results in overall, relatively uniform slowing of growth, energy inhibition during the period of rapid cell division will almost invariably affect organs differentially, leading to the states described above.

Agents that act by means of these mechanisms—that alter the "housekeeping" functions of cells—will act in a very time-dependent fashion, targeting different structures at different times in development (above). These agents will induce similar patterns of malformations at similar times of development: limbs on days 10 to 12, palate on day 12, and so on, depending on which structures are at the vulnerable stage. Equally interesting are agents that act only on certain structures, or only during a narrow window of development. Thalidomide is an obvious example. Despite the extremely high incidence of structural malformations, there is no evidence that thalidomide damages the developing brain. The incidence of mental retardation among thalidomide children is similar to the incidence in the general population (making allowance for the effects of some malformations, such as deafness, on cognitive development). In contrast, the herbicide nitrofen causes malformations of the heart, kidneys, and diaphragm in rodents, but does not induce limb malformations, even if given at the time when limb buds are known to be sensitive to damage. Clearly, such structure-specific teratogens do not act through a universal mechanism, such as killing cells, altering metabolism, or preventing cell division, but through some process unique to the affected organ or structure.

Specificity may also be expressed between species. Ethylene thiourea is a potent teratogen in rats, causing massive malformations of craniofacial structures at relatively low doses. In mice, tenfold the dose that devastates rat development causes only a low incidence of relatively minor malformations. One can assume that among the cascade of gene activations and inhibitions required to form the rodent head and face, certain processes differ between rats and mice. One of these, presumably, is affected by ethylene thiourea.

Identifying Human Developmental Toxicants

The ideal method for preventing congenital malformations requires a predictive animal model against which all chemicals will be tested before being marketed. Such a model does not now exist. Moreover, it is becoming clear that it may never exist, because of the inherent species specificity of developmental processes. Assuming good faith on the part of the producers and regulators, it is reasonable to substitute for this predictive model the use of the two species test system we now use. From such premarket assays, one can identify potent teratogens and keep them off the market if their benefit to the user is trivial (one more cold remedy) or minimize the exposure of pregnant women (radiation, antibiotics), or warn the user if the reason for exposure warrants continued use (anticonvulsives, anticoagulants). Even chemicals that are negative in the premarket screening should be monitored closely for several years after they are first marketed. Such monitoring would take the form of case reports for pharmaceuticals and of epidemiology for environmental and life-style chemicals. A central registry independent of the manufacturer is essential, however. This dual surveillance would provide early warning of unexpected side effects (including teratogenicity) and might prevent a sizable fraction of untoward effects associated with xenobiotics.

In practice, all teratogens but two have been identified by "alert clinicians": doctors in practice who recognize a common cause for a cluster of malformations. From Gregg's identification of rubella-induced cataracts to the 1983 identification of Acutane (an antiacne medication derived from vitamin A), the "alert clinician" has been the most effective warning system we have. Only smoking (1970s) and the anticonvulsive valproic acid (1970s) were identified by surveillance. In the case of smoking, the evidence was purely epidemiological; in the case of valproic acid, a case-control monitoring of pregnancies at risk identified the teratogen.

For a syndrome to be identified by an "alert clinician" requires not only an astute practitioner who is willing to press the case against apathy and harassment by defenders of the status quo, but also requires the existence of distinguishing features, either of the malformation syndrome or of the population. For example, the thalidomide malformation of phocomelia (seal limbs) occurs so rarely in the absence of thalidomide that a pediatrician would not expect to see more than one case (if any) in a lifetime. When several such phocomelics were born in each of many practices in one year, the suspicion of a new problem was easy. Nonetheless, initial suspicions of etiology focused on genetics, viral diseases, and fallout from nuclear bomb tests. Moreover, thalidomide, like rubella, produced a very high incidence of malformations among fetuses exposed during the sensitive period. This minimized the confusion about causation that results from the occurrence of significant numbers of normal exposed infants. Nonetheless, it took one year from the meeting at which a German pediatrician first described two phocomelic infants in his practice to the meeting at which Lenz identified Contergan (thalidomide) as the responsible agent.

Another route to alert clinicians is the occurrence of malformations among a closely monitored, unusual population: for example, women taking anticoagulants. Although anticoagulants are commonly administered after heart attacks, their administration to women of childbearing age, who have an extremely low incidence of heart attacks, generally signifies severe heart disease. Ordinarily, one does not anticipate the coexistence of severe heart disease and pregnancy; when it occurs, the pregnancy is closely monitored for both the mother's and the baby's safety. When the same syndrome appears in even five women of such a closely observed subgroup, suspicion can quickly focus on the one (or several) drugs to which all have been exposed. This was the situation in which the teratogenic effects of the anticoagulant warfarin were identified. Moreover, since warfarin is prescribed to very few women of childbearing age at any one time, there was no cohort of women already pregnant when warfarin was identified. In the case of thalidomide, hundreds of affected babies were born for as long as 8 months after the drug was removed from the market.

In contrast to these situations is that seen with fetal alcohol syndrome: the population at risk (women who drink "too much") is poorly defined and even more difficult to ascertain. The agent is widely distributed, making control groups difficult to find. Finally, the syndrome consists of undramatic physical symptoms and mental retardation, in varying degree, depending on the pregnancy. Even though fetal alcohol syndrome was also identified by clinicians, it was only identified in the 1970s, more than 10 years after thalidomide. A medical school professor, shortly after the thalidomide tragedy, asked his class what would have happened had thalidomide only decreased the children's IQ by 10 points. We now know the answer: nothing. The action would not have been recognized. Yet now that the fetal alcohol syndrome has been identified, it is recognized as a major contributor to birth defects and to mental retardation in particular, with as much as 5 to 10% of all French malformations attributed to alcohol.

The most recently identified human teratogen, the retinoid Acutane, is unusual because it is prescribed to healthy young adults at toxic doses (most people experience side effects at therapeutically effective doses) for acne, a skin problem that not only occurs most commonly during the childbearing years but may actually be aggravated by pregnancy. It causes severe malformations in about one-third of exposed fetuses. Acutane does not represent a failure of the postthalidomide testing protocols, however: its teratogenicity was identified before it was licensed. But it is the first medication to control severe, disfiguring acne. As such, it is overwhelmingly beneficial to teenagers and others with this "harmless" but physically and psychologically scarring problem. Moreover, many acne sufferers are male: there is no risk of pregnancy. Therefore, prescription of Acutane, but only by dermatologists, and only in women using reliable methods of contraception, was approved by the FDA. The result has been a small but continuing number of malformed infants and an unknown number of therapeutic abortions.

REPRODUCTIVE AND TERATOLOGY TESTING

Viewpoints

Much of the work done in developmental toxicology is basic research, motivated by interest in fundamental developmental processes, and using xenobiotics as tools to probe normal development. Such work does not lend itself to fixed protocols, since the outcome of one experiment may suggest new questions. As a result, sample sizes tend to be small, and experimental conditions often vary. Many of the factors considered critical in regulatory studies are unimportant, and such studies are often very frustrating to regulators. On the other hand, weakness in sample size and in uniformity of experimental conditions in such studies is offset by thoughtful consideration of oddities, unexpected results, and offbeat explanations for ordinary events. As a result, basic research often provides the first clue that a chemical is of interest to regulators. As an example, the herbicide nitrofen (2,4-dichlorophenyl 4'-nitrophenyl ether) was marketed for 10 years without concern for its developmental toxicity. The premarketing teratology assay was negative, because nitrofen causes neither prenatal mortality nor maternal morbidity at doses that cause 50% postnatal mortality. Thus even though premarketing screening identified a high incidence of postnatal mortality, regulatory action was not triggered. Only when Costlow and Manson became interested in the mechanism of postnatal nitrofen mortality did regulators act.[14]

In contrast to basic research, regulatory research is typically governed by practical considerations. It ranges from screening chemicals according to formal rules stipulated by regulatory agencies such as EPA and FDA, to basic research on questions of interest to regulators. In most cases there is a strong emphasis on developing methodology and/or assuring effective protocols. In the event that the work is being done as part of the premarket testing of a regulated compound, protocols are determined by the appropriate regulatory agency in the country in which the chemical is to be marketed. Note that although contract laboratories do (almost) exclusively contract work, universities, government laboratories, and chemical companies do various combinations of the basic and regulatory work. Teratology *testing* is essentially contract work.

Teratology Testing

Teratology testing is the process by which a chemical is evaluated for its potential to cause developmental toxicity in humans and/or animals. The principles that underlie teratology testing are those of Karnofsky's law: *Any chemical, given at the right time in the right dose to the right species, will cause malformations.* The corollary also holds: *Any chemical, even if given at the right time, in the right dose, will be nonteratogenic in SOME species.* The classical examples are thalidomide, a remarkably potent teratogen in humans but not in rodents or chickens; and cortisone, a potent teratogen in mice and rats but not in humans. One might say that regulatory teratology assays are *designed* with thalidomide in mind and that the test results are *interpreted* with cortisone in mind. That is, the regulations are based on the fear that one might miss a potent human teratogen, but the regulatory decision takes note of the information that over 700 animal teratogens, but only 30 or so human teratogens, are known.

Assay Systems.   Mammals are used for human safety studies. Rats, mice, and/or rabbits are the most commonly used species. For single-generation segment II assays (see below), protocols require testing in two species to minimize false negatives. Only one of the two species may be a rodent. Primates may be used in cases where exposure of pregnant women is expected or probable. The cost of primate studies means that very few animals are used per assay, and the results are often of dubious value. Nonmammalian species are not acceptable for evaluating the safety of a chemical for pregnant women. Unfortunately, relatively little money is available for reproductive and teratology testing of chemicals to assure the safety of wild animals or birds.

There is no nonmammalian developmental assay that is as reliable and

[14] R. D. Costlow and J. M. Manson, *Toxicology* 20:209–227, 1981; J. M. Manson, *Environmental Health Perspectives* 70:137–147, 1986.

also cheaper, easier, or less labor intensive than in vivo mammalian assays. Some in vitro assays are extremely valuable for investigating mechanisms of action of developmental toxicants. In particular, assays using either whole embryo culture, organ culture, or tissue culture can be used to discriminate between maternal and fetal components of toxicity. For these expensive and labor-intensive techniques to be worth their cost, it is desirable to have considerable information about the probable target and plausible mechanism of action of the agent.

In vitro assays that are heavily promoted include use of chicks, axolotls, *Hydra,* cell culture systems, organ culture systems, and tumor-cell binding assay. Because there is no way of extrapolating accurately from mammalian to human teratogenesis, evaluation of in vitro assays requires either the use of known human teratogens and known human nonteratogens, or comparison with mammalian in vivo assays. None of the in vitro assays reliably predicts results of either the nonhuman mammalian assays or the human data. Moreover, based on our growing knowledge of developmental biology, it is becoming very clear that modeling development of one species by use of another is very unlikely to be fully reliable, while the attempt to use paraembryonic processes such as regeneration to model embryogenesis is hopeless. Thus despite the extreme tedium and high cost, mammalian in vivo assays remain the best test for human safety studies.[15]

Sample Size.   An adequate sample size is approximately 20 *pregnant* females per dose, at least in controls. Since some agents prevent pregnancy, and others cause pregnancy termination, statistically significant differences in pregnancy rates may be indicative of reproductive or developmental toxicity.

Dose Selection.   Three doses per compound are normally used. One of the doses should cause maternal morbidity, and one should cause no apparent effect on dam or fetuses. The intermediate dose is spaced logarithmically between the high and low doses. *Controls* must be concurrent controls; historical controls are not sufficient, although they are extremely valuable in identifying anomalous results in concurrent controls. It is possible that a marked decrease in fertility of treated animals is masked by an unrelated fertility problem in controls. The experimenter's own knowledge of reproductive performance of common strains should identify phenomenally poor re-

---

[15] This is in stark contrast to the highly successful use of chicks and amphibians to study basic developmental processes. The reason for the difference is that the genetic elements of development are highly conserved, so it is even possible to correlate human with insect homeobox-containing genes. Most developmental toxicants, however, act at the level of cytoplasmic processes, especially on quasi-continuous processes with dichotomous outcomes. These processes appear to be as variable between species as the genetic sequences are conserved. Even at the genetic level, there must (by definition) be differences among species. Many of these differences are fixed during development. Thus it is not really illogical that control of development should also be highly species specific.

productive performance in the controls. This assumes, of course, that an experienced and unbiased investigator reviews safety studies.

Route of Administration. Ideally, the route of administration is the same as the route of exposure in humans. In practice, gavage is the most common route of exposure. Inhalation exposures are difficult and may cause severe stress in animals.[16] Except for drugs that will be so administered, intraperitoneal, intramuscular, and intravenous exposures are not often used in safety assays, although they are often used in basic research.

Types of Assays

Multigeneration Test. A multigeneration assay is appropriate if long-term exposure to, or bioaccumulation of, an agent is anticipated (e.g., food additives, pesticides). It should be required for any agent used therapeutically during pregnancy, as shown by DES. Figure 9.6 shows one protocol for a multigeneration assay. As a general summary, the multigeneration assay requires that the compound be administered to the parental or F_0 generation, beginning at 30 to 40 days of age for rats. Administration is usually in feed. Animals are mated after about 60 days on the diet. Surviving offspring of the first litters (F_{1a}) are necropsied[17] at weaning, and the parents are kept until second litters (F_{1b}) are weaned. Then the parental generation is necropsied. *Randomly* selected F_{1b} pups are kept on the test diet and mated at about 100 days of age; first litters (F_{2a}) are necropsied at weaning. The F_{1b} are necropsied when the F_{2a} litters are weaned. A three-generation study carries the procedure through one more iteration. Variations on the sequence of necropsies exist. The endpoints of such a multigeneration assay are:

· Fertility index, or (pregnancies/mating) $\times$ 100
· Gestation index, or liveborn/litter (or liveborn/total born)
· Sex ratio, at age 1, 4, 7, 14, and 21 days of age
· Weaning index, or (number alive day 21)/(number at 4 days)[18]
· Growth index, or mean weights of two generations of male and female offspring at 1, 4, 7, 14, and 21 days of age

The strengths of a multigeneration assay are that it identifies transplacental carcinogens (e.g., DES) and second-generation reproductive agents (e.g.,

---

[16] Since stress is itself a teratogen in rodents, any protocol that stresses the pregnant female must be controlled by treating a separate group of animals in the same way, but without administering the test compound.

[17] "Necropsy" includes all necessary postmortem evaluations, from gross dissection to histopathology.

[18] In many studies, litters are culled to specific size, often eight, on day 4, to avoid the effects of different litter sizes on weight gain. In behavioral teratology, it is considered necessary, since litter size affects development of pups. Many teratologists think it is a serious error to cull litters when physical development is being examined.

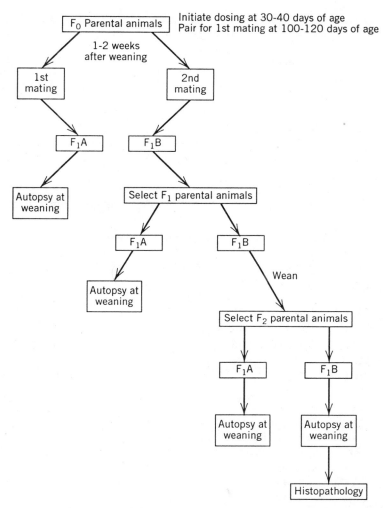

Figure 9.6 Protocol for a multigeneration assay using rats. (From J. M. Manson, H. Zenick, and R. D. Costlow, in *Principles and Methods of Toxicology,* A. W. Hayes, ed., Raven Press, New York, 1982, p. 153.)

the aromatase inhibitor fenarimol), bioaccumulative compounds (e.g., mirex), and postnatally lethal agents (e.g., nitrofen). The major weakness of the multigeneration assay is that it is expensive, costing over $500,000 for each species. It is also tedious and labor intensive. Because two generations of animals must be maintained and bred, it also takes a long time to obtain results.

One-Generation Three-Segment Assay. The one-generation assay attempts to overcome the long elapsed time required by the multigeneration assay by evaluating reproduction, prenatal toxicity, and postnatal toxicity separately. In theory it is possible to conduct the three segments in parallel, shortening the time elapsed before a result is available.

Segment 1: General Fertility and Reproductive Performance.   Ten males are dosed for 60 days, and 20 females for 14 days, before mating, and dosing continues throughout the study. Two females are mated with each male. This allows one female to be killed in midgestation, while the other is allowed to carry her litter through term to weaning. The endpoints include information on pregnancy incidence, implantation rate, pre- and postnatal survival, and pre- and postnatal growth rates.

The major strength of this segment is that it measures gonadal function of both sexes, as well as estrus cycle, mating behavior, conception rates, and early stages of development. It can also be used to identify postnatal effects on lactation and behavior, postnatally apparent malformations, and growth retardation. The major weakness of the assay is that if a problem occurs, localizing it will take another round of experiments. This is also true of the multigeneration assays, however.

Segment II: Teratology Segment.   U.S. guidelines require one rodent and one nonrodent species. Rats are the most common rodent species; the nonrodent species is usually the rabbit. Females are mated and treated during organogenesis (if the day the semen plug is seen is day 0 of gestation, treatment is on days 6 to 15 inclusive in rats and mice, days 6 to 18 in rabbits). At least two doses, one of which causes maternal toxicity and one which does not, must be used. Concurrent controls are required and should be treated with the vehicle in order to equalize stress effects between treated and control animals. All females are killed before parturition, and net maternal weight gain (excluding the uterus and its contents) is determined. Half the pups are examined externally, for soft-tissue malformations and half are examined for skeletal abnormalities.

The major weakness of the teratology assay is that there is no certainty that the results can be extrapolated to humans and considerable evidence that interspecies differences are significant. Therefore, both positive and negative results can be attacked—and are—either in regulatory proceedings or in torts. On the other hand, there has not been another thalidomide disaster since these requirements were instituted.[19]

---

[19] The one major human teratogen that has been introduced to commerce since teratology testing became obligatory are vitamin A analogs (e.g., Acutane) whose developmental toxicity was expected on the basis of numerous animal studies demonstrating that both hyper- and hypovitaminosis A results in developmental toxicity. Acutane was also identified as a teratogen in testing. Given the recent identification of retinoic acid as a *morphogen*, its profound and universal developmental influence is hardly surprising.

Segment III: Perinatal and Postnatal Segment.   For the investigation of perinatal and postnatal toxicity, 20 females per dose are treated from the last quarter of gestation through lactation to weaning. In the United States, two treated groups plus concurrent controls are required. In the United Kingdom, three treated groups plus concurrent controls are required.

The major strengths of the perinatal/postnatal segment are that it identifies postnatal toxicity, including effects on lactation. It also identifies malformations that are not prenatally apparent, delayed maternal toxicity, and behavioral toxicity in weanlings. This third assay can actually be combined effectively with either the two-generation or the segment I assays. It is uniquely useful in identifying "other effects" of chemicals that prevent implantation or alter mating behavior. In theory, it could also identify effects that would be masked by mortality if exposure occurs before implantation or during organogenesis. The major weaknesses of the perinatal/postnatal assay is that it is quite limited, both in the effects identified and in the dosing regimen. Alone, it would not identify either structural teratogens or agents affecting male fertility, implantation, or mating behavior (to name a few). In combination, it can be replaced by modifications of other assays. Thus it becomes a burdensome addition to other reproductive and developmental toxicity assays.

Summary of Testing

No single assay can adequately predict developmental toxicity, much less the full spectrum of reproductive and developmental toxicity. On the other hand, if the several assays required by regulatory agencies are adhered to, and if a modicum of intelligence is applied to the analysis of all of the data obtained, adequate warning of reproductive toxicity can be obtained. Even thalidomide, if evaluated in the three-segment reproductive assay, would have prevented reproduction in rats because it prevents implantation. Thus a warning would have sounded even though thalidomide does not cause *structural malformations* in rodents. If, on the other hand, the rules are followed blindly and interpreted without thought, atypical compounds can slip through any regulatory net.

An additional consideration in developmental toxicology is the significance of transplacental carcinogenesis. This phenomenon has been well documented in animal studies and there is no reason to consider that human fetuses are immune. The first human example was the 1973 identification of the transplacental carcinogenicity of diethylstilbestrol (DES).[20] In 1990, a study identified paternal employment at the British nuclear plant in Sellafield as a risk factor in the occurrence of childhood leukemia. The Sellafield results

---

[20] The DES case history is presented in Chapter 12.

must still be confirmed, since "clusters," unrelated to exposure, can and do occur in epidemiologic studies, but similar data exist linking parental employment in the petrochemical industry to childhood brain cancer.

A final consideration in developmental toxicology should be the occurrence of prenatally induced defects (other than cancer) that are not apparent until long after birth. In addition to its low-level induction of adenocarcinoma, which affects approximately 0.1% of the female offspring of exposed mothers (or 0.05% of total offspring), DES causes abnormalities of the reproductive system in over 50% of both females and males exposed in utero. Because the abnormalities are subtle and because they are not readily apparent until after puberty, these prenatally induced anomalies cannot be identified in early childhood, as are most "birth" defects. Several chemicals are known to induce malformations in animals that are only recognizable postnatally. In at least one case, death occurs at weaning. These examples raise the possibility that prenatal exposure to toxicants (including perhaps maternal stress) will cause deleterious effects that are manifest only in adulthood, and perhaps in senescence.

REPRODUCTIVE AND DEVELOPMENTAL TOXICITY IN NATURAL POPULATIONS

The ability to reproduce is essential to the survival of a species. Successful reproduction is so complex a process, and so many physiological parameters must be synchronized for its success, that reproductive failure is often the first identifiable result of toxic exposure in wild animals. Producing a fertilized ovum requires competent physiology and synchronized hormonal processes in a male and a female. Development of a fertilized ovum even to independent life is an equally complex process requiring, at the least, integrity of the embryo's genetic control systems and adequate support from the external environment, whether that be the egg with its nutrient stores or the maternal organism and placenta.

The basic principles of developmental toxicity do not differ between humans and other species, but practical considerations do. Efforts to prevent human developmental toxicology emphasize identification of potential developmental toxicants, preferably in advance of human exposure. The major obstacle to identifying human developmental toxicants is the difficulty of extrapolating between species. Most of the chemicals of interest are pharmaceutical or life-style chemicals, many of which are rapidly degraded in the organism and few of which are released to the environment. Few environmental pollutants other than lead and methyl mercury have been shown to perturb human development.[21]

---

[21] PCBs may cause cognitive deficits in children who are heavily exposed by their mothers' consumption of contaminated fish, but the data are still disputed.

In contrast, ecotoxicologists must often attempt to determine which pollutants (among many chemicals present in the ecosystem in varying amounts) are causing observed reproductive failure in a wild population. Extrapolating between species may pose problems, but the target species is usually available for field studies and often for laboratory experiments. The chemicals of interest to ecotoxicologists include pesticides, metals, and long-lived lipophilic pollutants such as hexachlorobenzene and the PCBs. Identification of the causative chemical rarely results in mitigation of the toxicity, since the agents are widely dispersed and very persistent. Even if a chemical is banned (as DDT was) because it adversely affects wild species, its levels in an ecosystem may decline only slowly.

Finally, there is a very practical difference in the scope of studies carried out in the laboratory and in the field. Laboratory experiments are usually designed to investigate either *reproductive toxicology,* the failure of adult organisms to reproduce, or *developmental toxicology,* the failure of the progeny to develop normally. This distinction is eventually made in field studies as well, but it is rarely incorporated into the design of ecotoxicologic studies as it is in the regulatory protocols for human developmental toxicity. Ecological studies more often begin with an uncontrolled experiment in progress, work backwards to identify the cause, and only then identify the facet of reproduction and/or development that was affected.

The basic principles of developmental toxicology nonetheless remain valid. First, a developmental toxicant must have access to the embryo or fetus. In mammalian species, developing embryos are exposed to chemicals that cross the maternal placenta and also exposed to chemicals that have been deposited in maternal fat and are secreted in milk. The embryos of nonmammalian species, in which a significant portion of embryonic development typically takes place outside the dam, are exposed to chemicals secreted into egg yolks before ovulation or secreted into egg whites during laying. Nonmammalian embryos are also exposed to ambient pollutants directly, without maternal mediation, once the eggs are laid. Thus, developing fish and amphibians are exposed to any chemical in the water; avian and reptilian eggs are exposed to chemical sprays and gaseous air pollutants to which the shell is permeable.

Second, environmental factors interact with the embryonic genotype in determining response to environmental insult. This interaction occurs within species (one hatchling in a brood may be malformed by PCBs while its sibling is normal; embryonic sensitivity to metals varies between populations of fish) and between species (one avian species is no better and no worse a model for another than the rat is for the human). The existence of variability among wild species is extremely important and must be considered in designing experiments. Initial laboratory investigations of the decline of avian species used chickens, domestic ducks, or quail and found no disruption of egg laying or hatching at levels of DDT far above those seen in the wild. It was soon established that these species are quite insensitive to DDT-induced eggshell thinning. Nonetheless, investigators continued to use chickens to study mechanisms of eggshell thinning.

Third, the embryonic sensitivity varies with stage of development. Since so many environmental pollutants are present in an ecosystem for decades, with consequent chronic exposure of the dams, the timing of exposure is often less critical than in human or laboratory exposures. Moreover, even if a chemical is present only transiently, exposure of the nonmammalian dam may lead to deposition of the chemical in the unfertilized egg. Lipophilic substances are especially prone to accumulate in the egg yolk, exposing the developing organism to potentially toxic levels even after maternal exposure has ended. For degradable chemicals like modern insecticides, however, the developmental toxicity will be highly dependent on the developmental stage at the time of spraying.

Finally, any chemical will act on the embryo by one of several mechanisms, and the manifestations of developmental toxicity in all phyla may include embryonic death, growth retardation, structural malformation, or functional deficit.

The data for different types of natural populations differ greatly, depending on the rationale motivating the various studies. Broadly speaking, there are three rationales for carrying out developmental toxicity studies in species other than those accepted as screens for human effects: to develop cheaper or quicker (so-called "in vitro") assays to predict developmental toxicity in humans, to identify developmental toxicity in the species under study, and to model developmental toxicity in a related but less accessible species. All of these rationales have influenced, to a greater or lesser degree, studies on invertebrates, fish, amphibians, and wild birds. Studies on wild mammals have generally been motivated by intrinsic interest in the species or because it is a model for an even less accessible species (e.g., seals, which are relatively small and plentiful, are often used as models for all marine mammals, including the very large and relatively rare great whales.)

Invertebrates

Reproduction, development, and regeneration in invertebrates have most often been examined as indicators of general ecotoxicity, rather than as developmental endpoints. Aquatic species have been studied most often. Some marine species are of economic importance, and pollution of coastal waters threatens the shellfish industry.

In a few cases, regeneration (*Hydra,* planaria) or development (crickets) of invertebrates has been suggested as the basis of assays for human developmental toxicity. Most of the agents used in developing such assays are pharmaceuticals rather than environmental pollutants, selected on the basis of known human or mammalian effects. Thus these studies have little relevance for ecological developmental toxicology. Both the planaria assay and the cricket assay have also been suggested as assays for environmental toxicity and show considerable promise in this sphere. The cricket assay is especially interesting since it gives an indication of the bioavailability of soil pollutants.

Fish

Development in fish has been studied because of the economic importance of many species of fish and because fish are indicators of aquatic toxicity. Recently the zebra fish (*Brachydanio rerio*) has been used as a model species for developmental biology.

The developmental consequences of exposing fish to chemicals must be interpreted very carefully. For example, some chemicals cause a temporary developmental delay in fish. These developmental delays have been considered evidence of developmental toxicity, by analogy with the mammalian paradigm, but several studies suggest that normal development can resume after such delays. Another difference between fish and mammals is that piscine malformations tend to be rather nonspecific. Quite dissimilar toxicants often produce the same malformations, suggesting that the developmental stage, rather than the chemical, determines the malformation. Some investigators do hypothesize that chemically induced malformations in fish are due to a generalized stress response acting on metabolism. Other investigators think that the occurrence of a very few similar malformation syndromes results from identification of only the dominant or most dramatic malformation, that subtle differences in phenotype are missed, and that the most severely deformed embryos may not survive to be examined.[22] The most significant difference in response between developmental toxicity in fish and mammals may be that fish embryos are most sensitive to structural malformation soon after fertilization and before gastrulation, whereas the period of maximum sensitivity in mammalian embryos occurs after gastrulation. A second critical period in fish development is the time at which the fry have absorbed the last of the yolk, at which time they are most likely to die from chemical exposure.

Chemicals that are known to affect fish development deleteriously include solvents, oil dispersants, certain pesticides, and metals. Tolerance to metals exists in populations of fish from polluted waters, but it is specific to the polluting metal, not developmental tolerance to metals in general.

Amphibians

Amphibians have proven to be useful tools in both classical and modern embryology. The African clawed toad (*Xenopus laevis*) has also been sug-

---

[22] By analogy, many older papers dealing with cortisone-induced malformations in mice stress the specificity of the defect caused, implying that cleft palate is the only malformation seen. In some strains, however, there is a significant incidence of cortisone-induced umbilical hernia; in other strains, open eyelid is as common as cleft palate.

gested as a model for mammalian teratogenicity. Amphibian development is also used to identify overall aquatic toxicity, and the chemicals used for such studies tend to be highly relevant to ecological problems. Moreover, in the past five years it has become apparent that amphibian species are in decline all over the world. It is generally agreed that there is no single cause for the decline of so many different species and populations, but it is quite possible that acidification or other aquatic pollution interferes with development of immature stages. It is too early to know whether one species can be used to investigate the worldwide decline in amphibians, but the only species for which a database is being developed is *Xenopus laevis,* through the FETAX assay.

Xenobiotics known to affect development in amphibians include methyl mercury (5 to 30 ppb), high levels of selenium, the fungicides dichlone and chloranil, and benzo[a]pyrene.

Birds

Birds have proven important in all aspects of developmental biology and developmental toxicology. Chick embryos are excellent experimental subjects for embryologic investigation, since they are easily obtained, can be maintained outside the egg for at least 24 hours, and are large enough for easy manipulation. Avian reproductive failures have been the most dramatic consequence of environmental pollution by persistent chemicals. The effects of DDT in thinning eggshells and of PCBs in causing malformations in birds have been documented repeatedly (see Chapter 5). More recently, selenium emerged as a developmental toxicant in the wildlife refuge at Kesterson (see Chapter 5). In contrast to the database for invertebrate and poikilothermic species, many chemicals have been tested in more than one species of bird. Chickens, ducks, and quail have been used most often in laboratory studies, but correlations between pollutants in eggs and reproductive problems in wild birds have been carried out in numerous species ranging from gulls to cormorants and ospreys. It is quite clear that there is as much variability among avian as among mammalian species in response to reproductive and developmental toxicants.

Besides chlorinated hydrocarbon insecticides, PCBs, and selenium, many or most organophosphate insecticides (OPs) are potent teratogens in birds. The avian teratogenicity of organophosphates is in striking contrast to their inactivity in mammalian development.[23] In birds, two groups of malformations are seen. One type, including wry neck and other axial skeletal defects, is due to acetylcholinesterase inhibition (see Chapter 8). A second group of

---

[23] It has been suggested that potential OP insecticides comprise half a million chemicals, so it is inevitable that one or another OP insecticide will affect mammalian development. But enough data are available to state that the primary mechanism of OP toxicity in adult animals, the inhibition of acetylcholinesterase, does not cause developmental toxicity in mammals.

malformations, including parrot beak and abnormal feathering, is due to inhibition of kynurenine formamidase. These latter malformations are not caused by all OP insecticides and can be mitigated by administration of nicotinamide (in the laboratory).

Birds are not usually exposed to enough methyl mercury for toxic effects to be apparent. Lead, on the other hand, appears to be less toxic to avian than to mammalian development.

Mammals

Little information is available about the effects of xenobiotics on wild terrestrial mammals. The easily studied mammals (rodents, rabbits) are more often targets for extermination than preservation. Those species that are most likely to be affected by persistent pollutants are the predators at the top of the food chain (wolves, bears). Few in number and with long gestation periods, these species are both expensive and difficult to study. Moreover, so much of their decline can be explained by habitat destruction that reproductive failure consequent to chemical exposures seems to be a minor factor. Wild mink, however, have almost certainly been affected by PCBs, to which they are extremely sensitive. Ranch mink pups whose dams were fed Lake Michigan fish containing 10 to 15 ppm PCBs die of pre- and postnatal PCB accumulation. Herbivorous mammals do not appear to be at major risk from environmental pollutants, although local kills from pesticide spraying of forage cannot be ruled out. Insectivores, especially bats, are adversely affected by insecticide spraying. In many parts of the United States, the combination of habitat disturbance and insecticide exposure endangers local populations of bats, although there are no data suggesting reproductive or developmental toxicity as the cause.

Reproduction in aquatic mammals is thought to be adversely affected by chlorinated hydrocarbons. The major hazards appear to be PCBs and possibly DDT. In seals, both abnormalities of the female reproductive system and premature births were closely correlated with levels of PCBs in the females' blubber. Elevated levels of hexachlorobenzene and dieldrin have also been associated with reproductive failures. Milk is apparently a major source of xenobiotic elimination in marine mammals, so offspring are contaminated postnatally as well as prenatally.

10

GENETIC TOXICOLOGY

Bottle of mine, it's you I've always wanted!
Bottle of mine, why was I ever decanted?
    Skies are blue inside of you
    The weather's always fine;
For
There ain't no bottle in all the world
Like that dear little bottle of mine!

—Aldous Huxley[1]

INTRODUCTION

It has been said in all seriousness that the organism is the gene's way of making another gene. On a more mundane level, the information contained in the genome is fundamental to all aspects of development, of structure, and of function of the organism. Genes determine eye color and height, the structure of enzymes and of cytoskeletal proteins, the working of our kidneys, and the cholesterol level in our blood vessels. Thus genetics is fundamental to all biology. But classical genetics, which depends heavily on an analysis of phenotypes and the matings that produce them, has been largely superseded by molecular genetics. Rather than identifying a phenotype, we can analyze proteins to find the variant and analyze its function. We can now identify the alteration in the genome that causes the variation in the protein. And we are finding that more and more human health problems, from birth defects to cancer and aging, depend on gene expression. Modern genetics—molecular genetics—is thus fundamental to all aspects of toxicology.

But not all problems are at the molecular level. We do not marry molecules, but people; we do not walk genotypes, we walk dogs. In all too many cases, even if we know that a syndrome is entirely or predominantly genetic, we cannot yet identify the molecular basis of a phenotype. Even when the molecular basis of a disease is elucidated (as it has been for cystic fibrosis and retinoblastoma), we may still need to treat the symptoms because we cannot yet precisely alter the genome. We must also consider the person who has the symptoms.

Moreover, many diseases are multifactorial: the result of interactions

---

[1] Aldous Huxley, 1932; *Brave New World,* Perennial Library, Harper & Row, New York, 1978, p. 52.

between genes and environmental factors. Examples include not only birth defects such as cleft lip, but susceptibility to heart attacks and to cancer. What is the probability that the child of a woman with cleft lip also has cleft lip? What are the odds that I die of a heart attack at the same age as my father? We may need the ability to predict these multifactorial events far more than we need to know the name of the molecular trigger.

Finally, we are often interested in the effects of genetic alterations on populations, especially when ecosystem effects are considered. It may be critical to know whether a given gene will become widespread in a population, or die out; whether a threatened species has sufficient genetic plasticity to survive. Such information can be gained from population genetics. Thus there is still a need for classical genetics and for classical genetic analyses.

The summary that follows is designed to provide the basic information about genetics that might be needed in ordinary life.

GENETICS: A QUICK LOOK

Biochemistry

Chemical reactions in living organisms are catalyzed by special proteins called *enzymes*. Enzymes can carry out specific cellular reactions that might otherwise not take place at the temperature, pH, and substrate concentrations present in the cell. Enzymes catalyze and control most cellular reactions, including the synthesis of new molecules or the breakdown of existing molecules. Special reactions include those that occur when light energy strikes the retina, leading to initiation of a nerve impulse, which is interpreted by the organism as vision; reactions leading to the contraction of proteins that make up muscle; or reactions that transport individual molecules across membranes, leading to uptake of nutrient or excretion of waste products. Proteins also have a wide variety of structural forms, including the keratin in hair and fingernails, the transparent lens of the eye, and the oxygen-carrying hemoglobin. Proteins are made up of a linear chain of amino acids, and the structural and/or catalytic properties of the protein are determined by the sequence of amino acids. There are 23 common amino acids, so a protein 100 amino acids long could have 23^{100} possible unique structures.

Within the cell the information for the structure of all proteins is encoded in the genetic apparatus, primarily in the DNA (deoxyribonucleic acid) of the nucleus. Like proteins, DNA is a linear chain made up of different subunits. However, unlike protein, DNA is tens of thousands of units long; one molecule of DNA can have a length of several millimeters. Moreover, there are only four different units for DNA: the nucleotides adenosine (A), guanosine (G), cytidine (C), and thymidine (T).[2] The information for protein

---

[2] In ribonucleic acid (RNA), the thymidine is replaced by uridine (U). Other variations occur, notably the methylation of certain of the A and/or G bases. This methylation seems to provide one means of controlling gene expression.

structure is encoded in the sequence of bases in DNA. This sequence is interpreted by the cellular machinery in groups of three bases or triplets. In most cases, a triplet specifies one amino acid; however, some triplets specify the end of a protein rather than an amino acid. These are called *stop codons*. Some DNA sequences also serve controller functions, determining the amount of DNA that will be transcribed or the choice of sequences within a longer sequence that will actually be included in a specific messenger (alternative splicing). In normal eukaryotic cells there are two complete copies of the information for most proteins, one copy on each of two homologous chromosomes.

When a protein is synthesized, the specific region of DNA encoding the protein is copied into temporary, working copies of the DNA, called *messenger RNA* (mRNA). Many copies of mRNA can be made from each DNA sequence. The working copies of mRNA move from the nucleus of the cell, where DNA resides, to the cytoplasm where the cellular machinery that makes protein resides. In the cytoplasm, the mRNA is interpreted and the proteins are synthesized by linking together amino acids present there, in the proper sequence. Many molecules of the specified protein can be made from each molecule of mRNA.

Mutation, Selection, and Evolution

A mutation is an alteration of the base sequence in cellular DNA. Mutation results if DNA is lost (*deletion*), if there is a change in the sequence of bases in DNA (*substitution*), or if there are changes in the packaging of DNA in the chromosomes. If the base sequence of DNA is altered, most often it will decrease or abolish the ability of the product protein to carry out its function in the cell—assuming that the mutation has any influence at all. Mutations can be neutral (i.e., without effect on the cell); or, occasionally, a mutation can improve the overall functioning of the cell. The modern interpretation of the theory of evolution postulates that mutation is the source of cellular and species variation that is selected by changing environments to produce the evolution of species.[3]

Let us consider the effect of a new mutation in an organism. First, it is

---

[3] To understand how mutations can be mostly deleterious and yet form the basis for evolution, one must also know that many ancestral genes have clearly been duplicated and have then diverged. For example, the gene for the ancestral molecule of the muscle protein, myoglobin, duplicated. One of the duplicate genes was then "free" to mutate without affecting the ability of muscles to produce myoglobin. In due time, perhaps after several more duplications, the extra myoglobin gene mutated and became more like hemoglobin and found a new function as the oxygen-transporting molecule of the blood. There are now six types of myoglobin-derived molecules, five of them hemoglobins. Note that the original gene sequence may be strongly conserved (very similar between species) because its function is so essential to the organism that even minor changes are lethal.

Such duplicated gene families can be used to calculate genetic distances between species (and phyla) since the degree of divergence between homologous proteins is indicative of the number of generations of independent divergence.

critical whether the mutation occurs in a germ cell or in a somatic cell. Only if it occurs in a germ cell does the mutation have a chance of being part of the genetic material of the next generation. Only such germ cell mutations have the chance to enter the gene pool of the species, and that only if the organism in which the mutation occurs succeeds in propagating its offspring in the population. Naturally, the new mutation might influence the organism's chances of reproducing itself or alter the probability of the offsprings' survival. That is, the mutation might be *selected against* or *selected for,* and could be *eliminated* from or *fixed* in the population. If the mutation occurs in a somatic cell (a body cell, a nongerm cell), it will most probably have no noticeable influence on the organism. The chances of influence depend on the stage of development of the organism and on the specific gene and tissue in which the mutation occurs (see also Chapter 11, which deals with carcinogenesis and oncogenes).

Effects of Diploidy

All cells have two copies of the genetic information for most proteins in the cell, one copy on each of two homologous chromosomes. Only the information on the X and Y chromosomes in males is not homologous (i.e., is present as single copies). Therefore, if a mutation *in a somatic cell* results in the synthesis of a defective or nonfunctional protein (a *loss of function mutation*), a sufficient amount of normal protein can usually be made from the second, "good" copy of the gene on the homologous chromosome to provide the cell with the needed function. In somatic mutations this would be true even if the mutation occurs in a gene that is expressed. (This is the molecular basis for "recessive" mutations, terminology that will be used in the section on human genetics.) Finally, even if the mutation occurs in the only good copy of the information for an essential cellular protein, the ordinary consequences of such a mutation extend no further than the death of that one cell. Often, but not always, the cell is simply replaced. Some cells, such as nerve cells, may not be replaceable, but as long as relatively few cells die, the damage will probably not be noticeable.

In summary, adult mammals would seem to have an enormous capacity to sustain somatic genetic damage. The deleterious consequences of mutational damage to somatic cells are greatest in the immature stages of an organism. The consequences at any time are strongly influenced by the gene mutated, the genetic constitution of the homologous gene, the function of the gene product in the cell, and the function of the cell in the organism. Mutations in germ cells are special because if the mutated ovum or sperm is fertilized and becomes a viable organism, it will carry the mutation in all of its cells, and pass it on to 50% of its progeny. A mutation in a germ cell has the potential of affecting an entire species, because it contains the possibility of the mutation spreading into, and perhaps becoming "fixed" in, the gene pool of that species.

Theory of Mutation

A *mutation* is an abrupt, heritable change in the genetic information of a cell, resulting from a structural alteration of the genetic material of the cell, the DNA. DNA encodes all cellular functions in the linear sequence of bases attached to its sugar phosphate backbone. This sequence is interpreted in groups of three bases, or *triplets,* according to the *genetic code;* each triplet specifies one amino acid in the primary sequence of a protein. There may be more than one triplet for a given amino acid, and there are special triplets that specify polypeptide chain termination rather than an amino acid.

Several sorts of heritable microscopic alterations of DNA are recognized: *base substitutions, base additions,* or *deletions; frameshifts;* and *large deletions.* Substitution of one base for another changes one letter of the triplet code and can (but does not necessarily) result in introduction of a new amino acid in the resulting protein or in the termination of the polypeptide chain. Addition or removal (deletion) of one to a few bases may shift the reading frame of the triplet code. For instance, the genetic statement

GAGGAG**GAG**GAG**GAG**GAG**GAG**GAG

would read

GAG**NGA**GGA**GGA**GGA**GGA**GGA**GGAG

following insertion of any nucleotide, N (underlined). Note that a second addition or deletion can restore the proper reading frame:

GAGNNNGAG**GAG**GAG**GAG**GAG**GAG**GAG

If exactly three bases are added or deleted, the proper reading frame is maintained, but one amino acid is added or deleted. Frameshifts cause a very dramatic change downstream from the addition or deletion, since all subsequent triplets are read out of the correct frame. Although alteration of one to a few amino acids in a protein can lead to loss of functionality of the protein, many alterations are tolerated, that is, they do not destroy the function of the protein. Examples of a *conservative replacement* might include substitution of an amino acid with one of similar functionality (e.g., leucine for isoleucine) or might involve substitution of an amino acid in a part of the protein where almost any amino acid would serve correctly, for instance in a region of α-helix. Such replacements will create a new amino acid sequence, but the protein's function may be retained or only moderately impaired. Examples in the human include the hemoglobins C (moderately impaired function) and S (severe impairment of function), which are distinct alterations at the same locus of the hemoglobin gene. Frameshifts and large deletions may cause the absolute loss of protein function and frequently lead to loss of viability of the cell (see above).

Continuing the analogy of the DNA sequence as "the code of life," one can construct examples of mutation of English sentences:[4]

Mutation Type	Original Reading	New Reading
Addition	Add one to the pile	Add none to the pile
Deletion	Always step on your heel	Always step on your eel
Large deletion	That new technician's figures were terrible; she couldn't divide worth a damn. But the result that she had apparently obtained was so enticing that I took her in hand and went over it all again carefully.	That new technician's figure was so enticing that I took her in hand and went over it all again carefully.
Substitution	Go plant a seed	Go plant a weed Go plant a reed Go plant a deed Go plant a need Go plant a feed
Frameshift	Mine eyes have seen the glory	Mine yesh aves eent heg lory ("e" deleted)

It is clear from the examples that the usual result of random alteration of the letters of the sentence would be to create nonsense, although as most of the examples show, the alterations can produce some new sense that might subtly change the meaning of the sentence (*Go plant a seed* becomes *Go plant a weed*). Similarly, the random alteration of the base sequence of DNA is usually harmful: introduction of a new amino acid that destroys the activity of the protein. However, the change may be innocuous, such as a conservative amino acid replacement or a new codon for the same amino acid. In rare cases, the mutation may confer an advantage in some way.

---

[4] I am indebted for these examples to Jim Johnston, who taught "Toxic Substances in the Environment" before me, for these permutations. It may seem impolitic to give him credit for the example of the "large deletion" in these days of political correctness, but Jim's point was precisely that the intended meaning is wildly distorted by such alterations.

It is currently believed that most mutations arise from two sources: mistakes made during replication of normal DNA and mistakes made during the repair of damaged DNA. Of course, replication and repair are not mutually exclusive, since the replication apparatus may be involved in a repair process. Chemical and physical agents that can induce mutations—*mutagens*—are therefore agents that can damage DNA or that reduce the fidelity of DNA replication.

Among the types of repair mechanisms currently identified, several always appear to restore the damaged DNA to its original base sequence, at least to the limits of the biochemist's ability to detect errors. In bacteria, the best characterized, error-free systems include *photorepair* and *excision repair*. Error-free repair systems depend on the existence of a complementary strand of DNA containing the correct sequence: the damaged section is "clipped" out of the DNA, and the correct strand is then used as a template to recreate the complementary sequence on the damaged strand.[5] In the event that both helices are damaged, other repair systems, which do not require a complementary template, are called into play. These systems exhibit significant frequencies of error and are therefore called "error-prone" repair systems. The most intensively studied of these are the postreplication (also called SOS) repair systems, several of which may exist. These SOS systems are induced by damage to DNA, but they require functional *Rec A* and *Lex A* gene products, and they raise the rate of mutation in the cell when present, presumably by making mistakes in the repair of damaged DNA.

Mutagenesis

In the Ames assay (named for Bruce Ames, who recognized that 80% of known carcinogens were also known mutagens), the mutagenic activity of a chemical is assayed by observing its ability to induce mutations that restore to the cell a functional enzyme originally destroyed by a previous mutation. Such mutations are designated reversions. The Ames assay uses strains of *Salmonella typhimurium* which have lost a single enzymic step in the biosynthetic path for histidine. Such strains cannot make histidine from glucose but will grow on glucose if supplied with a small amount of histidine. If a new mutation in any of these bacteria restores the missing enzymic step, that cell can grow without histidine in the medium, and it will give rise to a colony on a plate of minimal glucose agar. The number of colonies that arises reflects the mutagenic activity of the chemical to which the bacteria were exposed.

---

[5] One hypothesis for the role of methylation is that methylation of the parental strands is a cue to the original, and therefore undamaged, sequence.

In mutagen screening systems such as the Ames assay, an effort has been made to reduce the ability of the tester strains correctly to repair damaged DNA. The excision repair system has been destroyed by the *Uvr B* mutation and the mistake frequency of the SOS repair system has been enhanced by introduction of genes carried on a plasmid, pKM101.[6] Because of these alterations, most damage to DNA must be repaired by the error-prone system, creating more mutations per unit of damage.

There are several ways that a mutagen can reverse the effect of the original mutation in the DNA and thereby create a functional protein in the cell (such mutations are called reversions).

1. It can restore the original base sequence.
2. It can restore the original reading frame.
3. It can create a new codon that restores the original amino acid or inserts a conservative replacement.

In addition to these mechanisms, restoration of a functional protein to the cell might result from a new mutation at an entirely different site. For example, an original mutation that created a termination signal in the middle of a gene for a protein (and thereby destroyed the protein's functionality) could be reverted by a new mutation in a transfer RNA (tRNA) that allowed the tRNA to bind to the ribosome at the termination signal and add its amino acid to the polypeptide chain. The new mutation is a reversion since it corrects the functional defect introduced by the original mutation. The tRNA mutation is also called a *suppressor* mutation, because its effect is to suppress the original mutation without actually restoring the original mutated base sequence. For the purposes of mutagen screening it is not very important whether the reversions that are detected result from a back mutation or a suppression; both events are mutations.

Although mutation has occurred for as long as life has existed on the earth, abrupt increases in the rate of mutation in a population are likely to have profoundly deleterious impacts. A population of organisms is able to tolerate a certain number of weak, mutated individuals without threat to the viability of the species. This can be called the genetic carrying capacity of the population. An increase in the number or amount of mutagens that organisms encounter during their reproductive life will result in a greater steady-state level of mutated members of the population. If enough organisms in the population are sufficiently weakened by mutation to impair either reproduction or the ability of the species to adapt to changes in its environment, the species may perish.

---

[6] Plasmids are semi-independent units of DNA carried by some bacteria. Genes on these plasmids contribute to the host bacterium's capabilities, such as resistance to antibiotics. Some plasmids replicate independently of their host, allowing tremendous numbers of plasmids to be harvested from a bacterial culture. This is an important technique in obtaining *recombinant DNA*.

In addition, there is strong evidence that somatic mutation is the initiating step in the formation of most tumors. The initial observation was made by Bruce Ames, who identified an extremely good correlation between *mutagens* and cancer-causing agents—*carcinogens*. In a survey of 175 known carcinogens, Ames found that 157 (90%) were mutagenic in his test system while fewer than 10% of about 100 noncarcinogens tested showed mutagenic activity. In light of these data, our recently acquired ability to manufacture and disseminate kilotons of almost innumerable organic chemicals—many of which are mutagens—poses the threat of an abrupt increase in mutation rates. Increases in deaths from cancer have been attributed to this cause (most vehemently by Bruce Ames), although both the existence of an increase and the importance of synthetic chemicals in carcinogenesis have been vigorously disputed (again, most vehemently by Bruce Ames). Regardless of the cause, it is now thought that between 60 and 90% of all human cancer is the result of somatic mutations that "decontrol" cell division.

Because detection of mutagens is considerably cheaper and quicker than detection of carcinogens, the correlation between mutagens and carcinogens offers hope of identifying carcinogens by using bacterial or cell-based assays rather than the slow and expensive mammalian carcinogenesis assays. A bacterial test for mutagenesis costs under $1000 and requires only a few days, in contrast to a carcinogen test involving rodents, which costs over $300,000 per species and requires about 2 years.

Developmental Genetics

Mammals begin as a single fertilized ovum, which divides and differentiates to produce the organized set of organs that make up the adult. The full complement of genetic information needed to direct the process of cellular growth and differentiation resides in the original fertilized ovum and, usually,[7] in every body cell. It is believed that most of the genetic information in the original cell is used during the developmental process. But much of that information is not needed by the adult, especially by the fully differentiated cells of the body. Thus although the cells in your finger have the genetic information to make peptide hormones such as insulin, they do not. The pancreas makes insulin but does not synthesize acetylcholinesterase, liver cells do not synthesize hemoglobin, kidney cells do not make fingernails, and so on. In the process of differentiation, much of the vast quantity of genetic information present in each cell is used in precise temporal sequences (see Chapter 9) and then "turned off" forever. Only those genes needed for ordinary cell functions ("housekeeping genes") and the very few genes encoding the information needed by the particular differentiated cell are expressed continually. The process of differentiation has been described as a

---

[7] There are species in which this is not true, and there are exceptions in individual cells in most species; for our purposes, the qualification is nitpicking.

progressive, irreversible turning off of the genetic information in an organism. Most of the DNA in any given cell is turned off and will never be expressed. The conventional wisdom was therefore, that most mutations that happen in a somatic cell will have no influence on the organism since they will probably occur in the "turned off" genes, which form the majority in any given cell. Such mutations are almost impossible to identify, since the gene product is not expressed and alterations cannot be identified. In the past decade, however, attention has focused on those events in which a somatic mutation either turns on a gene that promotes cell growth and/or cell replication, or turns off a gene that inhibits cell growth or cell division. The effects of a change of state in such *oncogenes* are considered in Chapter 11.

In addition to genes permanently turned off during development, there is a second group of genes in most cells that are only temporarily "turned off" but which can be expressed under certain circumstances. These genes are said to be "inducible." For instance, if you cut yourself, you may damage some of the basal cells that underlie the skin and thus damage that tissue. The synthesis of scar tissue, rich in collagenlike proteins, will be induced in the cells surrounding the damage. But when the wound is closed, the expression of the scar-forming genes stops. Of more interest to toxicologists are the enzymes that are induced—their synthesis turned on—when particular xenobiotics are present in the organism (e.g., the cytochromes P-450 in the liver and other organs).

It can be seen from this explanation that a *loss of function* mutation (i.e., a mutation that changes a functional into a nonfunctional protein) will have a much higher chance of being expressed if it occurs in an organism that is still developing and differentiating (e.g., a fetus or a very young child). This maximizes the chance that the mutation will be expressed. In an adult organism, the chances of mutation in an expressible gene are low. Should the mutation occur in a gene that is expressed, diploidy—the existence of a normal copy of the mutated gene—mitigates against serious deleterious consequences.

But what if a mutation causes a *gain of function*? Thirty years ago, when the greatest emphasis in applied genetics was on the protein products of genes, gain-of-function mutations were assumed to be rare and perhaps not very interesting. It is now known that major portions of the genome are devoted to regulating gene action: determining when and how much of a gene product will be produced in a given cell. A gain-of-function mutation results precisely when a switch that turns off a gene (whether at the appropriate moment during embryogenesis, or when the wound has healed over) is itself inactivated. The resulting overproduction of gene products can have disastrous consequences, particularly when the stimulated function promotes cell division. A class of genes that causes overgrowth was first recognized somewhat over a decade ago. Because of their close association with cancer, they were given the name *oncogenes*. Eventually, normal cellular counterparts to the oncogenes were identified and were found to be strongly associated with regulating cell growth and division. The normal counterparts were named *proto-oncogenes*. Although it has been noted that this is somewhat like calling a new car a "proto-wreck," the terminology seems fixed. Oncogenes, and their role in carcinogenesis, are discussed in Chapter 11.

HUMAN GENETICS

Despite the advances of the past decades, much of human genetics is still descriptive. Our knowledge is too often still limited to a description of a syndrome or disease, and some knowledge of the transmission of the syndrome between generations.

The basic unit of heredity is the *gene*, which can be defined in terms of the characteristic it confers (its *phenotype*, or appearance), in terms of its product (typically, a protein), or in terms of the sequence of nucleic acids that comprise its code in the genome. Physically, a gene is a sequence of nucleic acids [deoxyribonucleic acid (DNA)] with a specific location on a specific *chromosome*. A chromosome is, functionally, a linear array of genes. In human heredity, however, we have rarely had the information to characterize genes in terms of their DNA sequence (the first exception was the mutation that causes sickle-cell anemia) or of their location on a chromosome.[8]

Somewhat more often, but still quite rarely, human genes have been characterized by their products. The best examples are diseases that cause the accumulation of large quantities of single chemicals (e.g., phenylketonuria, Tay–Sachs and other lipid-storage diseases), or diseases that involve a failure to synthesize a normal cell constituent, which leads to the absence of a normal chemical (e.g., the inability to synthesize melanin results in albinism). In most cases, however, we see only the end result of a mutant gene, the *phenotype*.

At the level of the phenotype (but not at the biochemical or molecular level), a gene can be characterized as *dominant* if it confers its characteristic phenotype on individuals who have only one "dose" of the gene. A gene is phenotypically *recessive* if it confers its characteristic phenotype only on individuals having two copies of the same gene. The usual example given is that blue eyes are recessive to brown: the child of a blue-eyed and a brown-eyed parent is brown-eyed. Actually, multiple genes are involved in determining eye, hair, and skin color, so it is quite possible to violate this "rule." A much better example is the gene for Tay–Sachs disease. A child must receive one copy of the defective disease from *each* parent in order to develop the disease, which is lethal in infancy. Individuals with a single copy of the

[8] The localizing of all genes on the human genome is only now beginning to produce results, although the pace is increasing. One major (common) and a dozen minor (rare) mutations for cystic fibrosis were identified between 1988 and 1991, and the deletion that causes the dominant disease of retinoblastoma is now very well characterized. Genes for Huntington's chorea, amyotrophic lateral sclerosis, and possibly for familial Alzheimer's disease have been located.

Table 10.1 In the Parental Generation, F_0, $AA \times aa$ Produces F_1 Offspring, All of Which Are Aa

	A	A
a	Aa	Aa
a	Aa	Aa

Tay–Sachs gene are phenotypically normal.[9] In contrast, the dominant Huntington's chorea affects 50% of the children of affected individuals, because only one copy of the gene is needed to express the disease state.

Genes in higher organisms[10] are located on chromosomes, with each species having a characteristic number of chromosomes referred to as its *karyotype*. A normal human has 23 pairs of chromosomes: 22 *autosomal* chromosomes and one pair of sex chromosomes, either XX (female) or XY (male). Humans are therefore *diploid* (i.e., they have two complete sets of chromosomes). *Haploid* organisms have only one set of chromosomes, *triploid* organisms have three sets, *tetraploids* have four, and so on. Among plants, *polyploidy* is very common; many crops and garden flowers are derived from plants with far fewer chromosomes.[11] Often, polyploidy in plants results in larger plants with larger flowers or more seeds. Among animals, in contrast, deviations from diploidy almost invariably results in abnormal individuals. In diploid organisms, one member of each pair of chromosomes comes from the father, and one from the mother. Let us follow some genes through three generations: from grandparents (F_0), through parents (F_1), to offspring (F_2) (Table 10.1).

Assume that the grandparents (F_0) are AA and aa. If A is dominant and a recessive, the phenotypes of all offspring will be the dominant phenotype. This is seen in crosses between short and inbred[12] tall sweet peas: all the offspring are tall. It is also seen in mammalian matings between albino (aa)

[9] Tay–Sachs disease is also interesting because its occurrence in Caucasians is essentially limited to people descended from Polish Jews. The Tay–Sachs gene is carried by up to 1 in 30 American Jews and poses so serious a risk for this ethnic group that premarital (or at least pre-procreational) screening is strongly suggested.

[10] *Eukaryotes* are organisms with chromosomes; *prokaryotes* lack true chromosomes and carry their genetic information as "naked" DNA. All multicelled organisms are eukaryotes, as are many single-celled organisms. Bacteria are prokaryotes.

[11] In the case of some crop plants, such as maize, there have been enough alterations in the genome, and they occurred so long ago, that it is hard to identify the ancestral species (or how many there were) with certainty. In ornamental plants, polyploidy is often induced deliberately: tetraploid day lilies are one example, triploid marigolds another. Tetraploid flowers tend to be larger than their diploid counterparts. Triploids tend to be sterile, and the triploid marigold therefore blooms profusely instead of going to seed.

[12] Note that not all tall sweet peas are genotype AA; if some of the tall parents are really Aa, these ratios will not hold.

**Table 10.2 In the F₁ Generation, Hybrid Offspring
(*Aa*) Produce F₂ Offspring That Assort in the Ratio
AA : *Aa* : *aa* :: 1 : 2 : 1**

	A	a
A	AA	Aa
a	Aa	aa

and normally pigmented (*AA*) individuals: all the offspring have pigment. If neither *A* nor *a* is dominant, however, the offspring will be of an intermediate phenotype. For example, when some red-flowered plants are crossed with white-flowered plants, all of the F₁ flowers are pink. If the F₁ individuals are bred to each other, F₁ × F₁ = *Aa* × *Aa,* and one can depict the mating as in Table 10.2.

The frequency of *genotypes* in the offspring is *AA:Aa:aa* :: 1:2:1. The frequency of *phenotypes* will depend on the dominance relationship between *A* and *a,* however. If *A* and *a* represent red and white flowers and neither allele is dominant, the phenotypes will be 1 red : 2 pink to 1 white. If *A* is dominant and *a* recessive, the phenotypes will be 3 dominant to 1 recessive phenotype, which is seen in the ratio of 3 tall to 1 short in sweet peas, and in the statistical probability of 1 albino to 3 normal infants born to pigmented parents, both of whom are carriers (*Aa*).

The same analysis can be carried out for two genes or more genes, although the tally gets very complex very fast. For two genes, *A* and *B,* and their recessive alleles, *a* and *b,* the matings are *AA BB* × *aa bb* → *Aa Bb* offspring (Table 10.3). When the *Aa Bb* offspring are mated to each other, the ratios of *genotypes* among the offspring are shown in Table 10.4.

The frequencies of the possible *phenotypes* will differ depending on dominance relationships. One can consider the situation where *A* is dominant to *a* and *B* is dominant to *b,* in which case the phenotypic ratio is 9 double-dominant to 1 double-recessive, with 6 having the dominant phenotype for one gene and the recessive phenotype for the other (3 for each gene). If, on

**Table 10.3 Transmission of Two Alleles When
Parents Are Homozygous *AABB* and *aabb***

	AB	AB
ab	AaBb	AaBb
ab	AaBb	AaBb

Table 10.4 Segregation of Two Alleles When Both Parents Are Double Heterozygotes, *AaBb*

	AB	*Ab*	*aB*	*ab*
AB	*AABB*	*AABb*	*AaBB*	*AaBb*
Ab	*AABb*	*AAbb*	*AaBb*	*Aabb*
aB	*AaBB*	*AaBb*	*aaBB*	*aaBb*
ab	*AaBb*	*Aabb*	*aaBb*	*aabb*

1 *AA BB* : 2 *Aa BB* : 2 *AA Bb* : 4 *Aa Bb* : 1 *aa BB* : 1 *AA bb* : 2 *aa Bb* : 2 *Aa bb* : 1 *aa bb*.

the other hand, the heterozygote (hybrid) genotypes *Aa* and *Bb* are distinct from *AA* and *aa* (*BB* and *bb,* respectively); then each of the genotypes can be identified separately. In that case the phenotypes are identical to the genotypes. It is also possible that the alleles of one gene are dominant/recessive, while for the other gene, hybrids can be distinguished from either parent.

Obviously, the situations become rapidly more complicated as more genes are added. For many traits, such as height in humans or weight in cattle, several genes act on a single trait. If more than three genes act on a single trait, the phenotypes tend to look as if they were distributed continuously, especially if the traits are also influenced by environmental factors such as nutrition or illness. Scientists before Mendel studied such quantitative variants and were unable to decipher the patterns that resulted. It was Mendel's genius that he chose simple traits with a dichotomous outcome—tall or short, wrinkled or smooth—and succeeded in simplifying the pattern and in analyzing it.

Note that hybrid parents *randomly* transmit each gene of the pair. Actually, it is chromosomes that are randomly transmitted; as a first approximation, the genes on a chromosome are inherited together.[13] For a given chromosome in a parent, the probability of transmission to a given offspring is one-half (0.5). Chromosomes are transmitted independently: that is, for each parental pair, the selection of which chromosome enters the sperm (or egg) is random. One copy of each gene is received from each parent. Remembering that *the probability of independent events is the product of their individual*

---

[13] In actuality, crossing over between homologous chromosomes results in some lessening of linkage between genes on a single chromosome, but the analysis is beyond the scope of this summary. Similarly, there are examples of preferential transmission of one chromosome over another, but for practical purposes, the random transmission of the two members of a pair of chromosomes, and of the genes on them, holds.

probabilities, we can determine that the probability that *each* of n genes (where each gene is on a different chromosome) is transmitted to a single offspring is 0.5^n, or $\frac{1}{2} \times \frac{1}{2} \times \frac{1}{2} \times \frac{1}{2} \times \frac{1}{2} \cdots$.

Chromosomal Abnormalities

In addition to the occurrence of deletions, insertions, and substitutions that make up mutational events, the genome can be altered by the aberrant migration of entire chromosomes or parts of chromosomes. Such events are thought to be due to abnormal movement of chromosomes during cell divisions. The consequences are most serious if the aberrant migration occurs during the (meiotic) cell divisions that gametes undergo preceding fertilization, or during the first (mitotic) divisions of the embryo. The later the mishap occurs during development, the smaller the portion of cells that will be affected and the less serious the probable consequences.

The effects of abnormal migration of chromosomes are categorized by the consequence to chromosome number. If one member of a pair of chromosomes is missing, the organism is said to be *monosomic;* if an extra (third) chromosome is present, it is *trisomic*. In humans, individuals who are either monosomic or trisomic for even part of one autosomal chromosome are very often both physically and mentally abnormal. Missing or extra sex chromosomes are less disastrous and quite compatible with completely normal functioning and nearly normal appearance, although some combinations lead to sterility.

The catalog of chromosomes present in the organism is its *karyotype* (Figure 10.1). Each human chromosome can be distinguished from the others by its physical appearance (its length, its shape, and location of the centromere) and by the pattern of bands that appear when metaphase cells are appropriately stained. Each pair is numbered (in humans, from 1 to 22 for autosomes; X and Y for sex chromosomes) so that abnormal conditions can be described (e.g., trisomy 21, XXX, etc.).

Inheritable Genetic Diseases in Humans[14]

It is thought that genetic causes account for 0.5 to 2% of *all* human diseases. A plausible breakdown of inheritable genetic diseases and their frequency per 1000 live births follows.[15]

---

[14] Note that identification of the following entities requires trained professionals who can both construct a good family history and read a karyotype. The amateur's assumptions about the mode of inheritance for a given phenotype are usually wrong. Quick judgments can lead to tragedy, so do not take these examples as anything *but* examples.

[15] Taken from the unpublished report of the mutagenesis subcommittee of the Committee on Medical Aspects of Chemicals in Food and the Environment, the Department of Health and Social Security, British Government, October 1977.

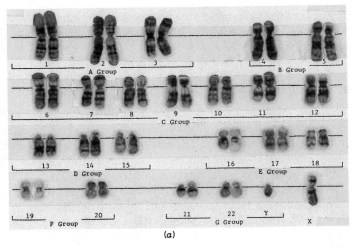

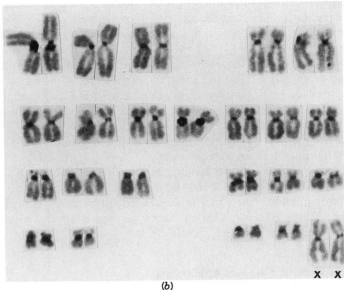

Figure 10.1 Human karyotypes. (*a*) Karyotype of a normal male child. (Courtesy of R. M. Fineman, Department of Pediatrics, University of Utah.) (*b*) Karyotype of a normal young woman. (Courtesy of A. D. Stock, Utah State University.)

· *Chromosome abnormalities* occur with a frequency of 6.9/1000 live births. Among defects more or less compatible with survival are:

An abnormal number of sex chromosomes. Examples include phenotypic males such as XYY or XXY (the latter is known as Klinefelter's syndrome); phenotypic females such as XXX and XO (the latter is also known as Turner's syndrome).

Autosomal trisomy. Trisomy of chromosome 21 is known as Down's syndrome or mongolism; trisomy 18, as Edward's syndrome, trisomy 13, as Patau's syndrome.

Triploidy occurs when the entire genome is present in triplicate. Triploid babies often, but not invariably, die before birth. Those that survive to birth usually die soon afterward.

Inversions occur when segments within a chromosome are reversed in direction. Translocations occur when two chromosomes swap pieces or when a piece of one chromosome attaches to another. Such chromosomal rearrangements need not produce abnormality in and of themselves, as long as all parts of the genome are present in the right dosage,[16] but when the chromosome number is reduced during meiosis, deletions or duplications may lead to abnormal chromosomes being present in the gametes, leading to chromosomal abnormalities in the next generation.

· Defects and diseases due to single genes of large effect include:

Autosomal dominant diseases, seen at a frequency of approximately 1.8 to 1.1 per 1000 births. Examples include *some* forms of childhood blindness, retinoblastoma, and Huntington's chorea.

Autosomal recessive diseases, which occur at approximately 2.2 to 2.5 per 1000 births. Examples include *some* forms of childhood blindness; cystic fibrosis, phenylketonuria, galactosemia, and the several mucopolysaccharoidoses.

Sex-linked diseases, which occur with a frequency of approximately 0.3 to 0.5 per 1000 births. Examples include hemophilia *a* and *b, some* forms of childhood blindness, Duchenne muscular dystrophy, and most color blindness.

· Congenital malformations with a sizable genetic contribution (approximately 50%) occur at approximately 20 per 1000 births. Examples include cleft lip with or without cleft palate,[17] polydactyly and syndactyly, malformations of the heart and great vessels, and congenital club foot.

· Complex disorders with a genetic component, of variable but not precisely known influence (perhaps 30% genetic?) occur at a rate of 7 to 10 per 1000 live births; examples include schizophrenia, epilepsy, and diabetes mellitus.[18]

---

[16] There is increasing evidence, however, that locational effects can have pronounced consequences.

[17] Isolated cleft palate (i.e., without a cleft lip), occurring in the absence of other malformations, appears to have a negligible genetic component. A woman who has had one child with cleft lip has a 10% probability of recurrence in the next pregnancy; a woman whose child has isolated cleft palate has the same risk as anyone in the population. If one has a child with a malformation, it is therefore essential to obtain the most accurate diagnosis possible, and preferably genetic counseling as well.

[18] The childhood form of diabetes mellitus is now thought to be an autoimmune response to a viral disease. What is inherited, therefore, is not diabetes itself, but the tendency to respond to one or more viruses with an autoimmune attack on the insulin-secreting cells of the pancreas.

These data suggest that the total genetic contribution to human diseases, at birth, is between 36 and 44 per 1000 live births. Put another way, it means that between 3.6 and 4.4% of live-born babies have a problem that is partially genetic. Among structural anomalies of chromosomes, approximately 20% arise de novo, while 80% are inherited. Most autosomal anomalies (especially trisomy) are lethal in early postnatal life: about 5% of all early neonatal deaths show such anomalies. Among stillborn children, 30-50% show such anomalies; about half of these are autosomal trisomy. If one examines spontaneously aborted fetuses (miscarriages), the incidence of chromosomal aberrations increases in frequency, and progressively more serious defects are seen as one looks at earlier and earlier miscarriages. Available data from spontaneous and induced abortions suggests that up to 80% of fertilized human eggs are lost before birth. In most cases the loss occurs before implantation—so early that the woman does not know that she was pregnant.[19]

POPULATION GENETICS

The frequency of a gene in a population depends on the relative fitness of the carrier organism, the back mutation rate, and the degree of inbreeding in the population.

Relative *Fitness* of Carrier Organism

Fitness is the relative ability of the organism to reproduce. Normal fitness is set at 1.0; total inability to reproduce, at 0. Genetically, it does not matter whether the failure to reproduce is due to death before maturity, to biological sterility, or to voluntary refusal to reproduce.

A mutant gene may affect (decrease) fitness and thereby be selected against in the population. The state of the gene in the organism (homozygous or heterozygous) may strongly affect this. For example, in the British population, a mutant gene determining hypercholesterolemia occurs in 0.1 to 0.5% of the population. Individuals who carry this gene produce very high levels of cholesterol, regardless of the cholesterol content of their diet. In persons who are heterozygous for the hypercholesterolemia gene, heart attacks typically occur by age 50 or 60. In persons homozygous for the hypercholesterolemia gene, the first heart attack typically occurs before age 20. Therefore, since almost all women and most men have completed their reproductive

---

[19] Given the evidence that an overwhelming number of early miscarriages consist of totally nonviable conceptuses, it is advisable for women who repeatedly miscarry to have their own and their partner's karyotype evaluated before heroic efforts to have children are instituted. The same precaution is advisable for women who "can't get pregnant," since they may be aborting too early to recognize pregnancy. Either the man or the woman can be the source of chromosomal abnormalities in the embryo. On the other hand, a *single* miscarriage is not evidence of such problems.

activity before age 50, fitness is strongly affected only in people who are homozygous for this gene. Other examples of genes that decrease fitness in homozygotes include autosomal recessives such as sickle-cell anemia and phenylketonuria (PKU).

Note that decreases in fitness need not be biological in origin. When the reproductive success (fitness) of children of people with Huntington's chorea is calculated, individuals who carry the Huntington's gene (and who usually develop the disease in middle age) have as many children as their sibs who do not carry the gene. Within Huntington's families, therefore, fitness does not differ between carriers and normal individuals. Compared to the general population, however, fitness in Huntington families is only 0.8. Since there is no difference between carriers and noncarriers, the decrease in offspring is not biological. Rather, the knowledge that they may develop the disease decreases the likelihood that children of Huntington's chorea victims will reproduce.

Social selection is also seen in neurofibromatosis, a dominant disease characterized by childhood onset of multiple tumors of the nervous system. Although not malignant, the tumors are moderately to severely disfiguring and can be crippling. The overall reproductive fitness of people with neurofibromatosis is approximately 0.2: they have one-fifth as many children as the average person. When one examines the reproductive capacity of *married* neurofibromatosis patients, on the other hand, fitness is seen to be >50%. Apparently, the disfiguring elements of the disease decrease the probability of finding a mate; those who find mates reproduce relatively well. Even the decreased reproductive rate among married people with neurofibromatosis may be social rather than biological: like possible carriers of the Huntington gene, those who carry the gene for neurofibromatosis may choose not to transmit it.

Back Mutation Rate

If the original mutation was a substitution within the DNA and not a deletion, there is a finite possibility that a second mutational event will return the gene to the normal state. Theoretically, the probability of a reversal is equal to the probability of the original mutation, and independent of it. The joint probability is therefore equal to the product of the individual probabilities, and the probability of both occurring in a single individual is vanishingly small. In very large populations over long periods of time, on the other hand, back mutations are significant.

Degree of Inbreeding of the Population

As discussed below, inbreeding increases the probability that an individual will be homozygous for a given gene, and therefore increases the probability of expressing a deleterious recessive. Most of the human data on the consequences of inbreeding are based either on clinical cases, in which a specific disease occurs in the offspring of consanguineous parents, or on epidemiological data, in which the relative fitness of progeny of consanguineous parents is compared with the fitness of progeny from outbred marriages. Since very few cultures encourage inbreeding closer than marriages of second cousins, relatively few studies on closer relationships exist, but there have been some. In one series of births resulting from admitted incest (father × daughter or brother × sister), at least 50% of the children were severely retarded mentally; several also had physical defects. The parents were essentially normal. None of the parents were mentally deficient, and only one had a physical malformation. Among children of survivors of Hiroshima and Nagasaki who married cousins, there was a significant increase in congenital malformations (48%), in stillbirths (25%), and in the postnatal death rate of offspring (35%) compared to the outbreeding survivors. The weight of surviving children at 9 months was decreased by 12% among children of consanguineous marriages.

Inbreeding in humans can be used to estimate the frequency of deleterious genes in the general population, as seen in Table 10.5. Preceding World War II, most rural Europeans not only lived and died within a few miles of their birthplace, but most of the population within a geographic area practiced the same religion. Records kept by the church as a by-product of recording births and marriages are remarkably complete, often go back for centuries, and provide an excellent data base for determining genetic relatedness. In a Swedish study, 16 to 28% of children of first-cousin marriages had a genetic disease. (The range is the result of uncertainty about how genetic some diseases are.) In contrast, only 4 to 6% of children whose parents were *not* related had a genetic disease. The difference, 12 to 22%, represents the increased risk of genetic disease due to an inbreeding coefficient of $\frac{1}{16}$, and suggests that each completely homozygous individual would have 16 times as many genetic diseases, or between 2 and 3.5. Since each such disease requires two alleles, the calculated number of deleterious genes per heterozygous individual is four to seven. All of these data suggest that each human

Table 10.5 Data from French Parish Records Illustrating Differences in Mortality Before Adulthood Between Children of Consanguineous and Unrelated Parents (Including Stillbirths)

Degree of Consanguineity	Mortality (fraction of births)
If parents are unrelated	0.12
If parents are cousins	0.25
Excess lethality from cousin marriage	0.13

Source: Data from J. Sutter and L. Tabah, 1958, presented by J. F. Crow and M. Kimura, *An Introduction to Population Genetics,* Harper & Row, New York, 1970, pp. 76–77.

carries about six deleterious genes. These results were not reproduced by larger studies in Japan, where first-cousin marriages are encouraged. Therefore, population geneticists suggest that when pedigrees are drawn from societies that frown on cousin marriages, the genetic consequences of inbreeding may be confounded with social or psychological causes (or consequences) of such inbreeding.

From Figure 10.2 it can be seen that $\frac{1}{16}$ of the genes from the *parents* of the affected offspring are identical due to the consanguineous relationship. That is, the fraction of each parental gamete that is homozygous due to the cousin relationship is $\frac{1}{16}$. Therefore, the number of lethal gene equivalents present in the offspring is $0.13 \times 16 \times 2 = 4$, the multiplication by 2 because there is one gamete from *each* parent. In other words, each of us (statistically) carries the equivalent of about four genes that would lead to stillbirth or death during early childhood *if we were homozygous* for any of these genes. In fact, there may actually be eight such genes, each of 50% effective lethality, or some other combination, to give the *equivalent* of four lethal genes. The Japanese data suggest that the number might be two rather than four.

How can it be that we *all* carry two to four lethal equivalent recessive genes? Why aren't these genes removed from the population by the failure of homozygous organisms to survive? To answer this question, we must look at the behavior of genes within a population.

Quantitative Inheritance: Hardy–Weinberg Principle

If two forms (alleles) of a gene, A_1 and A_2, are in a population in proportions p_1 and p_2, the frequency of the homozygote $A_1 A_1$ is p_1^2; the frequency of the homozygote $A_2 A_2$ is p_2^2; and the frequency of the heterozygote $A_1 A_2 = 2p_1 p_2$ (Table 10.6). Now, if we can only recognize the homozygote $A_2 A_2$ (e.g., if

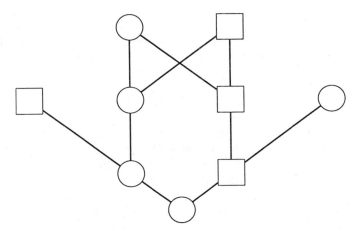

Figure 10.2 Pedigree of a first-cousin marriage.

Table 10.6 Hardy–Weinberg Law When Two Alleles, A_1 and A_2, Are Present in a Population at Frequency p_1 and p^2

	$p_1 A_1$	$p_2 A_2$
$p_1 A_1$	$A_1 A_1$ $(p_1)^2$	$A_1 A_2$ $(p_1 p_2)$
$p_2 A_2$	$A_2 A_1$ $(p_1 p_2)$	$A_2 A_2$ $(p_2)^2$

A_2 is the recessive gene for PKU) and the frequency of the homozygous recessive state (e.g., phenylketonuria) is 1/10,000, then

$$(p_2)^2 = \frac{1}{10,000}$$

$$p_2 = \sqrt{\frac{1}{10,000}} = \frac{1}{100}$$

The gene is present in 1% of the gene pool. Therefore, $p_1 = 1 - (1/100) = 99/100$ and the frequency of heterozygotes $A_1 A_2 = 2(1/100)(99/100) = 198/10,000 = 1/50$. (Note that there are two ways of becoming heterozygote—hence the 2 in the frequency calculation.)

About 2% of the population carries the gene. But only 1 person in 10,000 is homozygous. Only those carriers who produce an affected child can be recognized. The rest—some of whom married noncarriers, and some of whom, by chance, do not have an affected child—remain invisible. The child with the disease is the very small tip of a very large iceberg. It will therefore be impossible to remove the gene from the population by preventing affected (homozygous) individuals from reproducing.

It can be argued that Western civilization actually tends to promote an increase in frequency of deleterious genes:

1. By removing the "counterselective" effects of such genes. For example, children with phenylketonuria (PKU) used to be severely retarded; most were institutionalized. Now PKU infants are fed a special diet that is extremely low in phenylalanine; this diet reasonably effectively prevents retardation. The child therefore has a much greater chance of growing up normal and therefore of reproducing and passing on the PKU genes.

2. By exposing the population to more mutagenic agents, and so raising the mutation rate. This is speculative, because it has not been possible to demonstrate such an effect, but *if* many of the chemicals we make are mutagenic—and they are—and *if* they get to the genome, they should increase the incidence of mutations.[20]

3. By allowing selection of phenotypically normal fetuses and selective abortion of fetuses that would express a lethal disease. In the unselected situation, if *A* is the normal allele and *a* the deleterious mutant, the ratio of genotypes in the offspring is 1 *AA* to 2 *Aa* to 1 *aa*. The child who is *aa* dies or does not reproduce (the two results are genetically equivalent), and two abnormal genes are removed for each such affected infant of zero fitness. An even greater increase in population fitness occurs if parents of an affected child stop reproducing altogether, since this also prevents the birth of a number of heterozygous individuals. Selective abortion of affected fetuses, on the other hand, typically leads to the situation where the originally intended number of *phenotypically* normal children *is* born, since elective abortion is relatively untraumatic. The ratio of genotypes among surviving offspring is 1 *AA* to 2 *Aa*. Of the children born, two of three, or 67%, will carry the deleterious gene, and because replacement of the lost fetus is probable, there is a 0.67 probability that an extra *a* allele will enter the population.

Some good science fiction, and some bad science, has resulted from thinking about these ideas. It is nonetheless important to realize that a policy that in the short run minimizes costs (including both the burden to people who do not reproduce because of the severity of the defect they might transmit and the cost of affected children to parents and to society) may have the quite unintended consequence of increasing the genetic load of the population. What will be the cost of this policy to a specific future generation depends on a number of factors, including the sophistication of then-available medical care. For example, the cost of not preventing the spread of the genes for myopia is that many of us need corrective lenses. In our world, the cost of providing such lenses is not great enough to be a major factor in reproductive decisions. In contrast, the cost of caring for a retarded child in our "postindustrial" society is considered a major burden on families and on the society as a whole. In preindustrial societies, with their great need for manual labor, a "simpleton" was probably less of a parental tragedy than today, and certainly less of a cost to society. On the farm and in the shop, simpletons could earn their keep. Even further back, myopia was most probably a lethal phenotype in hunter-gatherer societies, where survival depended on seeing both predators and prey.

---

[20] If they reach the genome, such mutagens should also be carcinogens and should be increasing the incidence of cancer. See Chapter 11 for evidence on this topic.

Given that most mutations are deleterious, increases in the mutation rate due to the presence of synthetic mutagens in the environment would be a serious matter. It is well known that the rate of mutation differs between phyla. The spontaneous mutation rate is estimated to be between 1×10^{-6} and 1×10^{-5} mutations per site in lower eukaryotes (such as yeasts) and in bacteria; it is estimated to be 1×10^{-7} in higher eukaryotes. This means that an alteration of a *specific base pair within a gene* occurs once in every 10 million cells in humans (for example), and once in every 100,000 bacteria. Presumably the mutation rate is lower in higher organisms than in bacteria because it is evolutionarily advantageous to minimize the random changes that occur. There are, however, several repair systems in prokaryotes as well as in eukaryotes to prevent miscoding from being perpetuated.

What is the mutation rate in humans? Is it increasing as a result of the large numbers of chemicals being synthesized and released into the environment? There is no clear-cut answer to the latter question, because we cannot estimate the mutation rate accurately enough. Determination of mutation rates in humans is usually attempted by observation of the state of single genes and/or by observing the protein product of single genes. By using easily available cells, one can determine the frequency of genes in the population; by looking at differences in enzyme activities between large numbers of parents and children, one can determine how often a change occurs that is not (or seems not to be) due to assortment. Similarly, examination of enzyme activities in blood cells or cultured tissues allows an estimate of the incidence of mutational events in single cells, of which millions can be grown in a flask. Indirect estimates may come from a tally of the frequencies of genetic diseases of high visibility (e.g., frequency of hemophilia, PKU, or dominant diseases). Dominant lethal diseases such as retinoblastoma (no longer necessarily lethal) can be used to determine the *per locus* incidence of mutations, but it is not known whether this can be generalized across loci; one suspects not. The values for mutation/locus/generation vary; many assumptions must usually be made to calculate the rate from the observation. Problems of sample size often limit resolution. However, a currently accepted average rate for humans is 2×10^{-5} mutations per locus per generation.

11

CARCINOGENESIS

Statistics are like a bikini: what is revealed is interesting; what is concealed is crucial.

—A. R. Feinstein[1]

The great Ziggurat of Ur is one of the masterworks of early civilization. The ancient people who built these structures have long since vanished. This suggests that ziggurats may be dangerous to your health.

—T. Weller[2]

INTRODUCTION

A tumor is an example of misplaced growth: a localized failure in the mechanisms that control cell division in the organism. Tumors are often distinct from the surrounding tissue, separated from normal tissue by a membrane or capsule. On the surface of the organism, a tumor will be seen as a lump or raised area; internally, it may exert pressure on adjacent structures. Tumors occur on plants, where they are called *galls,* as well as in animals.

In animals, tumors are characterized by the degree to which they escape normal regulation of growth. Almost by definition, tumors exhibit *autonomous growth:* they do not grow in synchrony with the surrounding tissue or organ, but exceed the ordinary rates of cell division for their location, so that too many cells are present and the growth infringes on space allocated to other cells. Apparently, tumor cells are unresponsive to growth-stopping signals. There is certainly a range of such signals. A tumor may grow without regard to signals indicating cessation of growth, yet remain responsive to signals indicating growth is permitted (but not necessary). For instance, some breast tumors are still dependent on estrogens and will grow only as long as estrogen signals them to do so. Such tumors regress (shrink) following ovariectomy. Similarly, some prostate tumors are dependent on testosterone and regress following castration. Other tumors will grow even in the absence

---

[1] A. R. Feinstein, "Scientific standards in epidemiologic studies of the menace of daily life," *Science* 242:1257–1263, 1988.
[2] T. Weller, *Science Made Stupid,* Houghton Mifflin, Boston, 1985.

of growth-promoting signals. In the examples above, breast and prostate tumors that do not need sex hormones to continue growing would not regress following elimination of these hormones.

All tumors also exhibit the property of *transplantability,* although this is difficult to demonstrate outside the laboratory because the recipient organism must be genetically compatible. Inbred strains of mice, in which all animals are genetically identical to each other, provide the best evidence for transplantability of tumors. It is also possible to transplant tumors to different sites on the same animal. Finally, there are a few tragic examples in humans, in which tumors have been transmitted between close relatives. In most cases, however, tumors do not continue to grow if the new host is not genetically identical to the original host. From transplantation experiments, three important corollary conclusions can be drawn:

1. Progeny of a tumor cell *inherit* the tumor properties (i.e., they too grow autonomously).
2. Tumors are *clonal* in nature (i.e., each tumor is probably descended from a single transformed cell).
3. *Transformation* of a cell to the tumorous state is usually *irreversible.*

The quality of autonomous growth permits the cells of a tumor to continue growing if moved to a new site on the same host or to a new host of compatible genotype. Similarly, if most of the tumor cells are excised, the tumor will regrow from remnants. In some experiments it has been shown that as little as a single tumor cell can grow to a full-sized tumor. Only if the entire tumor is removed will recurrence be prevented. This has led to the assumption that a single transformed cell is sufficient to give rise to a tumor. Although this is clearly true for many tumors, the generality of this idea is not absolutely proven.

Many tumors remain localized and are considered *benign* since their removal ends their effect. Only if a benign tumor cannot be removed is it likely to be lethal. Although some benign tumors attain large size (uterine fibroids can weigh several pounds), they remain a part of the originating organ or tissue. They can nonetheless be lethal or crippling, typically because they "crowd out" essential cells—for example, by exerting pressure on the brain, trachea, or a critical blood vessel. From a tumor's point of view, the body is like a favorable growth medium. Since a tumor ignores the body's signals to stop growing while the rest of the body's cells obey, the tumor is *selected* in this environment and proliferates at the body's expense. Just as when bacteria are grown on a petri plate, growth ends only when the growth medium is exhausted. Even benign tumors can threaten the life of the organism if they grow to a size sufficient to compete with essential bodily organs for nutrients and removal of excretory products.

Benign tumors may also be medically intractable due to the sheer number of tumors present. Neurofibromatosis is a genetic disease in which many benign tumors form on nerves throughout the body. Even if each individual tumor is located where it can easily be removed, they occur in such numbers that removing them all becomes impossible. Some of them are also located along nerves that are too important to damage: an ever-present risk in neurological surgery. Similarly, benign tumors of the pituitary gland can be so intertwined with the brain that complete removal would be disastrous. Fortunately, the latter tumors grow slowly, producing symptoms during middle age. Removal of most of the tumor generally allows the victim to live a normal life well into old age before symptoms recur. Benign tumors may also be deadly if the social structure prevents removal: this is true of the poor in all countries without universal access to medical care, but especially in underdeveloped nations where few doctors are available to perform the necessary surgery.

Tumors that exhibit growth, transplantability, and the additional properties of *invasiveness* and *metastasis* are said to be *malignant*. A benign tumor grows as a coherent mass that may be adjacent to, but does not disrupt, normal tissues. Invasive tumors grow into normal tissues, destroying their function. Metastasis, the most insidious property of malignant tumors, is their capacity to shed individual cells or small groups of cells that migrate through the bloodstream or lymph and implant at a new site to form a new tumor. This seeding of tumors is called *metastasis*. Interestingly, tumors often show some specificity of implantation site of metastases. Metastases rarely occur in the heart, for example, and many cancers only form metastases in certain organs. Certain cancers metastasize while the original tumor is hardly detectable; others normally metastasize very late in the course of the disease. Thus each class of tumor tends to follow a pattern. Those that, like melanoma, have metastasis as an early part of the pattern tend to be strongly life threatening. Invasive, metastasizing tumors are also known as cancers.

Tumors take their name from the tissue and cell type in which they originate. Malignant tumors of epithelial cells are *carcinomas;* those arising in connective tissue and blood vessels are *sarcomas,* those arising in the blood forming cells are *leukemia,*[3] and those arising in lymph nodes are *lymphomas*. Tumors are also classified by the organ in which they originally arise. A *melanoma* is a cancer arising in melanin-producing cells; since melanin is the pigment in hair and skin, such tumors are strongly pigmented. A *squamous cell carcinoma of the larynx* would be a cancer arising in that part of the epithelium of the larynx that gives rise to the keratinized surface cells

---

[3] Although leukemias do not form distinct masses but are individual cells within the bloodstream, the rapidly dividing white blood cells fulfill all the other requirements for malignancy.

(squames); a *fibrosarcoma of the larynx* would be a cancer in connective tissue in the larynx; and so on.

CANCER AND ITS ORIGINS

Most of us have a strong interest in avoiding cancer. What puts us at risk of developing this disease? Can we do anything about it? What is the role of environmental chemicals? Several lines of evidence indicate that the incidence of cancer depends both on intrinsic factors, which are primarily genetic, and on environmental factors. Both categories have been demonstrated in humans as well as in animals.

Genetic predisposition has been most clearly demonstrated in inbred strains of mice, in which each animal is a genetic twin of all other animals in the strain. Both high- and low-cancer-incidence strains have been produced by selective breeding. Not only the incidence but the type of cancer is characteristic within strains: that is, some strains have a 98% incidence of mammary tumors in females by 1 year of age; others develop lung tumors but not mammary tumors; and so on.[4]

In humans, several diseases transmitted in simple genetic fashion carry an extraordinarily high incidence of cancer. Among these are retinoblastoma, Bloom's syndrome, xeroderma pigmentosum, and Down's syndrome or trisomy 21. Retinoblastoma, inherited as a Mendelian dominant, causes childhood eye tumors; the mechanism by which its oncogene acts is discussed below. Xeroderma pigmentosum is due to a failure of DNA repair mechanisms. Due to this inability to repair photodamage to DNA in skin, victims are extraordinarily sensitive to ultraviolet (UV) light. Skin cancers result from even minimal exposure to sun. Most victims have died young, but recent understanding of the DNA repair problem is leading to some improvement, mostly by fanatic avoidance of sun and meticulous removal of precancerous lesions.

More common than single-gene cancer syndromes are the predisposing syndromes. The Philadelphia chromosome is a chromosomal rearrangement very strongly associated with chronic myelogenous leukemia; it occurs in many leukemia victims and is very similar to the trisomy of Down's syndrome patients, who also have a very high risk of developing leukemia. In hereditary polyposis, the genetic predisposition is for an overgrowth of cells in the colon. People with hereditary polyposis are at high risk of colon cancer,

---

[4] Studying cancer was so important in the development of inbred strains of mice that most of the older strains are prone to one or more tumors. The exception is the C57BL strain, which was bred as a cancer-resistant control strain, not particularly susceptible to any of the major cancers. There are now several sublines of C57BL (e.g., C57BL/6 and C57BR). They are unusual among inbred mice not only for their low incidence of tumors but also because they are unrelated to most of the other inbred strains.

which arises in the polyps. People with many pigmented nevi (colored moles) run a very high risk of developing melanoma. In both pigmented nevi syndrome and hereditary polyposis, the genetic predisposition is for excessive growth by a particular cell type, a condition known as metaplasia. The cancer then arises out of these foci. The Li–Fraumeni syndrome, a familial high-cancer syndrome, causes very high incidence of several cancers, including colon and breast cancer, at an unusually young age.[5] In 1991 the genetic basis of the Li–Fraumeni syndrome was identified as inheritance of a mutation in a gene known as $p53$. The $p53$ gene has also been associated with nonfamilial cases of both breast and colon cancer. But whereas in "normal" people *both* alleles of $p53$ must mutate in a single somatic cell to deregulate cell division, Li–Fraumeni families already have one defective $p53$ gene. A single additional "hit" suffices to cause cancer.

Although relatively few cancers are as clearly genetic as the above, there is a tendency for cancers *of specific sites* to "run in families". If your mother or sister had breast cancer, your risk is double that of someone without a family history.[6] If a parent had lung cancer, you are (even more than most) foolish to smoke.

Despite the existence of clearly familial cancer syndromes, environmental factors are critical. Even with the very strong genetic predisposition to skin cancer that is seen in people with xeroderma pigmentosum, avoidance of the environmental trigger—UV light—reduces the risk of skin cancer. Epidemiological studies demonstrate clearly that ethnic differences in the incidence of cancer at specific sites are heavily determined by environmental factors, as can be seen for the high incidence of stomach cancer among Japanese in Japan. It might be argued that this is genetic in origin, but if one looks at cancer mortality in migrants, a different picture emerges. In Hawaii and on the U.S. mainland, where large numbers of Japanese emigrants have settled, their children have an incidence of stomach cancer that is intermediate between Japan's and the base level in the United States. The grandchildren of the emigrants have the low incidence of stomach cancer typical of Caucasian Americans. The same results can be seen for migrants from Europe to the United States, and for immigrants to Israel: the incidence of cancer changes within two generations to approximate that of the new compatriots.

Much of the credit (or blame) for such changes is given to dietary habits. Certain foods have been identified as potent carcinogens and are thought to

---

[5] In most people, doctors recommend screening for colon cancer beginning at age 50 or perhaps at age 40; women should have a baseline mammogram by age 40 and yearly mammograms beginning no later than age 50. In Li–Fraumeni families, most members have developed cancer by their fortieth year.

[6] This is one of the cancers that is increasing in the American population. The lifetime risk of breast cancer for American women is now 1 in 8, meaning that one woman in eight will develop clinical breast cancer in her lifetime. For women with a positive family history, the incidence is one in 5, or 20%. Nonetheless, roughly 90% of breast cancer occurs in women *without* a family history.

account for the high incidence of certain cancers in some populations and not in others.[7] Areas in Africa with high rates of moldy peanuts and rice due to infestations of the mold *Aspergillus flavus* also have high rates of liver cancer. In animals, aflatoxins made by *A. flavus* are a potent cause of liver cancer, providing further support for the cause-and-effect relationship. Similarly, a high incidence of stomach cancer is seen in countries like Finland, where large amounts of smoked fish are consumed. (Smoke is a potent source of the highly carcinogenic benzpyrenes.) Other specialty foods that contain carcinogens are fern shoots and oil of sassafras; Japanese who eat large quantities of bracken fern also have a high incidence of stomach cancer.

Certain cultural practices are also associated with a high risk of cancer. First and foremost is smoking, associated with lung cancer (Figure 11.1). Chewing tobacco is associated with oral cancer. Circumcision of males is associated with a *decreased* incidence of cervical cancer in their female partners, apparently due to better hygiene. Early sexual activity by women, especially with multiple partners, is associated with increased reproductive tract cancer in women, which is suspected to be due to the higher incidence of sexually transmitted genital warts.[8] On the other hand, childlessness, or delay of the first pregnancy until after age 30, is associated with increased breast cancer.

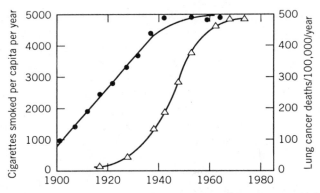

Figure 11.1 Relationship between cigarette consumption and lung cancer. (Adapted from J. Cairns, *Cancer and Society,* from Royal College of Physicians, *Smoking and Health Now,* Pitman Medical and Scientific Publishing Company, London 1971.)

---

[7] Certain foods may also act as anticarcinogens, notably those containing high levels of retinoids, and cole crops such as cabbage and broccoli. In some cases foods may contain both carcinogenic and anticarcinogenic chemicals, so the effects can be extremely complex. One suggestion, made by a nutritionist studying this subject, was to "eat anything you want, but not as much as you want."

[8] Warts are benign growths caused by viral infections. Although there is no evidence that most warts become malignant, the correlation between genital warts and cancer is high.

Synthetic Chemicals and Carcinogenesis

Finally, exposure to certain chemicals carries a high risk of cancer. The English physician Percival Pott was the first person to associate cancer with environmental exposure. He observed that chimney sweeps had a high incidence of scrotal cancer, and correctly attributed the disease to their heavy exposure to chimney soot. Today we know that the benzo[*a*]pyrene, found in all products of combustion, is the most important carcinogen in soot.

Although there is disagreement about how many synthetic chemicals are likely to contribute *significantly* to the incidence of human cancer, some chemicals are known human carcinogens. Industrial chemicals now known to cause cancer in humans include (but are not limited to) asbestos, vinyl chloride, β-napthylamine, benzidine, arsenic, and benzene. In epidemiological terms, the effects of chemicals can also be seen in occupational risks of cancer (Table 11.1). At the end of the chapter we discuss the controversies surrounding carcinogenesis assays.

Table 11.1 Occupations with Above-Average Incidence of Cancer

Occupation	Type of Cancer	Chemical Carcinogen[a]
Chemist	Brain, pancreas, lymphatic	?
Coal miner	Stomach	?
Farmers, sailors	Skin	UV light from sun
Foundry worker	Lung	Benzpyrene, metals
Leather worker	Bladder, mouth, larynx, pharynx	?
Metal miner	Lung	Cobalt, radon
Painters	Blood	Solvents
Petrochemicals	Brain, blood, lung, bone, stomach, esophagus, blood	Benzene
Printers	Lung, mouth, pharynx	Metals? solvents?[b]
Radiologist	Bone marrow	Gamma radiation[c]
Rubber industry	Bladder, blood, brain, lung, prostate, stomach	Organic volatiles
Textile industry	Nasal cavity, sinuses	Formaldehyde
Woodworkers	Lymphatic system	Terpenes, wood oils, formaldehyde

Source: Adapted from W. F. Allman, "We have nothing to fear," *Science* 85:41, October 1985.

[a] These chemicals are only named as *possible* agents. Other chemicals may be involved instead or in addition to those named.

[b] In the past decades, a revolution of printing has radically altered exposures. The source for this table does not indicate years of exposure, so either old agents (lead, metals) or new agents (organic inks, solvents) may be involved.

[c] Today, radiologists leave the room when giving x-rays, so this occupation should be safer now.

Table 11.2 Richard Doll's List of "Established Causes of Human Cancer"

Agent or Circumstance	Site of Cancer
Occupational Exposure	
Aromatic amines	
4-Aminodiphenyl	Bladder
Benzidine	Bladder
2-Naphthylamine	Bladder
Arsenic	Skin, lung
Asbestos	Lung, pleura, peritoneum
Benzene	Marrow
Bis(chloromethyl) ether	Lung
Cadmium	Prostate
Chromium	Lung
Furniture manufacture (hardwood)	Nasal sinuses
Ionizing radiations	Marrow and probably all other sites
Isopropyl alcohol manufacture	Nasal sinuses
Leather goods manufacture	Nasal sinuses
Mustard gas	Larynx, lung
Nickel	Nasal sinuses, lung
Polycyclic hydrocarbons	Skin, scrotum, lung
UV light	Skin, lip
Vinyl chloride	Liver (angiosarcoma)
Medical Exposure	
Alkylating agents	
Cyclophosphamide	Bladder
Melphalan	Marrow
Arsenic	Skin, lung
Busulfan	Marrow
Chlornaphazine	Bladder
Estrogens	
Unopposed	Endometrium
Transplacental (DES)	Vagina
Immunosuppressive drugs	Reticuloendothelial system
Ionizing radiations	Marrow and probably all other sites
Phenacetin	Kidney (pelvis)
Polycyclic hydrocarbons	Skin, scrotum, lung
Steroids	
Anabolic (oxymetholone)	Liver
Contraceptives	Liver (hamartoma)
Social Exposure	
Aflatoxin	Liver
Alcoholic drinks	Mouth, pharynx, larynx, esophagus, liver

Table 11.2 *(Continued)*

Agent or Circumstance	Site of Cancer
Chewing (betel, tobacco, lime)	Mouth
Overnutrition (causing obesity)	Endometrium, gallbladder
Parasites	
Schistosoma haematobium	Bladder
Chlonorchis sinensis	Liver (cholangioma)
Reproductive history	
Late age at first pregnancy	Breast
Zero or low parity	Ovary
Sexual promiscuity	Cervix uteri
Tobacco smoking	Mouth, pharynx, larynx, lung, esophagus, bladder
UV light	Skin, lip
Virus (hepatitis B)	Liver (hepatoma)

Source: R. Doll, "Established causes of human cancer," *Journal of the National Cancer Institute,* 66:123, 1981.

Truly environmental factors causing cancer include exposure to sunlight, which causes melanoma and other skin cancers; high levels of background radiation due to nearby uranium mines and their tailings or from radon in soil (see Chapter 4); and the so-called *urban factor,* the well-documented fact that when all known carcinogens are controlled for, cancer incidence is usually higher in cities than in rural areas. A minimalist's list[9] of environmental carcinogens in the broad sense, which encompasses occupational, pharmaceutical, and life-style exposures, is shown in Table 11.2. A suggestion of other causes is shown in the incidence of cancer by county (Figure 11.2), with its marked geographical variation in cancers of all sites (which are seen again and again, although not always in the same pattern, in the incidence of cancers of specific sites).

[9] This list was compiled in 1981 by the eminent statistician Richard Doll. It is quoted in N. Efron's *The Apocalyptics,* a virulent diatribe against those of us who worry about environmental causes of cancer. Efron calls this concern "a complex corruption of science and a prolonged deception of the public" and includes among the leading perpetrators not only Samuel Epstein (*The Politics of Cancer,* Sierra Club Books, San Francisco, 1978) and Paul Ehrlich (*The Population Bomb* Ballantine, New York, 1968), but also Rachel Carson (*Silent Spring,* Fawcett Crest, New York, 1962) and the Nobel laureates René Dubos and George Wald.

All Cancers Combined
White Males: 1970 - 1979

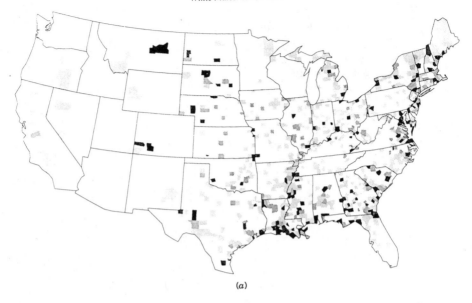

(a)

All Cancers Combined
Nonwhite Males: 1970 - 1979

(b)

Figure 11.2 Cancer mortality in the United States between 1970 and 1979, for all sites, by race and sex. Geographic areas for nonwhites are larger because of the smaller population. Note that cancer incidence for both sexes of all races is high in the industrial areas of the Northeast and the southern end of Lake Michigan. Additional high-cancer areas are seen in the oil-rich areas of Louisiana and Texas, where many petrochemical plants are located, and in the intensively farmed southern tip of Florida. Where male incidence is higher than the corresponding

All Cancers Combined
Nonwhite Females: 1970 - 1979

(c)

All Cancers Combined
White Females: 1970 - 1979

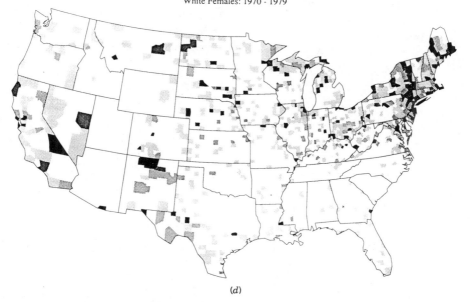

(d)

female incidence, occupational factors are implicated: given the latency period of most cancers, these statistics still reflect a time when relatively few women worked outside the household. Similar sex-related differences could result if an environmental factor increased the incidence of a common gender-specific cancer (e.g., prostate or breast cancer). (From W. B. Riggan et al., *U.S. Cancer Mortality Rates and Trends 1950–1979*, Vol. 4, *Maps*. EPA/600/1-83/015e. U.S. Environmental Protection Agency, Washington, D.C., 1987.)

MECHANISMS OF CARCINOGENESIS

Three kinds of studies have contributed to our understanding of cancer and its origins. The incidence of cancer within and between human populations provided some evidence about the factors that induce cells to escape the body's growth-regulating mechanisms. Animal studies, which allow manipulation of both environmental and genetic variables, provided data about the mechanisms by which cells escape growth regulation. Finally, cells in culture allow investigators to examine the mechanisms governing the longevity and growth patterns of individual cells, apart from the integrative capabilities of the whole organism. It is important to note that the principles of carcinogenesis hold across species. Although there are differences in the incidence of specific cancers within and between species, the process by which a normal cell turns into a malignant cell seems to be the same in all animals. Thus both controlled animal studies and observations of human epidemiology have contributed to the overall picture of carcinogenesis that developed between 1930 and 1980.

First and foremost, cancer has always been known as a disease of aging. The correlation between increasing age and the incidence of cancer, shown in Figure 11.3, is so strong that it must be accounted for by any hypothesis about mechanisms of carcinogenesis. Moreover, similar graphs can be drawn for all animal species, if one changes the age scale to reflect the fraction of the species' life span rather than the age in years. In laboratory mice, with a life span of approximately 2 years, the incidence of cancer becomes significant at about 1 year of age.

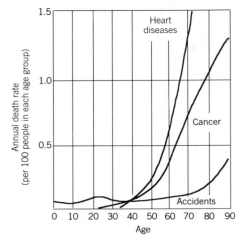

Figure 11.3 Causes of death at various ages. (Redrawn from U.S. Department of Public Health, *Vital Statistics of the United States,* Vol. 2, *Mortality.* U.S. Government Printing Office, Washington, D.C., 1968.)

Table 11.3 Age-Adjusted Death Rates for Various Cancers Observed among Male Japanese in Japan, Male Japanese Immigrants to California, and These Immigrants' Sons Born in California

Cancer Site	Death Rates (ratio to death rates in California whites)[a]		
	Japanese in Japan	Japanese Immigrants to California	Sons of Japanese Immigrants
Stomach	6.5	4.6	3.0
Liver	3.7	2.1	2.2
Colon	0.2	0.8	0.9
Prostate	0.1	0.5	1.0

Source: J. Cairns, *Cancer and Society,* W. H. Freeman, San Francisco, 1978.

[a] Ratio of the number of deaths that occurred divided by the number that would have been expected in a similar group of California Caucasians.

Second, when a common exposure unites a group of cancer victims, the time between the exposure and onset of the disease is often measured in decades in humans, and in months in rodents. These epidemiological data suggest that there is a long *latency* period between the initial event that triggers cancer and the actual appearance of a tumor.

A third peculiarity is that if one examines the incidence of cancer by specific type and site, different populations have markedly different incidence of certain cancers. If one looks at migrants, the incidence of a specific cancer shifts, in two or three generations, from the incidence of the old home's population to that of the new home's population (Tables 11.3 and 11.4). These data suggest that environmental factors are implicated in the origins of cancer.

Table 11.4 Overall Age-Adjusted Incidence Rate for All Forms of Cancer, among the Four Main Ethnic Groups in Israel in 1961–1965, Compared to the Rate in the United States

Group	Population	Annual Incidence per 100,000	
		Males	Females
Non-Jews[a]	140,000	179	93
Jews born in Israel	425,000	193	195
Jews born in Africa or Asia[b]	291,000	208	167
Jews born in the United States or Europe	352,000	294	313
U.S. population		300–400	300–400

Source: J. Cairns, *Cancer and Society,* W. H. Freeman, San Francisco, 1978.

[a] Mostly Arabs.

[b] Excludes Jews born in Israel.

The information that ethnic groups living in the same place differ in the incidence of cancer is consistent with hypotheses of a genetic basis for cancer, but can also be accounted for by *life-styles*—a term including diet, smoking, type of clothing, leisure activities, and less obvious variables. These factors may act within ethnic groups as they adapt to their new environment, or across an entire society. In the United States mortality due to several types of cancer has increased: notably lung cancer in both sexes, but also breast cancer in women, which has risen from an incidence of 1 in 20 a generation ago to 1 in 8 in 1993 (Figure 11.4). Mortality for some cancers (uterine cancer, colon cancer) has decreased as improved diagnostic methods prevent death; for other cancers, such as stomach cancer, the actual incidence has declined. It is thought that better methods of food preservation, especially the use of refrigeration instead of smoking or brining, may have caused this decline. Stomach cancer incidence remain high in countries where large quantities of smoked foods are eaten.

Animal studies on the mechanisms of carcinogenesis began in the 1930s. The first carcinogens to be identified were coal tar derivatives. These were initially painted on the skins of mice. Subsequently numerous chemicals that induced tumors in animals were identified. Some chemicals cause tumors at the site of administration (e.g., coal tars painted on the skin cause skin cancer), while others cause tumors in specific organs regardless of the site of administration (e.g., both vinyl chloride and aflatoxins cause liver cancer, even though the former is inhaled and the latter ingested in foods). Like humans, animals demonstrate a considerable latency period between exposure and the appearance of the first tumor. The latency period is a significant fraction of the life span of a species rather than a fixed interval of time. Thus, whereas the latency period in humans may be greater than 30 years, in rodents it is usually 1 to 2 years. Both the length of the latency period and the incidence of tumors can differ between chemicals and between animals of differing genotype, within and between species.

Inbred strains of mice, in which each animal is a genetic twin of all others in the strain, were developed in the 1930s for use in cancer research. Such strains were used to identify the role of viruses in the etiology of cancer. One dramatic example is the incidence of mammary tumors. The virus implicated in mammary cancers in mice is transmitted through the dam's milk and was known as the "milk factor" before the existence of viruses was recognized.[10] In female offspring of a sensitive strain, the incidence of mammary tumors can reach 95% by 1 year of age. Such a very high risk of developing particular types of cancer requires both a sensitive genotype *and* the appropriate viral infection, although either factor alone increases the incidence over uninfected, genetically resistant animals. This discovery of

[10] J. J. Bittner, "The milk influence in tumor formation," in *The Biology of the Laboratory Mouse,* 1941, reprinted by Dover Publications, New York, 1956.

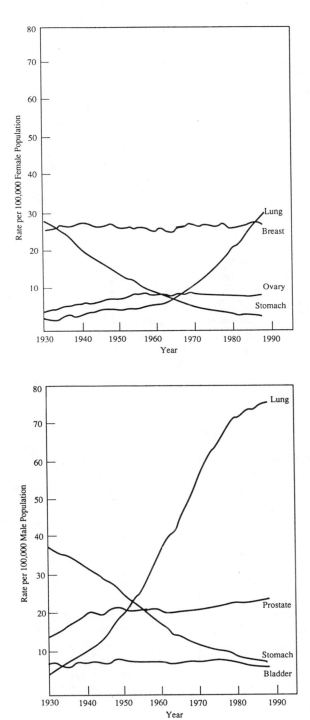

Figure 11.4 Changing incidence of several cancers in U.S. males and females. (Redrawn from *Cancer Statistics 1991*. American Cancer Society, Atlanta, Ga., 1991.)

the viral origins of animal cancers led to a massive effort to identify viruses in the etiology of human cancer: an effort that has only recently succeeded, albeit still indirectly. Until the appearance of AIDS (acquired immune deficiency syndrome), the Epstein–Barr virus was the only virus that had ever been consistently associated with a human tumor, and the association of Epstein–Barr antibodies with lymphoma was seen only in Africa.[11]

Differences in the latency period of tumor induction by different chemicals were partially explained by Berenblum, who showed that two distinct types of activity are necessary for induction of a tumor by a chemical. These activities are known as *initiation* and *promotion*. *Initiators* are chemicals that, given in increasing doses, increase the number of tumors per animal (and/or the number of animals producing tumors), but do not significantly shorten the latency period, which is the time between administration and the appearance of the first tumor. Moreover, the number of tumors produced depends on the total dose of initiator administered, almost without regard for how the doses are divided. Other chemicals shorten the latency period if they are administered after an the animal has been exposed to an initiator, but do not cause any tumors if administered alone or before administration of an initiator. Such latency-shortening chemicals also do not increase the number of tumors per animal. They are considered *promoters*. Promoters also reduce the minimum dose of an initiator that causes tumors. Despite the elegance of the model, things were never that simple. Of all chemicals with promoter activity, very few are pure promoters, and no chemical has ever been found that is a pure initiator (i.e., that produces tumors only when the animal is subsequently exposed to a promoter). Nonetheless, the activity of chemical carcinogens can be divided into those that show mostly promoting activities and those that show mostly initiating activity. In the absence of any mechanistic explanation for the carcinogenicity of chemicals, the model also provided a basis for categorizing the increasing numbers of chemicals that were shown to cause cancer in experimental animals. One might say that in 1970, the mechanisms of carcinogenesis were described by two black boxes labeled "initiation" and "promotion." The role of viruses in human carcinogenesis was doubted, although their role in animal cancers was well established.

In the early 1970s, Bruce Ames recognized a high degree of correspondence between lists of chemical carcinogens and of chemical mutagens. Besides providing the basis for the first in vitro assay for carcinogenesis, recognition of this correlation provided a framework for theories of carcinogenesis: that genetic changes in somatic cells are involved. Almost simultaneously, spectacular advances in genetics made it possible to identify specific gene sequences, analyze their products, and compare their occurrence across species and phyla. The presence of small fragments of genetic material could

---

[11] Elsewhere, Epstein–Barr virus usually causes mononucleosis.

be identified by using probes derived from the DNA of interest, which made it possible to demonstrate inserted viral genes without demonstrating infectious particles.

The correlation between mutagenic and carcinogenic chemicals was never claimed to be absolute, but early exceptions could be "explained" by metabolic differences between bacteria and mammals. For example, metals are inactive in the Ames assay. It is plausible to argue that bacteria have defenses against letting metals into the cell that mammalian cells do not have, and that access, rather than absence of interaction with DNA, accounted for negative results. As more extensive testing was carried out, entire classes of bacterial mutagens were found that cannot be shown to be carcinogenic in mammals.[12]

Equally important was the identification of known mammalian carcinogens that were simply not mutagenic in the Ames assay. A plausible rationale could be constructed in individual cases (for example, estrogen is a mammalian carcinogen, but inactive in the Ames assay, presumably because bacteria simply do not have receptors for hormones). However, many of the anomalous chemicals could not be shown to mutate mammalian DNA either. These are nongenotoxic carcinogens: chemicals that cause cancer without altering DNA.

Our understanding of the mechanisms by which such nongenotoxic chemicals induce cancer remains incomplete, but some mechanisms are now well-established. It is generally accepted that the initial step in carcinogenesis is mutagenic, but since the probability that a mutagenic event occurs in at least one cell in a tissue or organ is so very high (see the example of retinoblastoma, below) this is not a formidable obstacle to the subsequent action of nongenotoxic agents. The development of an actual tumor depends on division by the mutated cell at a rate exceeding the norm for that type of cell. For most cancers (other than childhood cancers like Burkitt's lymphoma, below), such facilitating steps are essential. They do not, however, all require DNA damage.

One very important carcinogenic mechanism that does not require changes in DNA is growth promotion. Once a cell has mutated, any environmental

---

[12] It is difficult, if not impossible, to *prove* that a bacterial mutagen is not a mammalian carcinogen since the argument can always be made that the sample size in the mammalian assay was not large enough, masking "real" increases in cancer (see Statistical Considerations under Animal Bioassays). But there are undoubtedly chemical mutagens that do not cause cancer in mammals. Caffeine, which is a mutagen in prokaryotes, yeasts, and fruit flies, is a possible candidate. There are almost certainly carcinogens that cause cancer in rats but not in humans. Arsenic is an example of the converse: a human carcinogen that causes cancer in rats only under unusual circumstances, if at all. The obvious reason why such differences occur between species is that many chemicals are procarcinogens that must be activated to become carcinogens. Since there are differences in metabolism between species, there will be differences in carcinogenic potential. Less certain is the existence of agents that alter DNA in one phylum or species, but not in another, even when access to DNA is facilitated.

inducement to more rapid cell division increases the number of mutated cells. Thus, carcinogens with estrogenic effects promote cancer in those tissues that are induced to grow by estrogen (providing that the necessary mutational change favoring unlimited growth has occurred in one of these target cells). The initial tumor therefore consists of a cluster of cells, progeny of a single mutated cell, all sharing the initial mutation. In the absence of (excess) estrogen, the mutated cell might have divided normally until the individual died of other causes. The carcinogenic role of estrogen is that it promotes the growth permitted by the initial mutation.

A second way for agents that do not directly interact with DNA to cause cancer is by producing free oxygen radicals, which can interact with many cell constituents, including both membranes and DNA. A third possibility is the induction of one or more cytochrome P-450 isozymes (see Chapter 7). Since P-450 enzymes also metabolize endogenous chemicals, inducing one or more isozymes may result in complex changes in the balance of cellular constituents.

Thus, by 1990, initiation had been redefined as activation of an oncogene or the suppression of an "anti-oncogene," and over 30 alterations in DNA that correspond to such activation or inactivation had been demonstrated. Numerous oncogenes had been characterized. The role of viruses had been identified and their elusive role in carcinogenesis explained (below). Even the seemingly nongenetic mechanisms by which hormones and nonmutagenic chemicals such as asbestos cause cancer had been explained plausibly in terms of gene activation and growth promotion. The modern synthesis of carcinogenic mechanisms is a triumphant demonstration of the interactions between various fields of biology.

CARCINOGENESIS: CURRENT MODELS

Background

The class of viruses called bacteriophages were identified and recognized as antibacterial agents in the 1930s.[13] In the 1960s bacteriophage (phage) proved enormously valuable in elucidating the principles of molecular genetics. For our purposes, the critical item is the ability of viruses to use the host's DNA to produce viral nucleic acids and viral proteins, essentially turning the bacterial cell (or other host) into a factory for the synthesis of new viral particles. Ordinarily, viruses force this synthesis of viral products until the cell lyses (explodes), killing the cell and releasing large numbers of new viruses into the environs. Such a virus infection is said to be *virulent*. It not only kills the cell involved, but also provides a means for the transmission

---

[13] As illustrated by Sinclair Lewis's novel *Arrowsmith*, whose hero attempts to cure bubonic plague with bacteriophage.

of genetic material between cells, since the virus can accidentally transcribe some of the bacterial DNA onto its own genome. The new viruses then carry this bit of bacterial information along to the next cell they infect. In virulent infections, this is irrelevant, since an infected cell dies, and the inserted material dies with it. In some cases, however, a virus will insert its genome into the host's genome, and replicate only when the host genome does. This is a *latent* virus infection. Perhaps the most familiar latent virus infections in humans are those due to *Herpes:* cold sores (*Herpes simplex I*), genital herpes (*Herpes simplex II*), and shingles (*Herpes zoster*). The characteristics of a latent virus infection include the incorporation of a single copy of the viral genome into the host genome, with occasional episodes of symptoms due to cytolysis when the virus changes to a mode of rapid replication.

It has been suggested that viruses, which cannot reproduce themselves without utilizing a host's genome, are either escaped organelles from higher organisms, or that many of the organelles in higher plants and animals (eukaryotes) are actually captured viruses. In any case, many bacteria contain latent bacteriophages. Latent viral infections can occur with or without the inclusion of extra genetic material into the new host. Such a latent infection would be quite inapparent, and quite harmless to the host cell, unless some trigger prompted the virus to turn virulent. The cell might divide repeatedly during such a latent infection, producing numerous virus-contaminated genomes. Triggers for induction of a virulent phase from a latent phase include a second viral infection (e.g., "cold sores" are so named because they tend to recur with colds or other illness), ultraviolet light, and stress.

In the 1950s, the development of molecular genetics began with the incontrovertible identification of DNA as the genetic material,[14] and the elucidation of its structure by Watson and Crick.[15] In the 1960s, the development of cell and tissue culture led to discoveries about cell growth, including the rather frustrating finding that whereas cells derived from cancers would grow and multiply essentially indefinitely, normal or nonmalignant cells become *senescent:* such cell lines replicated for several generations, but eventually died out. Only if the culture became *transformed*—that is, if the cultured cells acquired the characteristics of cultured tumor cells—would the line survive. Transformation includes changes in the shape of individual cells and in the aggregation patterns of cell clusters. Transformed cells, replaced in genetically compatible animals, cause tumors. Moreover, transformation very often includes the loss of part of the chromosome complement (i.e., there are distinct alterations in the karyotype of transformed cells).

In the 1970s came Ames's correlation between mutagenicity and carcino-

---

[14] Although many people contributed to this work, two of the critical papers are those of O. T. Avery et al. *Journal of Experimental Medicine* 79:137–158, 1944; and A. D. Hershey and M. Chase, *Journal of General Physiology* 36:39–56, 1952.

[15] The structure was announced by J. D. Watson and F. H. C. Crick in *Nature* 171:737–738 (1953).

genicity. This was the final concept needed to permit the development of a cohesive theory of the mechanisms of carcinogenesis. Today, with the identification of *oncogenes and their normal counterparts,* the pieces of the puzzle are falling in place. *Oncogenes* are normally occurring genes with ordinary functions in growth, differentiation, or metabolism, that are changed in the process of carcinogenesis. Over 20 oncogenes are known, and the number is increasing rapidly. Not all oncogenes are activated by the same mechanism. To date, the mechanisms by which oncogenes alter from their normal to their carcinogenic role include amplification, mutation, and changes in location.

Amplification means that the oncogene is somehow duplicated so that there are several copies present in the cell. An increase in the number of gene copies present in a cell and coding for the same protein may suffice to trigger a malignant transformation. The *ras* oncogene is perhaps typical of oncogenes that act by increasing production of a gene product. It is found in all organisms from yeasts to humans: an example of conservative evolution that suggests it is critical to cell function. In yeasts it is essential to the production of cyclic AMP (adenosine monophosphate), a major and ubiquitous constituent of metabolism in all organisms. The protein produced by the *ras* gene in yeasts can be identified and isolated. If *ras* protein is injected into animal cells in culture, the injected cells take on the shape typical of malignant cells and proceed to divide rapidly. The changes stop after 24 hours, the time it takes to exhaust the *ras* protein. This strongly suggests that despite its normal role in cell metabolism, *ras* protein can alter metabolic rate and/or differentiation if it is present in excess. One way of inserting duplicate copies of genes into cells is by retrovirus[16] infection, but this is still speculative. No mechanism for such gene copying has been proven yet, although it is becoming clear that temporary amplification of genes occurs normally.

Locational changes of the gene within the genome can also lead to escape from regulation, so that whereas in the normal location only normal levels of protein synthesis occur, in the new location, multiples of that level might be made. An example of increased gene product due to locational change and found concurrently with malignancy is seen in Burkitt's lymphoma. In the case of Burkitt's lymphoma, the translocation involves movement of a gene into the domain of an antibody-producing gene.[17] Understanding why a locational change leads to deregulated growth requires some knowledge of the chromosomal environment.

An antibody molecule consists of two heavy and two light chains, each consisting of a variable and a constant region. (Actually, there are 10 "constant" regions for the heavy chain and two for the light chains.) The former

---

[16] A retrovirus is a virus that uses RNA as its genetic material but can generate DNA from that RNA using host cell enzymes. Retroviruses include the HIV virus that causes AIDS.

[17] Taken primarily from C. M. Croce and G. Klein, "Chromosome translocations and human cancer," *Scientific American* 252:54–60, 1985.

are not of concern to us, but the two light-chain variants are designated kappa (κ) and lambda (λ). The genetic code for the constant region of the heavy chains is found on chromosome 14. Chromosome 2 encodes the light-chain κ region, and chromosome 22 encodes the light-chain λ region. The variable portion of the chains determines antibody specificity and is responsible for the enormous diversity of antigen/antibody response possible to the organism.

In Burkitt's lymphoma, malignant cells consistently display a translocation between chromosome 8 and one of the chromosomes encoding antibody chains. Seventy-five percent of the time the translocation is between chromosomes 8 and 14, 16% of the time it is between chromosomes 8 and 22, and 9% of the time between chromosomes 8 and 2. Moreover, it can be shown that the translocation occurs at precisely that portion of chromosome 14 encoding for the heavy chain of the antibody molecule. On chromosome 8, at the site of the "Burkitt translocation," is the *c-myc* locus, a human oncogene very similar in base sequence to an avian gene (*v-myc*) that is implicated in a B-cell lymphoma in chickens. It can also be shown that the *c-myc* gene is translocated to a site adjacent to the antibody loci in malignant Burkitt's lymphoma cells. There is *no* alteration in the *c-myc* gene sequence that is associated with malignancy, only the spatial relationship between *c-myc* and the antibody loci changes. The shifting of *c-myc* next to an antibody gene will result in an increase of *c-myc* protein production if, and only if, the host cells are antibody-producing cells. Lymph glands produce antibody, and Burkitt's lymphoma occurs in lymph glands. If the translocated chromosomes are present in fibroblasts (which do not produce antibody), the translocation does not result in increased *c-myc* protein production. Finally, there is no alteration in the *c-myc* protein as a result of the translocation of the gene. Only the quantity, not the structure, of the protein changes.

Note that these results do not imply that the *c-myc* gene is always translocated to one of the antibody-producing chromosomes; only the consequences of those translocations that *do* involve antibody loci are selected. It is obvious, however, that the significance of the translocation depends on the level of antibody production in the particular person, the suggestion being that the production of *c-myc* protein (whatever its function is) at such high levels leads to the unregulated division of the cells resulting in Burkitt's lymphoma. Because of the epidemiology of Burkitt's lymphoma, it has been suggested that the high levels of antibody being produced are related to the high incidence of malaria in these regions. Finally, the molecular evidence that suggests that this particular transformation of normal cells to malignant cells is a single-stage process resulting in cancer fits with the epidemiology of Burkitt's lymphoma: it is a childhood cancer, clearly lacking the long latency periods (often over 30 years) of cancers such as asbestos-induced mesothelioma or lung cancer.

Point mutations have also been implicated in malignant transformations. The NIH3T3 cell line, derived from a human bladder carcinoma, produces a protein that differs in a single amino acid from a protein produced by normal (nonmalignant) human bladder cells. The alteration is in the twelfth amino acid from the N-terminal end of the protein, normally glycine, which is altered to valine. The function of the protein in normal cells is unknown, but the gene and protein are structurally similar to the *ras* protein, which is associated with cancer in rats that are infected with the pertinent virus. The same quantity of protein is produced by normal and malignant cells. Thus, unlike the situation in Burkitt's lymphoma, there is no alteration in the regulation of synthesis. The alteration of the structure apparently suffices to distinguish normal from malignant cells. However, the NIH3T3 cell line originated from fully malignant cells. It is not known what other alterations occurred in this line prior to the removal of the tumor, and it is quite possible that there are numerous other differences between NIH3T3 and normal bladder cells. It is possible, for example, that the alteration in the *ras*-like protein represents a very late stage in tumorigenesis, and that one or more other genetic alterations occurred before the malignancy was expressed.[18]

Deletions have also been implicated in carcinogenesis. One good example is the hereditary human cancer of the eye, retinoblastoma. This is most often a childhood cancer, which can be cured by removing the affected eye (or eyes). In classical genetics, retinoblastoma is a dominant disease: 50% of the offspring of an affected person will inherit the disease. At the molecular level, retinoblastoma has recently been shown to be due to the deletion of a small portion of chromosome 13. For the cancer to occur, the critical segment must be deleted on both chromosomes: that is, the function of that segment of the chromosome must be totally nullified. Even the product of one copy of the gene suffices for normal regulation of cell division. Thus, at the molecular level, retinoblastoma is "recessive." Why, then, does inheritance of the syndrome follow the "dominant" pattern?

The answer seems to be that the deletion occurs with relatively high frequency. It is probable that each of us has at least one retinal cell containing the critical deletion on *one* chromosome, but as long as the *same* deletion does not recur in the *same* cell, no disease results. Because the probability that two deletions occur in the same cell is the product of their independent probabilities, a double deletion is highly unlikely—if the initial probability of a deletion is "one in a thousand" cells, or 1×10^3, the probability of the double deletion is $(1 \times 10^3) \times (1 \times 10^3)$, or "one in a million." The situation is quite different, however, for children who inherit one defective chromosome. In that case, every cell is already primed with one deletion, and a single deletion of the critical segment of chromosome 13, occurring in any cell in the retina, would cause retinoblastoma. The odds of that happening are very high, high enough that essentially every child who inherits such a defective chromosome 13 sooner or later has a second deletion and develops retinoblastoma. It should take longer for two "hits" to occur, and the likeli-

[18] The information is taken primarily from J. L. Marx, "Change in cancer gene pinpointed," *Science* 218:667, 1982.

hood of tumors forming in both eyes would be small. Sporadic cases of this disease—those without an affected parent—must be due to two mutational events affecting the same segment of homologous chromosomes. Moreover, it would be expected that two (presumably independent) mutational events occur much less frequently than one. Only an occasional child that has inherited normal chromosomes 13 would be unfortunate enough to have two deletions in a single cell, providing a "sporadic" case of retinoblastoma. The epidemiology of retinoblastoma is consistent with this hypothesis. Sporadic cases of retinoblastoma have a later age of onset, are more often unilateral, and produce fewer tumors per eye than do familial cases.[19]

Promotion and DNA Repair

The nature of promotion is still not fully incorporated into the oncogene theories. It is quite clear that many tumors require more than a single-step process to change from normal to tumor cells. Metaplasia—excessive growth of cells that remain fully differentiated, corresponding fully to the cell type of the tissue in which they are found—is often characteristic of precancerous lesions. It is hypothesized that these cells have been released from some growth controls but have not yet acquired some other characteristic required for tumor formation and/or for invasiveness. In the specialized case of pancreatic tumors in transgenic mice, it has been suggested that the additional factor is the ability to stimulate vascularization—to augment the blood flow to the metaplastic foci. Those foci that increased their blood supply subsequently became full-fledged tumors.

There is, of course, no reason to assume that all promoters act by the same mechanism. Evidence is also accumulating that some promoters are capable of damaging DNA.[20] In particular, the phorbol esters have been shown to cause oxygen radical release in phagocytes. Free radicals are well known to interact with nucleic acids, so that direct action on the phagocytic DNA would be plausible. More intriguing is the evidence that the damage is propagated by the transfer of the oxygen radicals to other cells, following the migration of the phagocytes from the site of phorbol ester application. The production of free radicals, the migration of stimulated phagocytes, and the transfer of the increased free oxygen radicals to other cells have been demonstrated; the consequences are speculative. It is also important to remember, however, that radicals can damage cell constituents other than nucleic acids, including membranes. Moreover, the carcinogenic action of estrogens can be explained by their growth-promoting activities once the

---

[19] Evidence linking the retinoblastoma gene to breast cancer and some lung cancers, and the increased susceptibility of retinoblastoma survivors to other cancers, suggest that this gene has a more general role in carcinogenesis. This link is described by J. L. Marx, *Science* 241:293–294, 1988.

[20] J. L. Marx, *Science* 219:158–159, 1983.

initial action of a mutagen has "permitted" the cell to replicate without control.

DNA repair is also implicated in carcinogenesis, primarily as a means of preventing the propagation of DNA damage that leads to cancer. The mechanisms have not been as elegantly elucidated as those for mutational events per se. Several human syndromes (fragile-X, Bloom's syndrome) have elements which suggest that DNA repair is faulty; these syndromes also carry a high risk of cancer. It is possible that for some cancers, inadequate repair corresponds to promotion, while common environmental factors such as ultraviolet light correspond to initiation in these syndromes. In other cancers, "promotion" may also require genomic alterations. Examples of probable sequences of events in multistage carcinogenesis are shown in Figure 11.5 for colon cancer, one of the most common cancers in the wealthy countries of North America and western Europe.

Recap: Carcinogenesis in Terms of Oncogenes

Having described the nature and action of oncogenes, we can now recapitulate the earlier observation on cancer and its causes in those terms.

Viruses cause cancer because viruses transport oncogenes. Such oncogenes may be normal parts of the viral genome, or genomic material picked up by the virus from earlier infections. Viruses need not be apparent in tumors, since the virus, having altered the genome, may well have been eliminated from the transformed cell line many divisions before the tumor develops. Alternatively, the virus may be so well integrated into the cell DNA that it is undetectable without the most modern technology.

Mutagens are carcinogens because alterations in DNA lead to base changes, deregulation, or amplification of oncogenes. These oncogenes have, in their unaltered state, normal and essential cellular functions, often relating to growth or cell division.

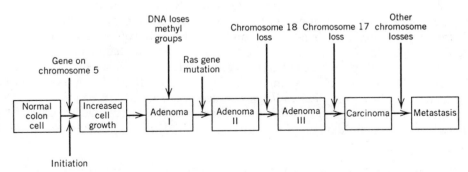

Figure 11.5 Multistage carcinogenesis: colon cancer. In the development of colon cancer, tumor suppressor genes are apparently lost from chromosomes 5, 18, and 17, and the *ras* oncogene is activated. Changes usually (but not invariably) occur in this sequence. (Redrawn from E. R. Fearon and B. Vogelstein, *Cell* 61:759–767, 1990.)

Latency between the exposure to a carcinogen and onset of tumor growth is due to at least a two-stage process of oncogene transformation or activation, corresponding to Berenblum's concepts of initiation and promotion. The number of stages required affects the frequency with which the full sequence of oncogenic events can occur in a single cell, since the frequency of the conjunction of two independent events equals the product of their separate frequencies. If each of the events is relatively rare, considerable time could be expected to elapse before they coincide in a single cell.

Transformation of cells in culture is associated with alterations in karyotype because it is the loss of regulatory loci that permits unlimited replication, which corresponds to malignancy in vivo; this also accounts for the ease of culture of tumor cells and for the tumorigenicity of established cell lines that are transplanted back to genetically compatible hosts. It does *not* explain why senescence occurs in normal (nonmalignant or untransformed) cells, nor does it explain how cells know when they are "old" and should die. The arguments favoring somatic mutation as the initiating step in cancer are:

1. Cancer arises abruptly and irreversibly.
2. The cancerous state is heritable *within a cell line*.
3. Cancers are clonal in nature.
4. Cancers are selected by their environment.
5. Many chemical carcinogens are reactive electrophiles or give rise to reactive electrophiles in the body.
6. Most chemical carcinogens react with DNA.
7. There is an excellent qualitative correlation between mutagens and carcinogens.

CASE HISTORY: BENZO[A]PYRENE

Benzo[a]pyrene (Figure 11.6) specifically, and polycyclic aromatic hydrocarbons (PAHs) in general, are ubiquitous contaminants of the human environment. Since they are the inevitable accompaniment of incomplete combustion, PAHs are found in crisply fried or broiled foods as well as in the smoke of incinerators, fireplaces, and cigarettes. It is therefore appropriate that PAHs were the first recognized environmental carcinogens. In eighteenth-century London, Sir Percival Pott (1714–1788) correctly traced the high incidence of scrotal cancer among chimney sweeps to the soot accumulated in their scrotal folds. Despite this early identification, PAHs probably remain the most prevalent of environmental carcinogens. They are also among the most thoroughly studied.

Formation and Occurrence.   Benzo[a]pyrene (BaP), as well as a large number of other polycyclic aromatic hydrocarbons (PAHs), are formed during the incomplete combustion of virtually all materials (pyrosynthesis). Important sources in the modern human environment include fires, coal combustion and conversion

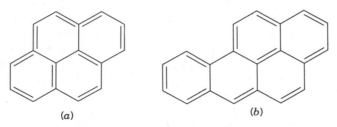

Figure 11.6 Structure of (*a*) pyrene and (*b*) benzo[*a*]pyrene.

industries, cigarette smoke, and internal combustion engine exhaust. The precise chemical mechanisms of pyrosynthesis are not known, but it appears that PAHs are built up from carbon units equivalent to acetylene, ethylene, and their free radicals. Formation typically occurs between 600 and 900°F; it has been suggested that tobacco smoke BaP is formed preferentially from plant terpinoids. PAHs are universally distributed in the environment. Airborne particulate BaP in New York City was recorded at levels up to 50 ng per 1000 m³ of air in 1972. A typical cigarette yields 25 ng BaP in its smoke. PAHs universally contaminate soils and sediments at levels between 100 and 1000 μg/kg, and levels up to 500 mg/kg soil have been recorded near oil refineries. PAH occurs in sediments that predate the Industrial Revolution and occur in all smoked foods, charcoal cooked foods, burnt toast, and so on. The estimated annual U.S. emission of BaP is 500 tons.

Physical and Chemical Properties.   PAH with four or more rings are virtually insoluble in water and poorly volatile. The lower homologs are more volatile and all are easily steam distilled. Many PAH are intensely fluorescent and readily photo-oxidize.

Biological Effects.   In 1776, Percival Pott reported the association of scrotal cancers with the occupation of chimney sweep, doubtless due to poor cleanliness and PAH in chimney soot. Some, but not all, PAHs induce cancer in experimental animals and humans. For example, pyrene is noncarcinogenic, whereas BaP is one of the most active carcinogens (Figure 11.7). All carcinogenic PAH are also mutagenic; noncarcinogenic PAH are nonmutagens. BaP carcinogenesis has been studied more intensely than that of any other PAH and is one of the best understood chemical carcinogens with respect to the structural basis of activity.

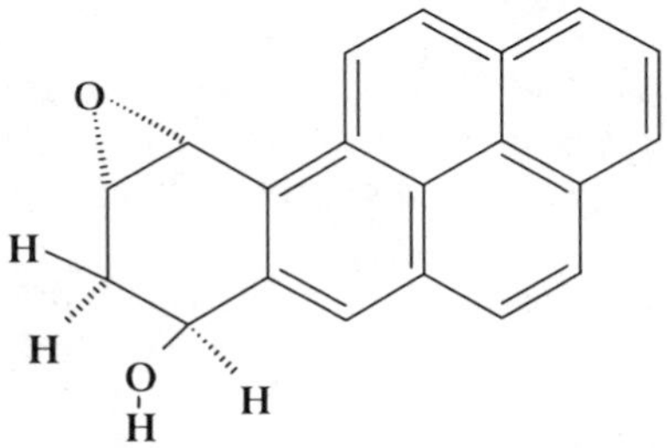

Figure 11.7 Proximate carcinogenic metabolite of benzo[*a*]pyrene.

Proximate Carcinogen. It is known that a three-step transformation of BaP is required to produce the proximal carcinogen, the (+)-benzo[a]pyrene-7,8-diol-9,10-epoxide. This compound attacks DNA. Other carcinogenic PAH all share the characteristic "bay region."

BaP is a complete carcinogen, containing both initiator and promoter properties, but the diol-epoxide mentioned above is active primarily as an initiator. The promoter activity of BaP must result from one or more of its other metabolites, over 30 of which are formed by mammals. A single dose of as little as 50 μg of BaP can induce tumors in 60% of treated mice if followed by promoter treatments. About 90% of the carcinogenic activity of tobacco smoke has been attributed to BaP alone. If for no other reason than its presence in cigarette smoke, this compound probably accounts for a large fraction of human cancers.

CARCINOGENESIS TESTING

Cancer is arguably the most dreaded disease in the developed world. A large fraction of all biomedical research is devoted to cancer: its causes, treatments, cures, and prevention. Similarly, a very large fraction of toxicologic research, and an even larger fraction of regulatory efforts, are devoted to identifying carcinogens. Last, but far from least, there are serious disagreements about the appropriate policy to pursue with respect to testing chemicals for carcinogenic potential and with respect to interpretation of the data. It is therefore worth considering the problems that are unique to carcinogenesis testing as well as the various solutions that have been proposed for these problems.

There are important conceptual differences between "classical" toxicology and genetic toxicology, including carcinogenesis. In classical toxicology it can be assumed both that toxic effects appear soon after exposure and that the intensity of induced pathology is directly correlated with the intensity of the exposure *in that individual.* Moreover, manifestations of toxicity are due to altered functional products and/or degenerative changes and/or cell death of target cells: that is, there are *terminal toxic effects on exposed cell populations.* The process of carcinogenesis differs from this classical model because the effects of the exposure are delayed, because the intensity of the effect in any one *individual* is independent of the intensity of that individual's exposure and it is the *probability of the effect* that increases with the intensity of the individual's exposure. That is, it is the frequency of the effects (e.g., tumor incidence) in the exposed population rather than their intensity in individuals that is correlated directly with the intensity of exposure. Finally, the manifestations of carcinogenicity are due to the proliferative response of a new, altered cell population, and the existence of such an altered population presupposes survival of the affected cells. Moreover, because the affected cell reproduces itself, the effects are *self-replicating.*

In practical terms, these differences mean that a carcinogen does not produce gradations of illness as exposure increases: all cancers, even the very easily treated common skin cancer, are lethal eventually if untreated.

Only the *probability* of the effect increases. For example, in houses built above geologic strata that allow a high percolation of radon, people are exposed to increased levels of ionizing radiation from the products of radon decay (see Chapter 4). When these houses are identified and the radon levels catalogued, one can make predictions of the probability that the inhabitants will get lung cancer. The probabilities range up to 1 in 10 (10%), equivalent to the risk of a pack-a-day smoker. But the *severity* of the disease for people living in these highly contaminated houses is no greater than it is for the 1 person in 10,000 who gets lung cancer from much lower levels of radon. Finally, removal of *most* of the problem cells is not enough, since *any* altered cells can reproduce and regenerate the tumor.

Moreover, the combination of delayed onset and low probability of disease in any one individual (for most carcinogens at common levels of exposure) means that one cannot identify the carcinogenic action of a substance for years or decades after the exposure. This makes it difficult to determine which chemicals are carcinogens. There are disadvantages to all testing methods: either time, or cost, or worry about their reliability makes them suspect.

Methods for Identifying Carcinogens

Epidemiology. This is the most reliable way to identify carcinogens, since it is based on human incidence of the disease. It is always retrospective and therefore usually 30 years after the fact in identifying carcinogens. This makes it somewhat impractical as a method of controlling the introduction of numerous new compounds, many of which will be produced at high volumes, unless emphasis on productivity far exceeds concern for human health. Epidemiological studies are also very expensive. Finally, although they can identify potent carcinogens to which clearly defined populations are exposed, retrospective epidemiological studies are not very good at detecting agents that affect less than 10% of exposed persons, those that are widely dispersed, or those for which exposure levels are difficult to determine.

A major controversy in toxicology is the extent to which synthetic chemicals are actually increasing the incidence of cancer in our time and our society. Extreme viewpoints abound, and it is almost impossible to avoid polemics. Among the factors that make the truth difficult to discern are the heterogeneity of causes of different cancers, the relative rarity of many kinds, and the very high levels of tobacco-associated lung cancer. If a cancer is rare, a large population must be screened to find enough cases for a *retrospective* study, in which people with a certain cancer are compared to people without cancer and/or different cancers to look for plausible causes. Most such studies can be criticized (are flawed, say those who disagree with the results) because of differences between controls and exposed individuals, because of uncertainty about actual levels of exposure, or even because present exposure levels are significantly lower than those in the study. The more reliable *prospective* study, in which populations exposed and unexposed to a suspect carcinogen are monitored until enough deaths occur to analyze differences between the groups, is prohibitively expensive for rare outcomes and may take decades to complete.

In addition, changes in many facets of our lives have accompanied the rapid increase in production of synthetic chemicals. Our diets, our exercise patterns, and our leisure activities have all changed. Each of these changes can interact with synthetic carcinogens. If, for example, smoking not only causes lung cancer but also produces mutagenic compounds that are excreted through the kidney, an increase in bladder cancer might result. This could be confused with a carcinogenic effect of, for example, an artificial sweetener introduced at the same time as smoking increased. Thus it is difficult to disentangle causes, especially 20 to 30 years later, which is a reasonable interval between exposure to a carcinogen and visible cancer.

Even when large numbers of people have been exposed to very high levels of potent carcinogens such as asbestos and cigarettes, consensus is hard to reach. In the case of the proven carcinogen asbestos, manufacturers have argued that asbestos is merely a "cocarcinogen" and that smoking is the actual cause of the lung cancers seen in asbestos workers. This argument was abandoned only when the occurrence of mesothelioma was shown to be uncorrelated to smoking. Cigarette manufacturers, in their turn, argued for decades that the association between lung cancer and smoking was spurious.[21] Since large sums of money, as well as reputations, ride on the decisions about the carcinogenicity of a product, this area of research will always be fraught with controversy.

In a scathing indictment of the conduct of epidemiological investigations, Feinstein[22] noted that many epidemiologists, rightly assuming that randomization—often considered essential in laboratory studies[23]—was not possible, also threw out all other basic elements of the scientific method.[24] In

---

[21] Note also that in 1993, the U.S. government still subsidizes tobacco growing within the United States and the Bush administration pressured foreign governments to accept imports of U.S. cigarettes. The Clinton policy has not been clearly stated yet.

[22] Feinstein, *Science* 242:1257–1263, 1988.

[23] Even in such studies, "randomization" often means little more than haphazard selection. For example, it is often assumed that to pick mice (or rats) out of a cage in no particular order is to "randomize" their assignment to dosage groups. In fact, more aggressive or healthier animals may be nearer the opening; sick animals might huddle in the dark corner. Alternatively, healthy animals might evade capture, while sick animals are easy to grab. In neither case are animals assigned randomly. To avoid manipulating tables of random numbers, it is often preferable to select animals systematically (e.g., so that weights are evenly distributed between treatments).

[24] Note that Feinstein's article dealt with the proliferation of studies showing that various "life-style" factors increase the risk of breast cancer. The same arguments can be applied to studies of the relationships between synthetic chemicals and cancer. It is somewhat disappointing, then, that few toxicologists apply the same standards to all epidemiological studies. There is too often an inverse correlation between how fervently a given writer believes in the epidemiological evidence for life-styles as the cause of cancer and synthetic chemicals as a cause for cancer, even though the same types of epidemiology support both.

particular, he notes that many epidemiological studies are characterized by the absence of an a priori hypothesis, by a multiplicity of analyses, and by modification of control groups in the course of the analyses (with the result that the study actually lacks a well-specified cohort). In addition, many epidemiological studies ignore dose–response quantification and neglect underlying differences in medical care. In some cases these flaws are difficult to overcome. For example, social drinking is more frequent in the upper middle class; at the same time, a higher incidence of self-reported, and therefore more often nonlethal, breast cancer may be due to better detection in this same upper middle class. It is possible that an observed correlation between moderate alcohol consumption and breast cancer is due to the association of moderate drinking with good medical care rather than with a causative role in breast cancer. Other flaws may be due to enthusiasm: for example, an association between reserpine and breast cancer in one epidemiological study may have been strengthened by removing from the control group precisely those patients with conditions for which reserpine might have been prescribed.

In sum, epidemiology is not a reliable tool for identifying human carcinogens, except under unusual circumstances. The inherent difficulty of studying oneself, as well as the difficulty of studying events that occurred decades ago, make it preferable to use epidemiology to confirm suspicions already raised by animal studies. Alternatively, one can use suspicions raised by epidemiological investigations to seek confirmation (and ideally, a mechanism of action) in animal studies.

Mammalian Bioassay. This is the oldest and best validated method of identifying carcinogens. It takes 3 years per compound, and the 1982 cost was approximately $500,000 per chemical. Its significance is usually disputed when the assay is weakly positive (e.g., saccharin). Because of its importance, the mammalian bioassay is discussed at length below.

Short-Term Assays. These assays, which include the Ames assay, sister chromatid exchange assay, and mouse micronucleus assay, are quick and therefore relatively cheap. They are also repeatable without exhausting the budget or personnel. Extrapolation is from observed mutagenicity (whether in bacteria, yeast, or corn) or chromosome damage (in eukaryotes such as yeast, or in mammalian cells) to (presumed) carcinogenicity. Further extrapolation is required between species if the Ames assay or other microbial assay is used. Use of these assays assumes that sensitivity to mutagenic activity can be predicted between species and that mammals and bacteria have the same detoxification systems. Use of human cells raises other problems, which are not yet fully delineated. In sum, the reliability of the short-term assays is uncertain: estimates of their reliability vary between 50% (chance) and 90% if only one short-term test is used. A reasonable, perhaps slightly optimistic guess is that one can identify carcinogens with 80 to 90% accuracy if a battery of short-term assays is used.

Procedures for Carcinogenesis Assays

Short-Term or In Vitro Assays.   The most common short-term test is the Ames assay, named for its inventor, Bruce Ames. It uses a strain of the bacterium *Salmonella typhimurium* that has been carefully altered to be highly sensitive to mutagens and unable to repair damage to its DNA. This sensitivity is incorporated in four ways:

1. The lipopolysaccharide coat of the bacteria is made more permeable via a mutation known as *rfa*, deep rough.
2. The UVRB gene, needed to repair damage caused by ultraviolet light, is deleted. UV light causes the formation of thymine dimers; these are ordinarily repaired by the function of the UVRB gene.[25]
3. The bacteria are *his⁻* mutants. *His⁺* is a gene that allows bacteria to synthesize the amino acid histidine from glucose and a nitrogen source. *His⁺* is the normal (or wild-type) gene: normal bacteria can make their own histidine. *His⁻* is a mutation that makes a bacterium dependent on an outside source of histidine—it can no longer synthesize its own. Such a cell is a mutant. A mutation from *His⁺* to *His⁻* is called a *forward mutation* because it is a change away from the normal or wild type. A *reverse mutation* takes a mutant back to the normal genotype. A bacterium that has gone from *His⁻* to *His⁺* is called a *revertant*.
4. The bacteria contain a plasmid, pkm10,[26] that carries ampicillin resistance and SOS repair genes. The SOS repair genes are a system of last resort to produce a functioning DNA molecule and save the cell when error-free repair systems break down. These SOS systems tend to result in an increased mutation rate due to errors (especially incorporation of adenine) during repair.

In the Ames assay, bacteria are cultured with and without the chemical to be tested, and with and without S-9, the microsomal fraction of mammalian liver that carries xenobiotic-detoxifying enzymes.[27] The S-9 fraction can transform promutagens into their active form, which would occur in vivo in the mammalian liver. In a standard assay, 5×10^8 cells are cultured in each tube or flask. The nutrient medium does not contain histidine, so only reverse mutants (*his⁺*) will grow. The number of such revertants in the treated (+chemical) flask, relative to the number in the control (−chemical) flask, will identify the mutagenic potency of the chemical. The difference between

---

[25] The human genetic disease xeroderma pigmentosum also prevents repair of UV damage and results in skin cancer at a very early age.

[26] Plasmids are DNA-containing, independently replicating organelles within bacteria. Some plasmids replicate only when the bacterium does; others reach high numbers inside single bacteria.

[27] These xenobiotic-metabolizing enzymes or *mixed-function oxidases* ordinarily function to render lipid-soluble chemicals more water soluble, since increased water solubility enhances excretion. In some cases, however, the initial stages of the biotransformation result in increased toxicity to the organism. One of the best examples of such activation is the P=S to P=O activation of organophosphorus esters (see Chapter 8).

the +S-9 and −S-9 flasks provides evidence that the mutagen is direct acting, if it is equally active in both, or needs to be activated, if it is inactive in the absence of S-9. *It is extremely important to include controls in every assay.*

Because the Ames assay uses bacteria, it is insensitive to chemicals against which bacteria have defenses. Metals, which do not ordinarily get into bacteria, are almost invariably negative in the Ames assay because they do not reach DNA. The Ames assay is also insensitive to carcinogens that target special molecules or organelles (receptors) that are not present in bacteria. Thus hormonal carcinogens are negative in the Ames assay: bacteria have no hormone receptors. Other short-term assays have been devised to overcome these shortcomings. Most depend on mammalian cells in culture (e.g., Chinese hamster ovary assay), but some can be adapted to in vivo dosing, followed by culture, or merely scoring, of the exposed cells (e.g., mouse micronucleus test).

All of the short-term assays depend on the assumption that carcinogenesis begins with, and is dependent on, mutational events. The particular DNA markers that are used include not only the reversions (due to deletions or substitutions) of the Ames assay, but sister chromatid exchange (Chinese hamster ovary assay) or chromosome breakage (mouse micronucleus assay). In reality, then, the short-term assays are excellent assays for mutagenicity and/or chromosome damage. As assays for carcinogenicity, they require some extrapolation.

Decisions about which form of the chemical (pure or technical grade) should be used are the same as for animal bioassays. In the case of lipophilic substances, serious consideration must be given to the vehicle used to solubilize the chemical, and a vehicle control must obviously be included in both the +S-9 and the −S-9 fractions.

Animal Bioassays

Animals.   The most commonly used species is the rat, followed by the mouse. There are several choices that must be made, notably of the species, of the strain within species, and whether the strain should be inbred, random bred, or uniform hybrid lines. Inbred strains are most readily available in mice. Rats and hamsters are the other two species most suited for cancer bioassay. There are fewer inbred lines in these species, but the same rules apply.

An *inbred strain* is one that has been bred for at least 20 generations by brother × sister (or parent × offspring) matings. Animals within such a strain can be presumed to be genetically identical in 99.5% of their genome.[28] The

---

[28] Note, however, that this genetic similarity begins to decrease *as soon as* brother × sister matings stop. Two colonies of an inbred strain that have been separated for 20 generations are recognized as genetically distinct sublines and should be so designated. Thus the inbred strains maintained at the Jackson Laboratory in Bar Harbor, Maine, all carry the designation "/J." If I were to buy inbred PL/J mice and inbreed them in my laboratory, after 20 generations of inbreeding they might be given the designation PL/J/UI.

relationship is that of identical human twins. Such strains may be highly sensitive but are very unlike the genetically highly outbred and heterozygous human population. Moreover, because many of the inbred strains of mice were selected for sensitivity to particular forms of cancer, a bias is introduced.[29] On the other hand, their use means that one can hold genetic variation to a minimum. *Random-bred animals* are generally derived from a small gene pool; even though their inbreeding is not defined, their similarity to human random breeding is very dubious.[30] However, if obtained from reliable breeders, they provide phenotypically similar animals on a genetic background selected for good laboratory behavior. *Hybrid animals,* if constructed by matings between two carefully chosen inbred lines, offer genetic identity between individuals with maximum heterozygosity at most loci. The National Cancer Institute's standard mouse line is a cross between the C57Bl/6 inbred strain, developed as a low-cancer line, and the unrelated C3H (high-cancer) line.

Route of exposure is best chosen to mimic human exposure, if feasible. Nevertheless, although the most common route of human exposure is probably by inhalation, the most common dosing regimens are oral or intraperitoneal. Inhalation studies are hazardous to technicians unless highly sophisticated equipment and extreme precautions are taken; other experimental exposures are often substituted.

Treatment Level. It is usually necessary to choose the maximum tolerated dose to minimize the size of treatment groups (see below). There are valid objections to this method. The most serious objection is that extremely high doses may overload enzyme systems and result in alternative metabolic pathways being utilized. The assumption then is that at doses closer to realistic levels of exposure, normal metabolic pathways would remove the chemical without producing carcinogenic intermediates. In this view, the overload doses lead to the use of alternative pathways of detoxification, and indirectly to long-range toxicity. The realistic significance of such a possibility is not known for carcinogenesis, but metabolic "shunts" are known. An

---

[29] This bias may be desirable. In the event that the chemical being tested is to be widely distributed, one may wish to investigate the risk to sensitive populations. Some inbred strains of mice, notably C57BL/6 and C57BL/10, were bred as controls for the cancer-sensitive strains and have a low spontaneous incidence of cancer. To use them might be to bias the assay against detection of weak carcinogens.

[30] Nor can it be assumed that all random-bred "strains" are genetically equivalent. Many or all of the large commercial breeders maintain closed colonies at each locality. There is no guarantee that 20 years later, the animals from North Carolina will have the same genetic makeup as those bred by the same company and designated by the same code but bred in New York. At recent meetings of the Society of Toxicology, there has been serious concern over the changing characteristics of a major producer's "outbred" rat. Litter sizes and body weight in this strain have increased, and male fertility has declined. Mean life span has decreased to such an extent that several carcinogenesis assays were compromised.

additional aspect of uncertainty in the use of high doses is that statistical extrapolation outside the range of doses used is a dubious proposition. Figure 11.8 illustrates possible effects of using different models to extrapolate from the high doses used in mammalian bioassays to the much lower levels typical of environmental exposures. Given three data points from a laboratory study, quite different risks are estimated using different models. If a one-hit model is used, dose response should be linear; two- or multihit models produce curves. Plot (a) illustrates the estimated lifetime risk of cancer from exposure to 100 mg/kg per day assuming a linear dose–response relationship without threshold: 10% incidence of cancer is equivalent to smoking a pack of cigarettes per day for 20 years. Plot (b) represents a linear dose–response relationship with threshold: there is no cancer risk below the threshold. Plot (c) illustrates a nonlinear extrapolation, assuming that there is no threshold for carcinogenesis. A small but definite risk of cancer remains. Plot (d) illustrates a nonlinear dose–response relationship with a threshold: risk of cancer is nonexistent below approximately 375 mg/kg per day, but risk rises very rapidly thereafter. Other models are possible. The question of appropriate models and of thresholds is fraught with tremendous controversy, since small differences in estimates of carcinogenicity translate into millions of dollars. The most reasonable opinion is that most cancers are multistage and that carcinogenic chemicals with thresholds are exceptions. Thus the most plausible model is a curvilinear model without threshold, as shown in plot (c). Note, however, that other curves could be used; guesses about underlying distributions are even more fuzzy.

Chemicals. It is necessary to choose between using the pure or the technical form. Use of the latter mimics workplace and environmental exposure; use of the former minimizes the possibility of being misled by active contaminants that vary between batches. The World Health Organization finally decided that both forms must be tested for acute toxicity, but the cost of including both forms in mammalian carcinogenesis assays is probably prohibitive. In either case it is necessary to monitor quality control between batches of a chemical, whether synthesized or bought, and also to monitor the condition of a chemical with time whether it remains on the shelf or is in the vehicle of administration.

Statistical Considerations. The probability of *identifying* a carcinogen from animal data depends on at least three factors: the size of treatment-induced effect, the incidence of effect in controls, and the number of animals in control and treated groups. Consider that the U.S. population is over 250,000,000 people. Therefore, an incidence of "one in a million"—our cheerful expression of vanishingly small odds—translates into 250 cases of whatever. This represents more cases than were seen in the "epidemic" of toxic shock syndrome. But to *detect* the effects of weak carcinogens, those that would cause "one case in a million per year," we would need to test thousands or millions of mice if we used environmental levels of exposure in our assays. Such "megamouse" experiments have been considered. But for purely practical financial reasons, the alternative has been selected: the National Cancer Institute tests use the highest feasible dose to minimize sample sizes and depends on statistics to extrapolate to the expected frequency of tumors at human exposure levels.

Extrapolation requires models. Alternatives include *one-hit* (linear) models with or without threshold, and *multistage* (curvilinear) models in which tumor incidence depends on the number of stages involved between normal cell and actual cancer. Here the curve is $y = x - pn$, where n is the number of stages, y the tumor incidence, and x the dose. These models are illustrated in Figure 11.8.

Thresholds.   The existence of a threshold implies that there is a safe level of exposure at which no toxic effects are found. The existence of thresholds is axiomatic in classical (acute) toxicology, as discussed in Chapter 7. The theory of somatic mutation as the initiating event in carcinogenesis implies that there is no threshold, since a single damaged DNA molecule could lead to a tumor and a single molecule of a mutagen suffices to damage that molecule of DNA. In practice, the probability that a single molecule actually reaches the DNA without prior metabolism *and* damages the DNA *and* that the damage is not repaired before subsequent steps in the carcinogenic process are triggered are so low as to be negligible for any given individual. More realistically, metabolism of mutagens present at very low levels might be too rapid for the compound to reach its target (DNA), and/or the damage is repaired, and/or the cell dies from other causes, and/or that cell is not the target of subsequent carcinogenic changes. Any or all of these assumptions lead to the existence of an *effective* threshold. However, across 250,000,000 people, even the "extremely low" probability can be expected to occur. Moreover, data from animals do not support the existence of thresholds, although such data cannot disprove their de facto occurrence.

Despite the wobbly theory, the regulatory significance of the existence or nonexistence of thresholds is enormous (cf. the DES case history in Chapter 12). This is an aspect of toxicology that we have not considered at length previously, although it was implicit in the case history on leptophos—that the data which toxicologists gather are used to shape public policy. These policy decisions are nominally based on the data but also take into account other factors, including cost, social policies, and the claims of competing interest groups about the benefits to be obtained from using or not using specific chemicals.

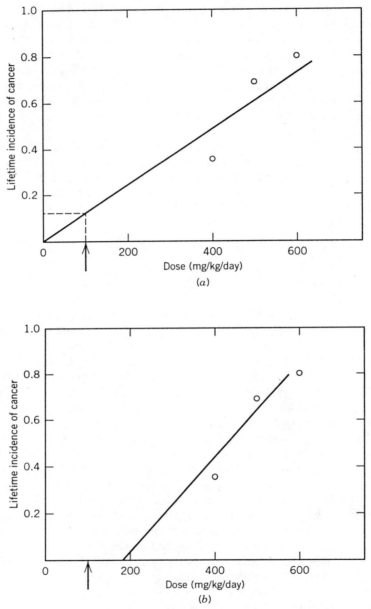

Figure 11.8 Implications of using different models to extrapolate data to lower levels of exposure: (*a*) linear, no threshold; (*b*) linear, threshold; (*c*) nonlinear, no threshold; (*d*) nonlinear, threshold. Arrows indicate level of human exposure.

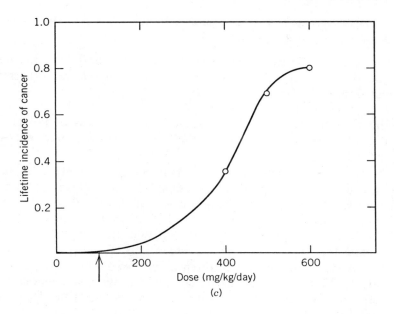

(c)

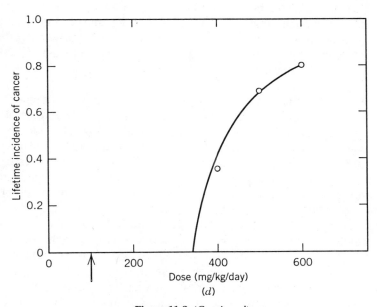

(d)

Figure 11.8 (Continued)

Policy Considerations in Carcinogenesis Testing:
Criticisms of Animal Bioassays

There is a serious controversy in the toxicology community about the validity of mammalian bioassays as predictors of carcinogenicity. The criticisms can be divided into a few broad categories.

1. Many of the models are too sensitive. The example used to substantiate this claim is, most frequently, the high incidence of liver cancers in inbred lines of mice.
2. Differences in metabolic pathways between species make extrapolation of data to humans invalid. For example, humans may be insensitive because the proximate carcinogen is a metabolite that is not formed in the human system; or the use of high doses may shunt metabolism into rarely used metabolic paths that include the formation of otherwise nonexistent carcinogenic metabolites.[31]
3. The route of administration often differs from human exposure; as a result, different and unrealistic quantities of the test compound reach target organs; for example, the lungs might detoxify an inhaled chemical that if given by mouth, causes stomach cancer.
4. The cost of false positives is too high. Many valuable drugs and/or industrial compounds may be kept off the market because of false labeling as carcinogens. The converse criticism is raised by quite different people, who cite the case histories for vinyl chloride and DES (see Chapter 12), among others.

Given the existing assays, a range of possible strategies can be used to identify (and subsequently to regulate) carcinogens. First and simplest, one can decide to do nothing unless there is epidemiological evidence that an agent causes cancer in humans, because "the proper study of mankind is man"[32]; and a chemical is a *human* carcinogen only if it causes cancer in humans. This would make all testing unnecessary. At the other extreme, one can simply ban all new chemicals because the risks are too great. These options are straw men, since it is obvious that our society is neither callous enough to repeat the asbestos saga deliberately (Chapter 4), nor so sated with progress that we would risk losing a new polyvinyl chloride. It is quite clear that for the foreseeable future, we will develop new chemicals—as plastics, as industrial intermediates, as pharmaceuticals, and as food additives—and that we will test those new chemicals for carcinogenic potential. The only real decisions are which tests are adequate and how much we are willing to spend on testing.

---

[31] It should be noted that the opposite argument—that there may be false negatives among the animal testing data—is rarely made by the same people who argue for the existence of false positives. And vice versa.

[32] Alexander Pope, *Essay on Man.*

The major options in carcinogenesis testing are animal bioa
so-called in vitro or short-term assays. Animal bioassays ai
labor intensive, and take 3 to 5 years before results are obi
reliability is questioned because there are far more animal carci
known human carcinogens. On the other hand, there is no cle
that any potent animal carcinogen is *not* a human carcinogen. It one looks
at the converse, the only proven human carcinogens that is not also an animal
carcinogen is arsenic. Thus it is plausible to argue that a positive result in
a mammalian carcinogenesis assay is strong evidence of human carcinogenesis. It is equally plausible to argue that there could be qualitative differences
between species, since metabolic differences between species can result in
formation of an active carcinogen in one species but not in another.[33] In
contrast to mammalian bioassays, in vitro assays are quick, relatively inexpensive, and of dubious reliability. Whereas the major objections to animal
bioassays are that they are prohibitively expensive, given that they do not
provide absolute certainty of human toxicity, the major objection to short-
term assays is that they are unreliable if not irrelevant.

Several policy options can therefore be considered. It is possible to carry
out animal bioassays on all new chemicals before they are marketed. This
is a very expensive option, but probably closest to the current policy for
pesticides, pharmaceuticals, and other consumer chemicals. It is extremely
expensive and cumbersome. Alternatively, one could carry out short-term
assays for all new chemicals and consider their results conclusive. There
are also two possible mixed strategies. If one chooses to maximize chemical
production, one can insist on short-term assays and consider a negative
result tantamount to a permit for the chemical to be marketed. If the short-
term assay is positive, the producer would have to carry out an animal
bioassay. Only if the chemical is positive in the animal bioassay as well as
in the short-term assay would the compound be banned. To minimize the
probability of marketing carcinogens, a positive result in a short-term assay
would be grounds to forbid marketing, while a negative result would require
a subsequent animal bioassay to be carried out. Only a negative mammalian
bioassay would constitute the go-ahead for production.

Which of these options is used depends on issues outside the laboratory.
Policy decisions may be swayed by considerations of economic "competi-
tiveness," which usually means that industry is given a greater leeway; or
by public concern, which when it reaches a crescendo generally results in
banning some chemical that has had the ill luck to be the "carcinogen of the
month." Some examples of the political processes involved in policy decisions are illustrated in the next chapters.

---

[33] No chemical is known to differ in this way between laboratory animals and humans. However,
human data—on carcinogenesis and on intermediate metabolites—are simply too sparse for
conclusions to be drawn. As more and more information becomes available on the often complex
sequence of genetic alterations that lead to actual cancer, the more likely it becomes that
qualitative as well as quantitative interspecies differences exist.

12

POLICY IN ENVIRONMENTAL TOXICOLOGY

The hand that signed the treaty bred a fever,
And famine grew, and locusts came;
Great is the hand that holds dominion over
Man by a scribbled name.

—Dylan Thomas[1]

INTRODUCTION

The scientific method theoretically requires a scientist to be dispassionate, impartial, and aloof from partisan issues. Although this is certainly critical in the formulation of hypotheses and the interpretation of data, it is ridiculous to assume that scientists can remain uncommitted about the *results* of their work, even when that work consists of laboratory experiments testing a hypothesis. The strong tendency for laboratory scientists nonetheless to remain apart from the political processes that determine environmental policy is to a considerable extent inevitable: a partisan advocate will not be believed by the uncommitted. Therefore, to retain credibility, it behooves the laboratory scientist to avoid being associated with specific regulatory viewpoints. It is essential, however, that scientific impartiality not become naiveté or ignorance. Regulatory decisions are not made in a vacuum. Laws are drafted and passed within a framework shaped by society and its elected leaders, but environmental laws often originate with data gathered in a laboratory. Regulations require interpretation of experimental results within a framework determined by the society that frames the laws and by the structure of the agency interpreting those laws. Demonstrating that a chemical has a toxic

---

[1] From the poem "The Hand That Signed the Paper," in *Major British Poets*, O. Williams, ed., Mentor Books, New York, 1963.

effect in laboratory animals is the beginning of the processes known as risk assessment and risk management.

Risk assessment is the determination of the degree of hazard posed by a chemical to the population of interest. Most often, this population is human. Occasionally, wild species are considered, but this is far more common for engineering projects, which may destroy significant segments of habitat. Relatively few environmental toxicants have *no* useful function—and those few, such as 2,3,7,8-tetrachlorodibenzodioxin, are unavoidable by-products of the synthesis of more useful chemicals. Therefore, every decision to ban a chemical carries a cost or a risk of its own. DDT stopped a typhus epidemic in Naples in 1944. Had a risk–benefit analysis been carried out prior to its use, the equation might have balanced the risk of cancer in exposed persons (let us say, 10 per million) against an estimated 50% risk of death from typhus in those who develop the disease. If 10% of the population had come down with typhus, 5% of the population would have died. Clearly, the risk of spraying everyone with DDT was less than the risk posed by a typhus epidemic.

Risk management is the policy for dealing with an estimated risk. If it is decided that the benefits of an acne treatment outweigh the risks of exposing fetuses to a teratogen, how shall we minimize the numbers of malformed babies? What restrictions can be placed on the drug to ensure that it is not taken by pregnant women? When it was realized that the insecticide endrin is extremely toxic to fish, risk management limited endrin use to the Great Plains, on the assumption that it would not reach water in the dry lands west of the Mississippi.

It is of course striking that when irreplaceable benefits are claimed for the continued production of a toxicant, the people at risk do not benefit, and the people who benefit are not at risk. Velsicol stood to profit from selling leptophos, but neither the doctor who dismissed nerve damage nor the executives who denied the occurrence of OPIDN were exposed to leptophos (see Chapter 8). When a representative of the Pear Council said that the fumigant DBCP was essential to continued growing of pears and suggested that people who did not want children might be recruited to be DBCP applicators, he could ignore the known carcinogenicity of this soil fumigant—he was not planning to apply it himself.

Leptophos, DDT, and DBCP are not sold in the United States today, and few products still contain asbestos. In many cases, however, we accept risk and unpleasant consequences: many cancer chemotherapeutic drugs are themselves carcinogenic, and most are given at debilitating doses. Those risks are acceptable because the alternative is death. A different problem is posed by *Acutane* (see Chapter 9), the antiacne medication that causes birth defects. More males than females suffer from disfiguring acne. Should they be denied a drug because if women take it while pregnant, the fetus has a 33% chance of being severely malformed?

The case histories presented so far have been selected because they represent clear-cut scientific data about the environmental or health effects of chemicals. The costs, or risks, have clearly outweighed the benefits. Some inkling of other possibilities could be seen, however, in the fireproofing qualities of asbestos and in the comparison between DDT and parathion. What if asbestos exposure of workers could be rigidly controlled and asbestos used only where it is not released into the air? Would the small risk of cancer due to very low level exposure outweigh the decreased risks of fire? What if the environmental damage due to DDT had been less devastating? What if insect resistance had not, in any case, made DDT use less attractive and less beneficial? Considering the many deaths due to parathion, would the benefits of DDT have outweighed its environmental costs? Even more intriguing: Was there a way to limit the use of DDT to public health uses, minimizing its release to the environment?

What follows is an excursion into the arena of policy decisions. Two of the cases present complex problems. In each, the chemical in question benefitted—or was thought to benefit—the same people it put at risk. Tris was used as a flame retardant, to protect children's clothing from catching fire; DES was thought to prevent miscarriages. In each of these cases, the particular governmental action that was eventually taken was approved by a significant number of scientists without a stake in either outcome. Since other scientists took the opposing view on each issue, even the often dilatory action of the government in the case could be justified. An argument could be made that the government agencies had succeeded in balancing risks and benefits among publics with different needs. In contrast, the third case illustrates the role of political agendas in deciding supposedly scientific issues. Formaldehyde is an extremely versatile industrial chemical that is used to provide cheap insulation for homes, to produce permanent press fabric, and as a glue in wood products. Its potent allergenic and significant carcinogenic properties also make it a hazard for many of the people exposed to it. The case histories of DES, Tris, and formaldehyde illustrate, each in its own way, that political considerations heavily influence regulatory decisions.

Introductory Glossary

Because agencies make policy decisions and presumably implement laws, the excursion begins with a brief glossary, designed to produce "acronym literacy": a summary description of the agencies and laws whose names litter discussions of environmental policy. Because the scope of agency activities is often expanded or altered, and because laws are frequently amended, superseded, and even repealed, only broad outlines are given.

Federal Agencies

EPA. The Environmental Protection Agency was established in 1970. The EPA is charged with regulating the quality of media in which toxic pollutants appear (i.e., air, water). Its authority derives from the Drinking Water Act, the Clean Air Act, and the Federal Water Pollution Control Act;

the Resource Recovery and Control Act or RCRA; the Federal Insecticide, Fungicide, and Rodenticide Act (FIFRA); the Toxic Substances Control Act (TSCA); and amendments to all of these. Because EPA was established well after food and drug laws were already in place, regulation of foods is split. EPA determines (as did the U.S. Department of Agriculture before it) what constitutes "safe" levels of pesticide residues in foods. But it is the Food and Drug Administration, not EPA, that is charged with keeping foods off the market if residues are too high.

OSHA. The Occupational Safety and Health Administration was also established in 1970, under the aegis of the Department of Labor, "to provide for the general welfare, to assure so far as possible every working man and woman in the nation safe and healthful working conditions and to preserve our human resources." OSHA sets interim standards when a substance in the workplace is identified as hazardous. By law, it must replace such interim standards by permanent exposure standards within 6 months. OSHA may also set emergency standards for severe hazards. To date, these regulations have been issued on a chemical-by-chemical basis. OSHA has sought to set generic standards for entire classes of chemicals, but this has been rejected in court. Generic standards would apply to classes of chemicals (e.g., carcinogens) rather than to individual chemicals (e.g., one set of regulations for vinyl chloride, a different set for benzene). In addition to regulating workers' exposure to toxic chemicals, OSHA is responsible for controlling physical hazards. These range from the dangers of repetitive motion syndrome (e.g., among telephone company employees who use computers all day), to assuring that fire doors are not locked shut, as in the North Carolina chicken-processing plant where over 20 workers died in 1991. This tremendous diversity of responsibility, plus a shortage of inspectors, has effectively crippled OSHA for most of its existence. OSHA is also a highly politicized agency whose effectiveness depends very much on the attitude of the current Administration in the White House.[2]

NIOSH. The National Institute for Occupational Safety and Health is the research arm of OSHA but is not under the Secretary of Labor. Instead, NIOSH is a subdivision of the National Institutes of Health (NIH), a subdivision of the cabinet-level Department of Health and Human Services (HHS). NIOSH has no regulatory power. It is officially the research arm of OSHA; however, there have always been administrative and intangible obstacles to OSHA and NIOSH acting in concert, since OSHA is in the Labor Department.

---

[2] It was OSHA that was alerted to the problems in the Velsicol plant that made leptophos. OSHA investigated the death, by cyanide poisoning, of a worker in a silver-reprocessing plant; their quick response in shutting the plant down may have saved several more lives. Unfortunately, it is very easy to ridicule OSHA's efforts, which also include such seemingly trivial regulations as the height of toilet seats.

FDA. The Food and Drug Administration is the oldest of health regulatory agencies. It is required to regulate exposures arising from the ingestion of food or drugs, or from exposure to cosmetics. The nature of its regulatory authority, and of triggers for regulatory actions, differ among foods, drugs, and cosmetics. In the case of drugs, recent decades have seen the tightening of standards to include not only safety but efficacy. More details about FDA are given below, under the laws that determine the scope of its power.

CPSC. The Consumer Product Safety Commission was established in 1972 to oversee consumer products. Earlier legislation that falls under its governance includes the Flammable Fabrics Act, the Federal Hazardous Substances Act, and the Poison Prevention Packaging Act. The CPSC cannot require premarket testing to demonstrate hazard but can force products off the market if risk is adequately demonstrated. CPSC banned Tris-treated sleepwear (case history below) as the result of a petition by the Environmental Defense Fund (below). CPSC has also been instrumental in identifying hazardous infant toys and furniture and played a role in the controversy over formaldehyde.

CDC. The Center for Disease Control is an arm of the Public Health Service. Much of its work is archival and epidemiological. CDC is also quite clinically oriented. It is not concerned primarily with toxic substances, but because it is called in when there is an unexplained outbreak of "disease," it is sometimes involved in toxicologic investigations. The "Annals of Medicine" in *The New Yorker* magazine draws heavily on CDC files[3] and has dealt with such cases as aspirin poisoning, rabies, and nitrite poisoning. Some excellent toxicological research has been done by staff members of the CDC.

ATSDR. The Agency for Toxic Substance and Disease Registry is a new agency, founded in the 1980s as part of the Center for Disease Control to provide information to the public on toxic and hazardous substances. In this it is succeeding, generating chemical-by-chemical summaries of toxic effects. These summaries can also be very useful to toxicologists. ATSDR is also charged with assessing the health risks posed by hazardous waste sites. Its success in this endeavor is not yet certain.

Laws

Federal Food, Drug and Cosmetics Act (and Amendments). This group of laws is the basic authorization for the FDA. The law sets different provisions for action with respect to food, to drugs, and to cosmetics. Food must not contain "any poisonous or deleterious substance which may render it injurious to health "—the General Safety Provision under which most food

---

[3] Many of these case histories by Berton Roueché have been collected into books (e.g., *The Medical Detectives*, Berton Roueché, Washington Square Press, New York, 1982).

additives (including red dyes 2 and A safrole, cyclamates, and carbon black) have been banned. A major source of controversy within FDA's authorizing legislation is the *Delaney clause,* included in the 1958 amendment to the Food and Drug Act. The Delaney clause, named for the Congressman who introduced it, states that "no additive shall be deemed safe if it is found to induce cancer when ingested in man or animal, or if it is found, *after tests which are appropriate for the evaluation of the safety of food additives* [emphasis added] to induce cancer in man or animals. . . ." The Delaney clause does not make any allowance for benefit–risk comparisons, but it applies only to *additives,* not to pollutants, pesticides in raw foods, or contaminants, and was specifically stated to exclude DES.

The Delaney clause has probably been criticized more vehemently and more often than any other legislation of interest to toxicologists. Nonetheless, it has been used only three times in 20 years: once to ban saccharin and before that to ban two obscure food packaging compounds. Critics rightly say that there is no scientific justification for the Delaney clause (why should one worry more about a food additive than about a naturally occurring secondary plant product if both are equally carcinogenic?). But the existence of the Delaney clause served, until recently, as a buffer between the FDA and political pressures to allow the scientifically equally dubious concept of "weak carcinogens" or "animal carcinogens" to be used to approve food additives.

The most recent onslaught against the Delaney clause is to redefine "carcinogen" to exclude agents that pose a less than one per million lifetime risk of cancer, which would be defined as a negligible risk. One point in its favor is that the new procedures would require uniformity in dealing with old and new chemicals, and also for raw and cooked products. At present, although a new food additive could not be approved if it has shown carcinogenicity, old additives that may pose serious risk remain in use because no one tests them properly, while carefully tested additives posing a very small risk are banned. The courts have rejected the redefinition of "carcinogen." Congress is now considering revision of the Delaney clause. If this were done, the present dichotomy, in which the Delaney clause is not applied to old, but only to new pesticides, could be remedied. It is estimated that there are 40 pesticides currently in use that leave residues in food and are carcinogenic. These 40 pesticides are thought to account for 90% of estimated food-transmitted cancer risk. Nonetheless, the Delaney clause has not been invoked for old compounds. On the other hand, the new plan would perpetuate case-by-case reviews, so the agencies could not characterize classes of chemicals. This means that the high cost of applying the regulations would not decrease, and the extension to old pesticides would remain doubtful.

Drugs. The FDA is charged with assuring that drugs are "safe." This is a relative designation, applied differently depending on use of the drug.[4]

[4] To give an extreme example, a drug used in cancer chemotherapy is allowed to have more side effects than an over-the-counter headache remedy.

The burden of proof of safety is on the manufacturer, who must provide FDA with enough data to demonstrate safety. Premarket testing can be, and is, required. Initially, the specific tests to be carried out were not prescribed by the FDA, which reserved the right to ask for additional information.[5] This smorgasbord approach has been superseded by more rigid guidelines. Nonetheless, it is advisable for a company to discuss its products with the FDA before "completing" safety evaluations, or expensive retesting may be required.

Recent amendments stipulate that drugs must not only be "safe" but also shown to be efficacious—that is, to do what they claim. This provision has altered the naming and formulation of numerous over-the-counter drugs and drug combinations. It has also provided FDA with the means to remove grossly fraudulent products from the market.

Cosmetics. In contrast to its mandate for determining drug safety, the FDA can act against cosmetics only if there is evidence of hazard. That is, the burden of proof is on the Agency to show that a given cosmetic poses a hazard. No premarket testing is required for cosmetics, although most manufacturers exhaustively test at least for acute effects. The Agency can sue in federal court to ban products that it considers dangerous (e.g., vinyl chloride in aerosols, chloroform in cosmetics).

FIFRA, FEPCA. The Federal Insecticide, Fungicide and Rodenticide Act (FIFRA) and the Federal Environmental Pesticide Control Act (FEPCA) provide for registration of pesticides (including herbicides, avicides, and slimicides as well as the classes named in FIFRA). Registration must be renewed every 5 years. The product must "perform its intended function without *unreasonable adverse effects on the environment*" (emphasis added). Premarket testing is required, and in recent years the EPA has stipulated which tests must be carried out. The burden of proof of safety is ostensibly on the producer.[6]

---

[5] It was the request by Frances Kelsey for additional information about the reproductive and developmental effects of thalidomide that delayed Merrill's application to market the drug in the United States. Dr. Kelsey has said that she was aware of the differences between adult and fetal drug metabolism due to work she had done previously. In any case, by the time the studies she requested were under way, thalidomide was known to cause malformations.

[6] Whether EPA carries out this mandate is disputed. An article in the *New York Times,* August 23, 1991, states: "E.P.A. failed to evaluate warnings on at least 10 dangerous pesticides." The article notes that in the past 3 years, manufacturers had "sent EPA 154 notices of adverse health or environmental effects for 75 different pesticides. Complete reviews of studies indicating adverse effects had not been conducted for 10 to 20 chemicals, a number that is changing as the agency proceeds with its investigation." The article goes on to note that "since 1988 the EPA has received 50,000 scientific studies from manufacturers to support the continued use of 420 major pesticide ingredients used in more than 10,000 pesticide products. The studies are being reviewed by 400 scientists and technical experts at the agency." To illustrate the complexity of the problem: The *Times* goes on to discuss the spill of an herbicide, metam sodium, which spilled into the Sacramento River, noting that studies showed that metam sodium causes birth defects of the brain and spinal cord in animals, and that the active teratogen is the metabolite methyl isothiocyanate, formed when metam sodium reacts with water.

If a hazard is detected after a pesticide has been registered, the *cancellation* of registration requires 30 days' notice to the registrant (manufacturer), who may request a hearing. The pesticide can be sold until final adjudication (often several years). *Suspension* can be ordered if the EPA considers the product to be an "imminent hazard." Suspension must be followed by cancellation proceedings, but the pesticide cannot be marketed during legal proceedings. RPAR (rebuttable presumption against registration, usually pronounced "ahrpahr") is EPA's announcement that there seems to be evidence against a pesticide. The subsequent gathering of information, hearings, and/ or legal proceedings can take years. Often RPARs just fade out, to be resurrected years later. (The neurotoxic OP insecticide EPN has been under RPAR three times in less than 20 years and has not been banned.) The Scientific Advisory Panel is a group of scientists outside the Agency that oversees EPA pesticide decisions and may recommend stricter labeling, the banning of a chemical, and so on.

TSCA.   The Toxic Substances Control Act (TSCA; say "Tosca") was passed in 1976 and is administered by EPA. Designed to fill the gap between EPA and FDA regulatory functions, it requires premarket testing of a chemical if it "may present an unreasonable risk of injury to health or the environment." The costs of testing are borne by the manufacturer.

RCRA.   The Resource Recovery and Conservation Act (RCRA; say "Rickra"), passed in 1976, authorizes EPA to regulate the disposal of hazardous wastes. The law defines hazardous wastes as solids or mixtures of solids that pose a hazard because of physical, chemical, or infectious characteristics. Radioactive wastes are specifically excluded from RCRA, since their regulation is the province of the NRC, the Nuclear Regulatory Commission [an acronymic descendant of the Atomic Energy Commission (AEC)]. RCRA refers specifically to solids, leaving the status of liquid wastes unclear, but EPA has been legislating their disposal as well. RCRA is, therefore, the federal legislation under which dumps are regulated (see Chapter 6). States may regulate hazardous waste disposal, but their plans must meet EPA standards.

Safe Drinking Water Act.   This is the 1974 law under which EPA regulates water quality. It states that water regulations shall protect health "to the extent feasible . . . taking cost into consideration. . . ." It authorizes EPA to set *recommended* (not mandatory) maximum levels for contaminants that have adverse effects on human health, and stipulates that these levels shall provide "adequate margins of safety."

Clean Air Act of 1990.   This amendment to the older Clean Air Act was several years in the making, due to compromises between legislators from auto-producing states and those from states with strong environmental lobbies. The bill was ultimately considered a workable compromise by both industry and environmental lobbies. Its import is not clear, however, because in the Bush Administration the executive branch, through the Office of Management and Budget, on occasion refused to provide the necessary money to implement the law. The Clinton Administration's attitude is not yet obvious.

Pressure Groups, Lobbyists, PACs

Environmental lobbying groups and political action committees (PACs) have profoundly influenced the actions of the federal government in environmental matters, both by engaging in traditional lobbying activities and by suing federal agencies for not carrying out their legal obligations. A few of the more influential organizations are identified below.

EDF.   The Environmental Defense Fund (EDF) is one of the most re-search oriented of the environmental lobbying groups. Their unofficial motto has been "sue the bastards." Similar environmental lobbying groups include the National Research Defense Council (NRDC), the Sierra Club, and the Audubon Society. In the early 1990s, EDF appalled some of its constituents (and delighted others) by collaborating in a waste-reduction effort with Mc-Donald's restaurant chain. The success of that collaboration is not yet clear, but it provides a way of changing behavior other than endless confrontations in court.

Friends of the Earth and GreenPeace.   These are somewhat more radical environmental groups, which do not use lawsuits and publicity to present their viewpoint, but which actively intervene in environmental cases. GreenPeace has plugged outflows from chemical plants into rivers, and placed its ship and its people between ships and populations of endangered species. The original *Rainbow Warrior,* GreenPeace's ship, was sunk in New Zealand by the French, who objected to her antinuclear activities.

National Wildlife Federation and the World Wildlife Federation.   These are among the most conservative of the environmental organizations. The National Wildlife Federation was one of the last environmental groups protesting the actions of Reagan's Interior secretary, James Watt. At the time (1981), the NWF president noted that their very large membership was predominantly Republican and had a median income of over $30,000.

Nature Conservancy.   This relatively uncontroversial organization is unusual in buying land rather than lobbying. They acquire ecologically important land by purchase or donation, and collaborate with all levels of government to maintain critical habitats, either by ceding the land to public agencies as wildlife preserves or by maintaining stewardship directly. An international effort is also in progress.

CASE HISTORY: DIETHYLSTILBESTROL

Diethylstilbestrol (DES; Figure 12.1) is a white, crystalline powder. It is odorless, almost insoluble in water, but readily soluble in alcohol, chloroform, ether, and fatty oils. It melts between 169 and 172°F. It is inexpensive to synthesize and considerably more stable than natural estrogens, both in organisms and in the environment. It was the first estrogen that could be taken orally. It was first synthesized in 1938 by E. C. Dodds and was approved for use in human medicine by the FDA in 1941.[7] In human medicine, the major use of DES for over 20 years was its administration to pregnant women, at levels of 1.5 to 150 mg/day, to control bleeding during pregnancy. It was also heavily used to treat symptoms of menopause, inoperable prostate and breast cancer,[8] and senile vaginitis. Finally, it gained favor as a postcoital contraceptive, at 25 to 50 mg/day for 5 days, around 1972.

Between 1945 and 1948, DES was shown to be an effective means to castrate young roosters chemically, to fatten chickens, and to increase muscle mass of sheep and cattle. In 1947 the FDA approved the use of DES in animal feed and in animal implants, for use as a growth promoter and to make caponettes in poultry. DES was thereafter extensively used in animal feed and implants as a growth promoter, in animal feed at a level of 5 to 20 mg/day for beef cattle, or as implants of up to 36 mg.[9]

In 1958, DES was shown to induce cancer in experimental animals. In the same year, the 1958 Food Additives Amendment to the Food and Drug Act began to require proof of safety for food additives. This amendment included the Delaney clause (see above), which stipulates that no chemical causing cancer in humans or in animal tests could be a food additive. Although drugs administered to food animals were included in the Delaney clause, another clause of the law specified that the Delaney clause does not apply if previous sanction had been granted. Therefore, DES could legally be used for medicine, and was legally manufactured for use in feed, if the manufacturer had preexisting approval. If, however, a factory burned down, the owners could not relocate and still make DES for animal feed, since that would constitute a "new" manufacturer approval.[10] In 1962, DES was specifically excepted from the Delaney clause of the Food and Drug Act by the Stilbestrol Amendment, with the provision that it was a permitted food additive

---

[7] The Food and Drug Act of 1938 required approval for safety but not effectiveness of all *new drugs*. A new drug was defined as any drug "not generally recognized as safe."

[8] Its efficacy in treating these cancers probably derives from the hormone responsiveness of certain cancers, as discussed in Chapter 11.

[9] There is the very plausible anecdote of the waiter who liked chicken necks and was given all of the restaurant's supply of this commodity. He developed severe symptoms of feminization due to the DES implants, which were located in the necks. More readily documented is the feminization and infertility of workers in plants manufacturing DES, which led to a lawsuit by their union, the Oil, Chemical and Atomic Workers' Union (OCAW) in 1966. An earlier lawsuit was brought by mink ranchers, because mink fed on chicken necks and heads became sterile.

[10] This was an actual case.

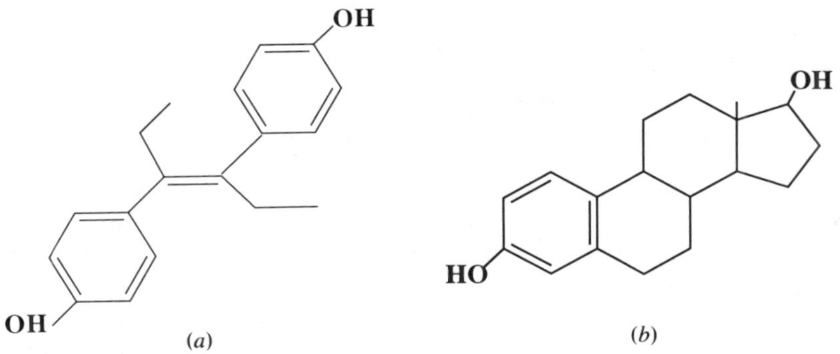

Figure 12.1 Structures of (*a*) DES and (*b*) estradiol-17β.

as long as no detectable residues were found in meat from treated animals. In 1959, new analytical methods showed that DES was present at levels of up to 50 ppb in poultry. However, its use permit in poultry was withdrawn on the basis of *drug safety* rather than by invoking the Delaney clause.

In 1971 the USDA began to monitor meats for DES and promptly found 0.5 to 2 ppb in approximately 1% of livers. No residues were found in muscle meat at that time. The FDA thereupon increased DES withdrawal time (the length of time between the last permitted dose of DES before slaughter) in cattle from 2 days to 7 days and demanded certification of 7-day periods.[11] In 1972 a radiotracer study showed that detectable DES or its metabolites were present even after a 7-day preslaughter withdrawal period. Based on this study and on evidence that DES is a human transplacental carcinogen (see below), the FDA then withdrew approval of DES use in other animal feed. Because FDA did not hold public hearings, however, industry sued. In January 1974, DES was allowed back on the market pending public hearings ordered by court. It remains in use in cattle feed to this day, although several European countries have refused entry to meat unless it can be certified as DES free.

Toxicology.   DES is rapidly metabolized by humans and animals at least to the extent of being conjugated. The major metabolites are excreted as glucuronides in the feces of animals or in the urine in humans. In animals about 95% of a single oral dose is excreted within 7 days. Because of its heavy use, but more because of the numbers of animals concentrated in feed lots, DES in its conjugated forms is present in feedlot wastes (animal feces) and has occasionally been detected in wastewater.

DES acts as an anabolic steroid in all of its effects. Estrogen-like activity accounts for its ability to increase the deposition of muscle tissue in cattle and sheep and to induce formation of caponettes (desexed males) in poultry. In humans,

---

[11] Since cattle feedlots keep careful track of prices for cattle and sell beef on very short notice if the price goes up, it is difficult for them to comply with the 7-day withdrawal period. Since FDA inspects a very small fraction of all meat, it is worth it for large producers to risk detection, although small producers probably cannot risk it.

DES promotes thickening, stratification, and cornification of vaginal tissue, induces growth of mammary gland ducts, and promotes endometrial growth. The same effect is certainly seen in animals; however, it is not of interest. On the basis of the observed human effects, one would characterize DES as a transplacental carcinogen and/or a teratocarcinogen (see below).

Teratogenesis. DES causes nonneoplastic abnormalities of the vagina (vaginal adenosis) and of the cervix in about 50% of the female offspring of mothers treated with DES during the first 17 weeks of pregnancy. There is an elevated incidence of epididymal cysts and abnormal semen in male offspring. Estimates of the incidence of *reproductive tract* abnormalities in prenatally exposed offspring are as high as 100%; a significant fraction of women with prenatal DES exposure have difficulty maintaining pregnancies to term.[12]

Transplacental Carcinogenesis. DES causes clear-cell adenocarcinoma of the vagina in 0.14 to 1.4 of 1000 female offspring of mothers treated during the first 18 weeks of pregnancy. This rare cancer had previously been seen only in women over the age of 50. Among DES children, the incidence of cancer is confined to girls between the ages of 10 and 25. A total of 244 cases had been recorded as of March 1976. Since DES was used during pregnancy until the year its transplacental carcinogenicity was identified, one must expect additional cases until approximately 1997. Despite the relative rarity of the cancer, it is important for young women who may have been exposed to DES in utero to be checked for symptoms beginning at puberty at the latest.

Mechanism of Toxicity. To date there is no evidence suggesting that DES acts other than as an estrogen. Estrogens, which are normally present in all mammals, induce tumors in experimental animals and induce vaginal adenosis when administered to newborn mice. It is not known whether natural estrogens can induce clear-cell adenocarcinoma if administered in doses equivalent to those used in human DES therapy. If DES acts only as an estrogen, and if this is also true of all of its metabolites, then a "safe" dose of DES must exist, since all mammals produce estrogens and must indeed have estrogen to develop normally.

History: Estrogens and Carcinogenesis. The first clue to the relationship between hormones and cancer was the observation by Beatson in 1898 that ovariectomy of women with breast cancer often helped recovery, or at least increased length of survival. In 1923, Allen and Doisy defined the ability of an agent to induce cornification of the vagina in ovariectomized rodents as the distinguishing mark of an estrogen. Between 1950 and 1962, numerous investigators demonstrated the induction of tumors by estrogens. The tumors produced included tumors of the breast, cervix, endometrium, ovary, pituitary, testicle, kidney, and bone marrow in one or more of five species of laboratory animals. However, attempts to induce tumors in monkeys failed.

[12] There is evidence that DES will also affect the grandchildren of women who took it while pregnant. Because the women exposed in utero have difficulty maintaining their pregnancies, the incidence of prematurity increases. Premature babies are more likely to have assorted developmental problems, including mental retardation, than are full-term babies.

In 1962, the Drug Amendment of the Food and Drug Act required proof of effectiveness as well as safety for drugs. However, Congress included the Stilbestrol Amendment, saying that FDA *must* approve all uses of DES in food animals as long as no residues were found in the meat. The precedent was the earlier amendment allowing the use of carcinogens in food animals if no residue is left in food. P. B. Hutt, the former chief council for FDA, called this amendment "a legal and scientific game of hide and seek," since it meant that the use of DES could continue only until techniques to detect small quantities of residues were invented rather than on any risk–benefit estimates.

In 1973, DES was conclusively shown to be a human carcinogen. Within a year, estrogen was shown to cause the precancerous state of *vaginal adenosis* when administered to laboratory mice. Nonetheless, for some years thereafter, DES continued to be used legally as a postcoital contraceptive." This required administration of very high doses for 5 days, making the uterus unreceptive to a fertilized egg (if fertilization had occurred) and preventing implantation.[13] R. Hertz, of George Washington University School of Medicine in St. Louis, MO, commented that "I know of no other pharmacological effect, so readily reproducible in such a wide variety of species, which has been generally regarded as inapplicable to man."

In animals the induction of tumors by estrogens depends on the presence of three critical factors: administration of a very high dose; genetic susceptibility (both high responders and unresponsive strains mice are known), *and* the presence of a viruslike factor that is transmitted through milk and that predisposes to tumorigenesis. The age of the animal at treatment and the duration of treatment are also important. The human conditions are obviously not as well defined. Most human data come from epidemiological studies, many of which may be badly flawed.[14]

Epidemiology of DES-Induced Cancer.   From 1941 to 1972, DES was used to control bleeding during pregnancy, regardless of earlier pregnancy history; if there was a previous history of pregnancy loss, DES was administered even in the absence of bleeding.[15] In some cases it was also prescribed on the theory that it would do no harm. Finally, many women have reported that they were not informed that hormones were being prescribed. It has been estimated that 1 to 2% of all women who were pregnant in the 1950s were given DES or other estrogens. The estimates of the total numbers receiving hormonal treatment range between 500,000 and 2,000,000.

[13] To avoid risks to the fetus, most college health services that used DES in this way required the women to sign a promise that if they became pregnant anyway (DES fails to prevent implantation in about 10% of cases), they would have a surgical abortion. They apparently did not consider that this short-term use of high levels of estrogen posed any risk to the woman, a hypothesis that has not been fully evaluated.

[14] A. R. Feinstein, "Scientific standards in epidemiologic studies of the menace of daily life.," *Science* 242:1257–1263, 1988.

[15] In the first months of pregnancy, many women experience some bleeding at the time they would normally menstruate. But since bleeding can also signal the beginning of a miscarriage, preventive measures were sought. Today, with a greater appreciation of the high frequency of inviable conceptuses, early miscarriages are not aggressively countered.

Between 1966 and 1969, *eight* cases of adenocarcinoma of the vagina were reported in young women between the ages of 15 and 22. Cancer of the reproductive tract is extremely rare in such young women, and clear-cell adenocarcinoma is rare even in women over 50, in whom it is normally seen. Thus the appearance of a cluster of eight cases sufficed to raise questions about the cause; DES was identified within months. As of February 1976, 244 cases of DES-associated clear-cell adenocarcinoma of the vagina had been identified. The pattern has been that cancer occurs only (although not necessarily) if exposure to DES or other estrogenlike chemicals occurred before the eighteenth week of pregnancy, suggesting the existence of a critical period. Dosage levels varied widely, ranging from 1.5 to 150 mg/day. Because of variations in the time during which DES was administered, total doses ranged from 135 to 18,200 mg. No consistent relationship between daily or total dose and the occurrence of tumors has been identified, other than that exposure must occur before the eighteenth week of gestation. The age of the girls at diagnosis of tumors ranged from 13 to 23 years.

Risk of Prenatal DES Exposure.   Overall, the risk of clear-cell adenocarcinoma is estimated at 0.14 to 1.4 per 1000 exposed females, but the risk of cancer differs considerably between studies. At the Mayo Clinic in Rochester, Minnesota, there were no cases of cancer among 818 prenatally exposed girls. Boston Hospital found 9 cases of cancer per 1000 exposed girls. Obviously, the details of exposure are important in some unknown way.

Nonneoplastic abnormalities affect up to 50% of the daughters of DES-treated mothers. These abnormalities range from vaginal ridges that have no known adverse effect to abnormalities that severely decrease the ability to carry a fetus to term. Despite the shock value provided by cancer occurring in young girls, it is almost certain that the less dramatic but more common reproductive abnormalities will cause a far greater load of suffering to the generation of "DES daughters" than the carcinomas. There has been no increase in genitourinary cancer observed in male offspring of women who received DES during pregnancy, but there is an increased incidence of epididymal cysts and abnormal semen analyses. There is some indication of decreased fertility in these men, but it is not well defined.

DES or other estrogenlike substances are routinely used for menopausal estrogen-replacement therapy. Recent studies suggest a decrease in uterine cancer incidence related to stopping the replacement therapy. Other studies suggest, however, that estrogen replacement therapy *at the very low doses that are now favored* does not increase the *overall* incidence of reproductive tract cancers in women because it slightly increases the risk of some types and decreases the risk of other types. It must be stressed that the effects of DES in postmenopausal women are totally consistent with the assumption that it acts solely as an estrogen. All side effects of DES must be expected, at corresponding levels of estrogenicity, from any other estrogen.

Continued Agricultural Use.   As might be expected, there has been considerable controversy about the continued use of DES as an animal feed supplement. It is estimated that banning the use of DES (or equivalent levels of other estrogens) would force farmers to spend $500,000,000 per year in additional feed to produce the same amount of meat, or 7,700,000,000 pounds of animal feed per year. It is not disputed that the continued use of DES in livestock feed means that there will be low-level residues, in the ppb range, in liver and perhaps occasionally in muscle meat. Opponents argue that any level of a carcinogen is hazardous (there is no threshold) and that exposure to DES-treated meat increases the risk of estrogen-induced cancer in the general population. Proponents of continued DES use argue that since DES acts purely as an estrogen, it is as safe as natural estrogens; and that because to develop and function normally, all humans need estrogens, there is *for estrogens and other hormonal carcinogens* an effective threshold consisting of normal levels of estrogens in the body. Only levels of DES that *significantly* increase the normal levels of estrogen would pose a carcinogenic risk, and the meat residues are orders of magnitude below any such increase.[16]

Moreover, there are natural estrogens in the diet. Examples include coumestrol, from alfalfa meal, which has 113 to 186 ppb DES-equivalent estrogenicity, and genistein, from soybean, clover, and forage crops, which has about $\frac{1}{40}$ of this potency. Plant estrogens also occur in unpolished rice, in vegetable oils, in soybeans, and in legumes. One study claims to show that the estrogenic activity of wheat germ is equal to 420 ppb DES, far higher than the 2 ppb DES found in "contaminated" liver. An extreme example of dietary estrogens is subterranean clover, a plant native to New Zealand, which is so estrogenic that it causes infertility in sheep and cattle that are pastured on it.

Summary. As is depressingly typical in these case histories, the estimates of benefit and risk focus on risks to the consumer and benefits to the beef producer. Totally neglected are the people who produce DES and other estrogens. There is evidence that feminization of workers is a common occupational hazard in hormone factories. This risk, and any increased risk of cancer due to occupational exposures, should also be factored into the risk–benefit equations for DES.

CASE HISTORY: TrisBP

History. The first U.S. standards for the flame retardancy of wearing apparel went into force in 1954. These regulations were the result of several incidents in which a spark touched brushed rayon sweaters, which ignited with nearly explosive speed. The 1954 regulations were designed to prevent the sale of such highly combustible apparel. This law was amended in 1967 to permit the regulation of a broad range of consumer products, including carpets, mattresses, upholstered furniture, tents, curtains, sleeping bags, and children's clothing. The object of these laws was to protect the public from "unreasonable risk" of fires leading to death, personal injury, or significant property damage. Under the amended law, the flammability of carpets and rugs was regulated beginning in 1971, of mattresses and mattress covers in 1973, and of children's sleepwear between 1972 and 1975. To meet the children's sleepwear standard, it was required that a bone-dry garment char less than 7 inches on its bottom edge when exposed to a gas flame for 3 seconds. The flame resistance had to last for at least 50 launderings. Similar standards were to be developed for all children's and adults' clothing, but infant and toddlers' clothing was given the highest priority and earliest deadline.

[16] By comparison, the normal premenopausal woman has approximately 800 μg of estrogen in her body at a given time.

Fire Retardants.   The fire retardants commonly used in clothing include organic chemicals containing bromine, chlorine, and phosphate esters. Among the more notorious flame retardants used in plastics and durable goods are the PBBs, PCBs, and mirex (see Chapter 5). In carpets, inorganics such as alumina and antimony oxide are used. The major chemical developed for use in children's sleepwear as the deadline approached was tris(2,3-dibromopropyl)phosphate (TrisBP; Figure 12.2). Tris was patented in 1951 and initially used as a flame retardant in foams and sheet acrylic. It was first used to treat polyester fabric in 1972, and thereafter was favored because it could produce the very soft, cuddly texture preferred for baby clothes, whereas many flame retardants adversely affect the feel of the cloth. It was also cheap to apply, since it could be applied at the same time as dyes and in the same bath. In practice, up to 10% of the weight of a child's pajamas was TrisBP, and up to half of that was "surface Tris": that is, it was not covalently bonded to the fabric. Thus for a 200-g garment, up to 6 g of TrisBP was readily removable from the surface. TrisBP was used primarily in 100% polyester, in polyester blends with triacetate, and in acetate fabrics.

Biological Properties of TrisBP and Related Compounds.   TrisBP and fabrics containing it are known to be moderately allergenic in humans and can cause delayed hypersensitization, the degree of sensitization being related to the amount of TrisBP in the fabric. Tris BP is also mutagenic. This is not surprising, since it is closely related to the well-known mutagens (and carcinogens) 1,2-dibromo-3-chloropropane (DBCP) and 1,2-dibromoethane. In fact, random samples of TrisBP manufactured for use in flame-retardant fabrics have been found to contain up to 0.05% of DBCP. TrisBP itself is a more potent mutagen than DBCP in the Ames assay. TrisBP is also carcinogenic when fed to mice and rats. Therefore, the absorption of surface TrisBP through the human skin is of major concern. In general, neutral compounds in contact with the human skin seem to be readily absorbed, and Tris is readily absorbed by rats wearing a gauze pad impregnated with it. Moreover, infants often suck their clothing, opening the possibility of direct ingestion of TrisBP. In sum, the data demonstrate that the chemical in question is a mutagen and a carcinogen; since it is absorbed through the skin, its use as clothing results in exposure of the wearer.

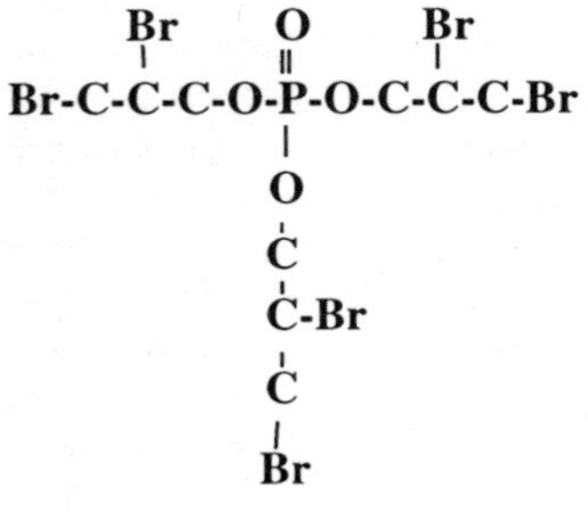

$$
\begin{array}{ccc}
\text{Br} & \text{O} & \text{Br} \\
| & \| & | \\
\end{array}
$$
Br-C-C-C-O-P-O-C-C-C-Br
$$
\begin{array}{c}
| \\
\text{O} \\
| \\
\text{C} \\
| \\
\text{C-Br} \\
| \\
\text{C} \\
| \\
\text{Br}
\end{array}
$$

Figure 12.2 Structure of TrisBP.

Risk–Benefit Assessment. The purpose of flame-retardancy standards is to prevent injury and death from fire. To determine the risk–benefit equation for Tris, it is necessary to estimate the risk of not using the chemical (i.e., the risk of fire-related injury in the absence of fire-retardant clothing).

In the United States, various estimates of annual burn injuries and deaths associated with textile-related burns range from a high of 200,000 to a low of 16,000 injuries; deaths may be as high as 4000 or as low as 500. If 20% of the injured or dead are children, 100 to 800 deaths might be prevented annually by the use of TrisBP in children's clothing. Based on the available mutagenicity and carcinogenicity data, it was estimated that up to 1.7% of all children who wear TrisBP pajamas for 1 year will get cancer as a result: that is, among 50,000,000 U.S. children, it is estimated that there would be 850,000 cancers.[17] This potential risk of cancer would have to be weighed against the alternative risk of up to 800 deaths per year.

Alternatives to Tris. Wool, vinyon, modacrylic, and polyvinyl chloride (matrix) fabrics are inherently flame-resistant fabrics that were in use in the 1970s and do not contain added flame retardants. Other fabrics, such as Trevira, have since been marketed. If the requirement for oven drying of wool were dropped, its natural flame-retardant properties would meet the most rigorous proposed flame retardancy standards.

Manufacture of self-extinguishing cigarettes should reduce wearing-apparel fires (and might also reduce forest fires). Similarly, self-extinguishing or child-proofed matches and cigarette lighters could reduce fires in children's clothing. Stronger public education in fire prevention is an obvious corollary.

Finally, the wisdom of fire retardancy standards for the very broad range of textile products currently under consideration should be examined in light of the hazard posed by the flame retardants they will require. To some extent this happened as a direct result of the publicity given to TrisBP. The parental public was presented choices phrased in dire terms by the headlines, but with no solid data behind them. Fire risks are uncertain within an order of magnitude. Cancer risk was by extrapolation from bacteria (the mammalian bioassay demonstrating actual carcinogenicity of Tris came later). A quite spontaneous boycott of all fire-retardant baby clothes led to the sprouting of labels denying the use of Tris on (for example) modacrylic "sleepers." The official ban by the CPSC came in April 1977, but by then very few stores still carried Tris-treated clothes. Manufacturers sought to export their stocks to Europe, which resisted vehemently. It is thought that most of the clothing and fabric was then sent to the Third World, but proof is unavailable. As a result of the outcry and its attendant losses, clothing manufacturers became more interested in developing inherently flame-retardant fabrics.

---

[17] Much of the information in this section comes from A. Blum and B. Ames, "Flame-retardant additives as possible cancer hazards," *Science* 195:17–23, 1977. The authors do not, however, predict an incidence of cancer among Tris-exposed children. Additional information came from *Controversial Chemicals: A Citizen's Guide,* P. Kruus and I. M. Valeriote, eds., Multiscience Publications, Montreal, Quebec, Canada, 1979.

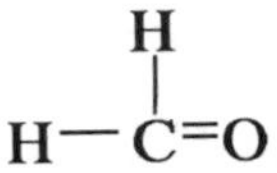

$$H - C = O$$
$$\overset{|}{H}$$

Figure 12.3 Structure of formaldehyde.

CASE HISTORY: FORMALDEHYDE[18]

Background.   Formaldehyde (Figure 12.3) is an industrial intermediate. It is also used heavily in a variety of applications, including embalming, fabric finishing, house insulation, and plywood and pressed wood manufacture. It is an irritant. Exposure leads to severe respiratory tract symptoms at rather low levels of exposure. It is also an allergen, and sensitivity can be severe enough to be life threatening at levels that are not irritating to unsensitized individuals. Finally, formaldehyde is also a mutagen and a carcinogen. The case history that follows demonstrates that the problems involved in risk assessment are not limited to determining the validity of scientific or epidemiological data, but also include political factors that may render suspect the decision reached by regulatory agencies.

Identification of Carcinogenicity.   In October 1979, the Chemical Industry Institute of Technology (CIIT), an industry-funded but autonomous research institute, informed EPA that formaldehyde was positive in a mammalian cancer assay: that is, it caused cancer in rats. Other regulatory agencies were also informed, because formaldehyde is used in products that fall under their control. To coordinate policy on formaldehyde, EPA, OSHA, CPSC, and other agencies then convened the Federal Panel on Formaldehyde to review the available data and assess the risk and consequent need (or absence of need) for regulatory action. In November 1980, the panel concluded that "formaldehyde should be presumed to pose a carcinogenic risk to humans." CPSC thereupon banned the use of urea–formaldehyde foam, the major product over which it had jurisdiction. Section 4(f) of TSCA, the Toxic Substances Control Act, specifies that once EPA receives evidence that there "*may be* a reasonable basis to conclude" that a chemical poses a *significant* cancer risk, the agency has 180 days to initiate action *or* to explain why such action should not occur.

Early EPA Response.   In March 1981, EPA determined that the applicable section of TSCA had indeed been triggered. This marked the beginning of the 180-day countdown. However, the Reagan Administration had taken office in January, and incoming EPA administrator Gorsuch delayed conclusions to allow for additional review of the data on formaldehyde. Eleven months later, in February 1982, the EPA's assistant administrator for toxic substances, John Todhunter, concluded in a memo that Section 4(f) had *not* been triggered and that there was in fact *no* basis to conclude that there might be a significant cancer risk from formaldehyde.

---

[18] Much of the information in this case history is taken from N. A. Ashford, C. W. Ryan, and C. C. Caldart, "Law and science in federal regulation of formaldehyde." *Science* 222:894–900, 1983.

The process by which this decision was reached is of interest. Like most facets of governmental activity, the framework of regulatory decision making has been shaped by the courts as a result of torts. The Washington, D.C. Circuit Court considers that agencies should take a hard look at relevant issues (data), in a way designed to negate the danger of acting in arbitrary or irrational ways, and make rational connections between the facts found and the choice (of regulatory action) made. For health risk determinations, decision making often requires not only evaluation of the technical data, but also consideration of science policy—that is, of a somewhat subjective series of guidelines that can be called on when the technical data are in dispute or do not lead to unequivocal conclusions. One can rephrase the obligations of the regulatory agencies as follows:

1. To evaluate technical data
2. To follow their own administrative procedures
3. To carry out their statutory (congressional) mandate

The Todhunter Memorandum. In contrast to the guidelines above, the Todhunter memorandum stating that formaldehyde did not trigger section 4(f) of TSCA relied on methodologically inadequate epidemiological studies,[19] assumes that the irritant properties of formaldehyde are identical with its carcinogenic action,[20] and also assumes both site and species specificity for formaldehyde carcinogenicity. That is, Todhunter assumed both that rats are *not* models for human carcinogenesis and, simultaneously, that sites not affected in rats would not be affected in humans either, regardless of differences in routes of exposure. He further assumed that workers *will* avoid irritating levels of formaldehyde and that consumers *will* be able to avoid irritating levels of formaldehyde, and concludes, without any supporting evidence, that homes insulated with urea–formaldehyde foam do *not* have higher formaldehyde levels than homes not so insulated.

One can start endless debates about the validity of epidemiological studies, since in almost all such studies there are potentially confounding factors that were not (or could not be) controlled. Among such confounding factors are actual exposure levels and duration of exposure; socioeconomic differences between exposed and control populations; exposure to other carcinogens in exposed or control groups; and differences in such variables as age, ethnicity, and diet. The methods for extrapolating data on carcinogenicity between species, coupled with the assumption that irritant and carcinogenic properties can be equated, may not be quite in the category of claiming that the earth is flat, but they are certainly minority views in the scientific community. Moreover, there is considerable evidence that workers put up with horrendous working conditions if the alternative is loss of a scarce job. Similarly, consumer exposure to formaldehyde insulation often occurs in conjunction with a purchased home. Few people have the resources to walk away from such a purchase because of seemingly minor irritations. Also, of course, if formaldehyde is carcinogenic, even nonirritating levels of exposure

---

[19] Ashford et al. (*Science* 222:894–900, 1983) consider the epidemiological studies inadequate. In defense of EPA, there does not ever seem to have been *any* epidemiological study that was not immediately characterized as "inadequate" (although less polite terms are often employed) by at least one other epidemiologist.

[20] In effect arguing that if there is no irritation, there is no risk of carcinogenesis.

could be hazardous. Finally, in the absence of data, it is clearly wishful thinking to assume that homes with urea–formaldehyde insulation do not have higher levels of airborne formaldehyde than houses without such insulation.[21]

Nevertheless, it is possible to come up with these conclusions, and, if the procedural process used had adhered to the rules required, and ordinarily followed by EPA, it would be difficult to overturn the decision. But were reasonable procedures used?

Background to the Todhunter Memorandum.   The Federal Advisory Committee Act, recognizing that federal agencies need to consult with outside groups, requires that "fair balance" among opposing viewpoints be provided. The record shows that EPA did not meet this requirement. First, in the summer of 1981, EPA officials had three meetings with representatives of the formaldehyde industry. One or two neutral scientists were included in these meetings, but most of the participants were selected by the Formaldehyde Institute, an industrial lobbying group. There were also private meetings between Todhunter and chemical industry representatives. Scientists from OSHA and CPSC, who were members of the Interagency Regulatory Liaison Group (ILRG), were *specifically* refused admittance to these EPA/industry meetings. Second, in October 1981, the Science Advisory Board of EPA (which is charged with overseeing EPA actions) recommended that EPA permit the National Academy of Sciences to review the formaldehyde data before issuing any decision. The Todhunter memorandum was published without such consultation. Third, the memorandum contradicted previous EPA positions. In 1979, the EPA Regulatory Council had issued guidelines that said:

1. Negative epidemiological studies will not be presumed to indicate that a substance is not a carcinogen.
2. Those sites or tissues that would be exposed by a different (untested) route of administration will be presumed to be at risk. For example, most chemicals are given to rats by stomach tube, which means that the liver is the organ with the most concentrated exposure. For chemical exposure by inhalation, the nasal passages and lungs are most heavily exposed. Therefore, if an agent is given orally to rats and causes liver cancer and does not produce cancer of the nasal passages, but human exposure is by inhalation, one must assume that nasal cancer is still a risk.
3. Negative bioassays in one species will not be used to detract from well-established positive evidence for other species.
4. A threshold level will *not* be assumed for carcinogenesis. That is, if a chemical causes cancer at high levels, it will be assumed that it causes cancer (albeit with lower frequency) at *any* level of exposure.
5. Positive results in one species, at one dose level, will suffice as evidence of carcinogenicity, assuming that the study is sound.
6. The *maximum* risk is to be estimated.

---

[21] By now there are, in fact, numerous studies showing the opposite, although one must give Todhunter the benefit of the doubt and suggest that the studies appearing after his memo were done in response to the labeling of formaldehyde as a carcinogen.

Todhunter, in his memorandum, essentially adopted the opposite position for each of these points. For example, rather than estimating the maximum risk, which would have included using the occurrence of benign tumors as an indication of possible future malignancies, he discounted their significance entirely. He also assumed that irritating levels of formaldehyde would in all cases be avoided, thus minimizing exposure levels. Finally, the Todhunter memo was not reviewed by scientists within or outside the EPA, despite Gorsuch's statement that her goal was to "improve the scientific basis" of EPA decision making. Todhunter concluded that "there *may be* human exposure situations . . . that *may not* present carcinogenic risk which is significant." On this basis, that there could be some situations in which exposure to formaldehyde did not increase the risk of cancer, he decided there was no need to regulate formaldehyde, even though the law [TSCA 4(f)] *explicitly* requires action if there *may be* a *reasonable basis* to conclude that a *risk* of harm exists. The qualifications in the law are all on the side of assuming maximum risk and regulating the "worst case." Todhunter set forth the best case and chose not to regulate on that basis.

Epilogue.   In 1987, seven years after initially dismissing the risk as trivial, the EPA finally acknowledged that formaldehyde is a "probable human carcinogen." The basis of the reversal was the same CIIT study that began the furor, plus human epidemiological data showing that high levels of exposure, usually occupational, are associated with an increased incidence of respiratory cancer. The human studies are—as usual—not conclusive in themselves. Combined with the CIIT animal data and with strong evidence in short-term assays that formaldehyde is a mutagen, the evidence for carcinogenicity has become extremely convincing.

The concession by EPA that formaldehyde is a carcinogen did not end the controversy, however. The National Research Defense Council protested EPA's implicit demand for human data (the epidemiology studies) because waiting for human data means that the Agency is not preventing cancer but only reacting to human cases. In contrast, the Formaldehyde Institute claims that the epidemiological data are flawed and that were formaldehyde a carcinogen, "I think you'd see people dropping left and right," whereas the spokesman claims that there are no "clusters of cancer in this industry."[22] A contrasting view was offered by Arthur Upton, a former head of the National Cancer Institute, who said of the original EPA decision that "if the carcinogenicity of formaldehyde is ignored, it would mean that no agent could be regarded as carcinogenic in the absence of human data.[23]

OVERVIEW

These three case histories illustrate the complexity of risk–benefit determination. The chemicals under consideration in all three cases are carcinogens: exposure increases the risk of cancer. Tris, however, is an effective flame

[22] E. Marshall, *Science* 236:381, 1987.
[23] S. S. Epstein, L. O. Brown, and C. Pope, *Hazardous Waste in America,* p. 348, Sierra Club Paperback Library, San Francisco, California, 1982.

retardant and protects the same population from fire that it exposes to risk of cancer. Tris provides an example of "straightforward" risk assessment. In theory, one simply counts the number of burns that Tris could prevent, sets it against the expected increase in cancer incidence, and determines which is the greater risk. The only uncertainties involved in the *risk assessment* are the actual risk of cancer, since there are no human data available, and the actual number of burns that would be prevented. Tris has been shown to cause cancer in rats, so a *relatively* simple extrapolation among mammals is involved in determining the cancer risk.

Risk management might pose some additional uncertainties. First, it must take into account the availability of alternative chemicals and of naturally fire-resistant fabrics. It is possible, after all, that a resolution is available involving neither risk of cancer nor of combustible fabrics. An additional consideration is the *perception of risk* by parents. Regardless of the relative risks from fire and cancer, parental decisions in buying clothes will be based on their perception of the risk of fire (which they can to some extent control) versus the risk of cancer someday (which is totally uncontrollable). The spontaneous avoidance of baby clothes suspected of being Tris-contaminated makes it clear that most American parents felt confident of their ability to keep fire from their children's clothes. Whether this confidence is realistic is a different question given the annual incidence of fabric-related burns in the United States. The problem of perception of risk, and the concomitant perception of one's ability to cope with a risk, is dealt with again in Chapter 13.

DES illustrates a different regulatory problem. DES caused cancer in a small fraction of prenatally exposed girls. Its ostensible function was to prevent miscarriages, but 20 years before the adenocarcinomas were recognized as DES induced, DES had been shown to be quite ineffective in preventing spontaneous abortions. There was no benefit to its use during gestation, and that use was discontinued as soon as the carcinogenic risk was identified.[24] Agricultural use, in contrast, seems to have benefits, in terms of reducing the costs of meat, without demonstrated risks to the consumer other than the miniscule levels of DES residues found primarily in beef liver. Given the hormonal nature of DES-induced cancer and the normal levels of estrogens in humans and in other foods, these residues cannot be said to increase cancer risk significantly. Is there a risk to users? Is that risk significant? And why is the risk to occupationally exposed populations rarely included in risk–benefit discussions?

Finally, formaldehyde is an example of the role of political agendas in risk–benefit determinations. Ironically, it provides an excellent example of genuine balancing of risks and benefits, since—again with the exception of

[24] This begs the question of why DES was used, and heavily used, for 20 years after it was shown to be ineffective. Obviously, good science does not necessarily triumph over heavy advertising. Current FDA regulations require proof of efficacy not only before a new drug can be marketed, but also for drugs already on the market. This may alleviate DES-type situations.

occupational exposure—the benefits accrue to the same people who incur the increased risk of cancer. Formaldehyde is a valuable chemical in all its guises. Risk–benefit analysis should have evaluated the existence of alternatives for each of its uses (e.g., in wood glues, in permanent-press fabrics, in insulation, in embalming) and the individual and total additional cost (if any) incurred by shifting to substitutes. These costs should have been weighed against the increased risk of cancer *from each type of use*. Then a reasonable answer might have emerged. Instead, an Administration that promised to decrease government regulation chose to do so by ignoring laws and established procedures, by denying its own agencies' scientists a voice, and by making patently false claims about the available data.

"When I use a word," Humpty Dumpty said in a rather scornful tone, "it means just what I choose it to mean, neither more nor less."

"The question is," said Alice, "whether you can make words mean so many different things."

"The question is," said Humpty Dumpty, "which is to be master."

—Lewis Carroll[25]

---

[25] In *Through the Looking Glass* by Lewis Carroll; from *The Annotated Alice*, edited by Martin Gardner, Penguin New American Library, New York, 1974.

13

SOCIAL ISSUES

Whether the State can loose and bind
  In heaven as well as on earth:
If it be wiser to kill mankind
  Before or after the birth—
These are matters of high concern
  Where State-kept schoolmen are;
But Holy State (we have lived to learn)
  Endeth in Holy War.

Whether The People be led by The Lord
  Or lured by the loudest throat:
If it be quicker to die by the sword
  Or cheaper to die by the vote—
These are things we have dealt with once
  (and they will not rise from the grave)
For Holy People, however it run,
  Endeth in wholly Slave.

—Rudyard Kipling[1]

INTRODUCTION

The decisions a society makes about the health of its people and its environment may have some basis in scientific information, but are ultimately determined by the values of the society. The sciences underlying environmental laws include not only biology and chemistry, but economics. The values are expressed in customs and in laws. Because both individuals and groups of individuals shape how a society functions, both psychology and sociology are essential to understanding the resolution of environmental conflicts, whether or not physical scientists think the issue is clear-cut or in doubt. Finally, the role of political processes is obviously critical, not only when laws are drafted but also when they are implemented, as was demonstrated by the formaldehyde controversy. The branch of political science that deals

[1] R. Kipling, "McDonough's Song," from the short story *Easy as ABC*.

with how policies are formulated and implemented is called *policy analysis*.

Of the social sciences that deal with environmental issues, economics is not only the most quantifiable, but also the most often applied. Economic arguments are used to justify most forms of pollution, both in the process of reaching decisions, which may involve complex calculations of competing costs and benefits, and in justifying the decision to the society.

ENVIRONMENTAL ECONOMICS[2]

It is often implied by "industry spokespersons" (some of them self-styled) that laws designed to protect the environment are inevitably bad for business. Environmentalists are portrayed as fuzzy-minded people who put a 1-inch fish like the snail darter ahead of human welfare.[3] This viewpoint is coupled with a shrill insistence that any new legislation to clean up the air or to protect groundwater or the health of workers is the reason that factories will close, jobs will be lost, the economy will suffer. Similarly, much animosity between economists and environmentalists is engendered by the environmentalists' perception that the only thing that matters to economists is money. This perception arises from the economists' emphasis on costs and benefits, expressed in monetary terms. To many noneconomists, placing a price on a life (whether of a migrant worker or of the last California condors) seems indecent. Despite this resentment, economic factors inevitably enter into environmental decisions when public policy is being determined. More accurately, economic considerations are intrinsic to many environmental decisions, since any transactions that transfer wealth are de facto economic in nature. Few things have clearer intrinsic value than life and health, followed closely by land and the right to use or manipulate it. In Western societies, economic activity is *market-based*. Therefore, some consideration must be given to the role of markets in determining prices, and to the more particular situation in which economic principles are applied to regulating toxic substances.

Free Markets

Much of Western (North American and western European) economic theory depends on the theoretical existence of *free markets*.[4] In this hypothetical free

---

[2] The information in this segment draws heavily on lectures given by Gary V. Johnson and Henry van Egteren when they were at the Institute for Environmental Studies. I have adapted their lectures, however, so only I am responsible for any errors.

[3] The snail darter, *Percina tanasi*, was threatened by the building of the Tellico Dam in Tennessee, as discussed in footnote 6 in Chapter 2.

[4] In contrasting ideology, but probably similar fidelity, the economic policies of the Soviet Union and its eastern European satellites between 1945 and 1990 were based on Marx and Engels' theory of dialectic materialism.

market, perfectly informed buyers and sellers agree on mutually satisfactory prices for goods. Perfectly competitive markets exist if *and only if:*

1. Firms produce a homogeneous commodity and there are no advantages to selling to one consumer rather than another. The free market does not permit the existence of kickbacks. Moreover, there must not be a real difference between different brands of the same commodity. A Ford and a Honda must be essentially the same.
2. Both sellers and buyers are sufficiently numerous that each transaction is small relative to the aggregate volume of transactions. There may not be any monopolies or near monopolies.
3. Both buyers and sellers have perfect information about prices. A buyer knows what every other producer's price is for the given item, and a producer knows what other buyers are offering, so every transaction is fully informed.
4. Both sides must also maximize their own advantage: the seller his profit and the buyer his "well-being."
5. Entry into and exit from the market is, in the long run, free.
6. Price is the *only* signal that producers and consumers use to allocate goods in a perfectly competitive market.

The classical example of a free market is of the farmers at a market, each hawking wares, with customers going from one to the other, selecting the best price–quality combination. Such a market results in efficient allocation but not necessarily in an equitable or socially desirable allocation. If, for example, there has been a bumper crop, the price may drop below the cost at which farmers can support themselves. If, on the other hand, there is a famine, the few farmers with grain can beggar everyone else, and those rich enough to meet their prices can feast while the less fortunate starve. A more modern example is seen in wage differentials. Women are typically paid less than men for comparable work (in a recent court case, "comparable work" was the traditionally female job of nurse versus the traditionally male job of security guard). It may be efficient to pay women "what the market will bear"; nevertheless, it may not be socially desirable. In the context of equity issues, it must be remembered that there is a strong link between money and power. If equity is not observed, efficiency can lead to a very undesirable society. Nevertheless, equity issues fall outside the realm of economics.

Even in the absence of equity issues, the free market is an idealized construct that rarely occurs in the real world. The sum of the necessary conditions is unlikely to be fulfilled very often. For many products, there are few producers. The small number of automobile firms that existed before the Asian producers began to import cars come to mind, but other industries are equally oligarchic. Fewer than 10 firms handle all wheat marketing for the entire United States. Even when many sellers exist, only a few may be accessible to a given consumer. Although the United States still has a large number of airlines, most cities are served by two or three carriers, if they are lucky. In 1992, one airline flew from Champaign, Illinois to St. Louis; two different airlines to Chicago; and a fourth carrier flew to a hub in Pittsburgh. Thus despite the extreme competition between airlines on the national level, a buyer's selection was severely limited by destination. A seller rarely has access to all consumers (can your local stereo store sell to people in Flagstaff?), and a buyer's access is similarly restricted (do you know what a Sony Discman sells for in San Diego?).

That perfectly competitive free markets rarely exist does not lessen their value for formulating theory, any more than population genetics suffers because the constraints of the Hardy–Weinberg law (Chapter 10) are rarely met in real populations—which, at the minimum, rarely approach infinite size.

Pareto Optima

The Pareto principle states that a Pareto optimum is reached when goods and services are allocated so that no one group of individuals can be made better off without at least one other person being worse off. That is, at the *Pareto optimum,* each individual (or group) is as well off as it can be *without hurting someone else.* In the simplest example of a Pareto optimum, consider the situation in which A has bread and B has wine. An omniscient dictator can reallocate these two products so that A and B each has as much bread and wine as they want. The same goal can be achieved by a market transaction: A and B negotiate until they come up with a fair price for wine in terms of bread, and vice versa. Note that neither type of allocation requires *equal* division of the bread and the wine. In a larger society run by the omniscient dictator, production of all goods is set up so that just the right amount of each product is produced so that (a) everyone can get as much as they want of it and (b) none is left over.[5] Any deviation from either (a) or (b) means that either a producer or a user could improve his or her situation, which would not be a Pareto optimum.

---

[5] In light of the recent collapse of the managed economies of the Soviet Union and its allies, the importance of omniscience becomes obvious. It is very difficult for dictators or their committees to determine how much to produce. When they are wrong, severe dislocations result. An employee of a farm cooperative in eastern Germany reports that in the year before the Communist regime collapsed, his farm was informed that their allocated 10 truckloads of pesticides would be ready, and must be picked up, at a certain factory on a certain day. The farm had only one truck; the trip was too long to make more than one trip per day. That did not matter: the factory had a quota to fulfill. So the narrator drove the single truck to the factory on the appointed day, loaded up, drove to a nearby river, and dumped the truckload. He repeated the process with eight other loads and took the tenth load back to the farm. His chilling punch line was: "And I was not the only one who did that."

General Competitive Equilibrium

Although Pareto optimality does not depend on the existence of a market, *perfectly competitive markets generate Pareto optimal outcomes*. For markets to generate Pareto optimal results requires that (c) producers pursue profit-maximizing behavior and (d) consumers practice welfare optimizing behavior. This requires, of course, that producers *know* what will maximize profits, and consumers *know* what will optimize their well-being, and also that both sides have *absolutely free access to the other*. Perfectly competitive markets satisfy our conditions (c) and (d), and this efficiency condition is what people argue for when they argue for *free markets*.

Despite its theoretical charm, the Pareto principle is not a good guide for decision making. There are an infinite number of optimal Pareto allocations for any allocation of goods. Consider a variant of the initial transaction, involving two people and two items, bread and wine. If one person begins with all the bread and all the wine and wishes to keep it all, a Pareto optimum exists if *A* has all and *B* none, since *A* would be "worse off" if he had less than all of the goods. Almost any other pattern dictated by human greed is also possible.[6] In short, the particular Pareto outcome that is achieved depends on many conditions other than availability of a free market.

One of the additional constraints is the distribution of income. Although noneconomists who are proponents of free markets often imply that free market efficiency—in which the product is sold with no consideration other than price—also requires that the distribution of wealth should not be subjected to any constraints, *this is not part of the economic theory of free markets*. It is acceptable to reallocate income (by taxation, for example) to achieve desirable social ends. Pareto outcomes are achieved by free markets, but the theory says nothing about how *income* should be distributed.

Public Policy

The object of public policy, in simplest terms, is the *increase of public well-being*, always recognizing that resources are limited and that any decision creates losers as well as winners. In fact, economists prefer to talk of maximizing the social well-being of the *publics* (plural rather than singular) because one group's good is not necessarily identical with another group's

---

[6] In fact, one can argue that there are famines in the world precisely because food is efficiently distributed by Pareto's principle. Several of the conditions of free markets are met: both producers and consumers know what they want, the producers know what prices are, world over. Except when the United States and Russia negotiate a grain deal, individual transactions are not a significant portion of the total. While the starving people of sub-Saharan Africa do not have access to all sellers of grain, that is irrelevant as long as they cannot *afford* the grain anyway. Other buyers, notably feedlots, can. Producers and other sellers want more money than the hungry can afford, so the grain goes to meat producers, who can afford the price because their (rich) customers will just pay a bit more for meat.

well-being. In the case of food distribution in times of famine, the sellers' optimum price may mean that many people starve. The determination of public policy therefore requires decisions on how to distribute benefits, how to define well-being, and whether public or private action is the best means to the desired ends.

To measure the objective, in what is called *political economy* or *social welfare theory,* one constructs a *social welfare function.* This concept derives from the work of Jeremy Bentham, an eighteenth-century philosopher who substituted *utility* and *disutility* for the emotion-laden concepts of *good* and *bad.* The social welfare function is a theoretical construct that serves as a framework or model around which to construct experiments or indices. Its correlate is the *impossibility theorem,* which states that it is impossible to construct a social welfare function that is valid in the real world.[7]

The primary obstacle to a realistic social welfare function is that one cannot measure the *intensity* of satisfaction. For example, the GNP (gross national product) measures the total dollar value of all production in the United States. In theory, if the GNP increases, "everyone" is better off. This is not necessarily true in the real world.[8] Moreover, the benefit of more money may matter more to you than to me: even if everyone achieved the same dollar increase in income, different people get a different degree of satisfaction out of the raise. Similarly, one person gets far more satisfaction out of a new car than does someone else, even though both pay the same price. Thus the economist is handicapped because only the *efficiency* of allocation can be measured; the *equity* cannot. Economics, like sociology, is the study of *aggregate* behavior, and therefore cannot determine the individual response. Interpersonal comparisons are impossible, and price is the measure of "wanting." The number of units bought and the price one is willing to pay are the closest measure of the degree of liking that can be achieved. If you and I each buy a new car at the same price, we are assumed to obtain the same "benefit."

In short, economic theory states that it is possible to achieve *efficient* allocation of goods by the use of free markets. Economic theory also makes allowance for *incommensurables,* which are things that cannot be converted to dollar values even though they can be quantified (such as a child's life, or the extinction of a species) and for *intangibles,* which can neither be converted to dollars nor even quantified (such as the pleasure of being alone on a hillside). Moreover, most (but not quite all) economists would agree that *equitable social goals* can be achieved by free markets only if these are coupled with income allocation *by nonmarket means.*

---

[7] The existence of the impossibility theorem suggests that economists have a better grasp of reality than those toxicologists who think their models—QSAR, PBPK, Hydra, or FETAX—will actually explain reality.

[8] There is abundant evidence that in the past 10 to 15 years, the poor have really been getting poorer, even though the rich have been getting richer. This reallocation of income is the result of the Reagan policy of cutting taxes for the highest tax bracket.

Characteristics of Goods in Free Markets

For markets to be free, goods must have certain characteristics. *Rivalry* means that when you buy an item, you are reducing the supply available to others. Breathing, in contrast, does not noticeably reduce the air supply. There is no rivalry for available air.[9] *Excludability* means that those who do not pay do not derive benefits from a transaction. If I buy a pair of shoes, I pay and I benefit. In contrast, many pollution control devices violate the concept of excludability. If a paper mill decreases the quantity of toxic effluent it emits into a river, everyone downstream benefits, even if they still buy cheaper paper from another company that pollutes more. Without rivalry and excludability, there cannot be an efficient private market, *since there is no way of making people pay for what they get.* Nor is it economically efficient to make polluters drink dirty water (even assuming one could), since water pollution does not reduce the amount of available water. Still, clean water is a public good. To maintain it—to keep it from being depleted—will require incorporating into the cost of products, the cost of keeping water clean. However, if a single corporation installs pollution control devices, it incurs costs and must price its goods higher than must the unscrupulous company that does not install such devices. In a free market, therefore, the "good guy" will be at an economic disadvantage and will go out of business. Only government intervention *at the level of regulating effluents* can deal with such issues.

Rivalry and excludability mean that the consumer of a specific item incurs all the costs of purchase and derives all the benefits of consumption. If I buy and eat an apple, I am the only one who pays and the only one who benefits. If the apple was sprayed with a hazardous pesticide and the worker spraying it became ill, I probably do not pay for the hospitalization of the worker who sprayed the pesticide. So although I derive all the benefits of eating the apple, I do not pay all the costs of producing it—society does so, either as worker's compensation, as welfare, as payments to hospitals for handling charity cases, or as a surcharge on some hospital bills because the government did not reimburse the hospital for charity care. Similarly, the producer does not incur the cost of the poisoning and need not consider that cost in setting the price of the apple. Such *externalities* skew the price of goods. To remedy the problem, the externality must be brought into the price of the item. If the worker's illness were covered by worker's compensation or by an employer-funded health plan, the employer would include the cost of the insurance premium in the price of the apples. The consumer would pay a higher price for the fruit. The externality would be eliminated. If full costs were internalized, the grower who neglects workers' health would pay higher insurance premiums, offsetting the savings derived from the neglect.

---

[9] This is not to say that one is satisfied with the quality of the air, just that it is not significantly diminished by use as it would be in a submerged submarine whose air tanks were empty.

Externalities

Externalities are defined as costs (or benefits) that are not reflected in the price of a good. Use of markets to allocate goods assumes that all externalities are included in the cost of the transaction (i.e., that all the costs are in the price). To illustrate, consider the national defense: everyone benefits, whether or not they contribute. Similarly, in a union shop, the wage settlement won by the union for its members also accrues to those who do not pay union dues. Similarly, homeowners who plant an exquisite garden in their front yard pay the entire cost of the garden; without taking extra steps, they cannot prevent passersby from benefiting (extra steps would include fences, and might well entail extra cost). In these cases, consumption by one (group or individual) does not prevent consumption by others; the costs are borne by a smaller group than the benefits. In the same way, the cost of certain transactions is borne by a larger group than the beneficiaries. For example, introduction of a pesticide on a crop usually neglects externalities. The cost of the pesticide to the farmer rarely includes the cost to the environment in terms of nontarget species injured or killed; the elimination of hedgerows by herbicides deprives the passerby of the pleasure of songbirds; downstream populations pay the cost of extra water treatment to remove pesticide residues; and so on.

The existence of externalities leads to inefficient and non-Pareto outcomes even if a perfectly competitive market exists, since *price is the only signal that producers and consumers use to allocate goods in a perfectly competitive market*. Moreover, the effects of externalities are aggravated when dealing with common property resources such as air. The producer who pollutes air to minimize production costs (scrubbers cost money) benefits, while the cost is borne by the whole of society. Similarly, ocean resources—fish, whales—are overexploited because there is no benefit to the fisherman if he does *not* catch a fish: *if he does not, someone else will*. A lack of well-defined property rights leads to overexploitation from society's viewpoint.[10]

Solutions to Problems of Externalities.   Economists recognize three major ways in which externalities can be included in the price of goods.

1. A tax can be imposed on the producer (firm, industry) that generates the externality. This attaches a cost to a formerly free input and allows the producer to decide if the cost is worth it (is it cheaper to clean up the factory or to pay the tax for polluting?). Obviously, the tax must be set to (a) cover the cost to society if the producer pays rather than cleaning up, or (b) be high enough that the producer cleans up. It must also be carefully designed so that the people who actually produce the problem also pay the tax.

---

[10] Thus Hardin's "Tragedy of the Commons" is the same as the economist's "lack of well-defined property rights [that] leads to over-exploitation from society's viewpoint."

2. A quota, or standard, can be set for the amount of pollution that each unit may produce. An example would be the setting of emission standards to minimize air pollution which decree that each chimney may emit only x ppm of a given chemical (or of all chemicals). Emission standards must set draconian penalties (e.g., shutting down the factory) for exceeding allowable levels of emission, since it is otherwise profitable for the producer to pay the fine and continue polluting.

3. Create well-defined property rights. This at least eliminates the problem of common resources and gives each producer a stake in preserving the resource.[11] In the case of air pollution, one can combine the creation of standards for a large area (the air above Chicago shall not have more than x ppm of pollutant y) with the "rights" of each factory to contribute a certain amount. Then Zyzygy Inc., which is allowed to emit 0.05 ppm ozone into its air but which has just discovered a new process that emits no ozone, can *sell* its right to release 0.05 ppm ozone to its neighbor, United Gaskets, for $1 million a year, because United Gaskets cannot afford to clean up its process, which emits 0.05 ppm *more* ozone than allowed.

Note that any of these approaches can be used equally well to maintain the *status quo,* decrease pollution levels, or increase pollution levels. The endpoint should be defined by collaboration among a number of disciplines. In the case of air pollution, such input should come from toxicologists (how bad *is* ozone for Chicagoans' health?), engineers (can we decrease ozone emissions by x ppm, and how much does it cost?); economists (how many companies will go out of business if we insist that they decrease ozone emissions by x ppm?); and politicians (will the voters stand for the increased costs of preserving their lungs, and do I risk my political career on this?). Which effect occurs depends on the political will and is to a large part outside the domain of both the economist and the toxicologist.

Applications of Economics to Environmental Regulation: The Use of Laboratory Data outside the Laboratory

Toxicologists spend large amounts of time, energy, and money identifying the toxic effects of chemicals that have been, are, or may be, released into the environment. Although some of this research is undoubtedly "basic science" with no immediate applicability, much of it is intended to influence

[11] If not very carefully regulated, it may also allow the rich to pollute the environment of the poor, by trading, for example, food for pollution. This technique is used every time a manufacturer threatens to close a plant if pollution controls are imposed. Note, however, that the threat is not necessarily phony. If other producers—in Asia, for instance—are not faced with the same pollution control requirements as those faced by American manufacturers, they have an advantage that may genuinely drive American manufacturers to choose between continued pollution and closing the factory.

public policy. This is the information identified above as "input . . . supplied by a number of disciplines." How are data used in a policy context? What do laboratory data actually mean to economists? Such questions are outside the topic of toxicology as a discipline but are critical in deciding environmental policy. Since so much is said about the related issues of *risk assessment* and *risk management,* let us consider the uses of data outside the laboratory.

In practice, toxicity data used by economists come from two sources: epidemiology, and model systems. Epidemiologic data are generally drawn from human data, unless the case of a particularly desirable species such as the California condor or the black-footed ferret is publicized. The use of model systems involves translating laboratory data on toxicity (e.g., the LD_{50}) into the significance of the given effect to the environment (flora, fauna) and to human health. Computer models are also used, but are derived, however distantly, either from epidemiologic or from laboratory data.

The most common measure of toxicity is the number of deaths per unit of population. Two things must be stipulated about this measure: first, the *unit* may differ, depending on the type of effect being studied. For example, cancer deaths are often recorded in cases per thousand of population; acute poisoning deaths might be measured per hundred workers but per million of general population. Second, although for shorthand we will talk about "deaths," the same analyses can be made for *any* deleterious effect, such as illness or loss of reproductive capacity. Moreover, the process is the same regardless of the species of interest. It is harder to put dollar values on some effects than on others: if acid from a chimney ruins the paint on my car, I can find out the exact cost of repainting it and even add in the cost of renting a car while mine is being painted. If I get emphysema from the air pollution, how do I count the cost? Moreover, some effects are intrinsically harder to measure than others. If 10% of married couples who want children fail to conceive, how can one couple prove that their infertility is due to pesticide drift from aerial spraying of nearby forests?[12]

Once a decision is made that the correlations in the data can be accepted as cause and effect, the appropriate public policy must be determined. Economists approach this subject (not unreasonably) from economic principles, just as biologists tend to look at their laboratory data. Economists are perfectly aware that broader social principles must also be considered, but these are often difficult to evaluate in an economic framework. The object of public policy, in simplest terms, is the *increase of public well-being,* always recognizing that resources are limited and that any decision creates losers as well as winners. The determination of public policy therefore requires

[12] This was the issue raised by the people of Alsea, Oregon, who observed that the incidence of miscarriages in their community rose after the Forest Service sprayed nearby. The epidemiological study done by U.S. EPA compared miscarriage rates between Alsea and a similar community that was not subjected to aerial spraying, and supported the suspicion of Alseans. But because the EPA study selected cases and controls differently, the study was thrown out.

decisions on how to distribute benefits, how to define well-being, and whether public or private action is the best means to the desired ends.

It is always necessary to identify the target population accurately. For the introduction of a pesticide into central Illinois agriculture, the target population is not necessarily everyone in Illinois. The population may be restricted to farmers (e.g., acute poisoning) or it may extend to the population of New Orleans (e.g., drinking water contamination) or of the nation (e.g., contamination of meat by pesticides used on corn). Similar definitions are necessary for nonhuman target populations.

Basic Concepts

In addition to the theoretical economic concepts discussed above, risk–benefit calculations involve some decisions. First, one must decide how risk estimates are generated. Second, one must generate a metric: that is, data must be quantitative and useable. Finally, one must decide what constitutes the *economic perspective*. This is obviously too vague a concept for implementation; the objective is usually a matter of reducing risk. Since policy tends to be highly anthropomorphic, the risks reduced are most commonly those to human health, although some recognition of ecosystem welfare is included.

In political economy or social welfare theory, one constructs a *social welfare function* (see above) to measure the objective. This theoretical construct serves as a framework around which to construct experiments or indices. Remember that the social welfare function is actually a fiction, inasmuch as the *impossibility theorem* states that it is impossible to construct a social welfare function that is valid in the real world, since one cannot measure the intensity of satisfaction. Thus the economist is handicapped; only the efficiency of allocation can be measured; the equity cannot.[13]

Even though it is known that the theory will never explain specific real-world transactions accurately, the concept of the social welfare function serves a valuable role by providing a theoretical framework for data evaluation. If, as is frequently suspected, one group attempts to manipulate data for its own benefit, the existence of a theoretical framework provides impartial ways of assessing such manipulation. It should also be noted that although economists (like laboratory scientists) tend to think of themselves as analysts, they often function as policymakers. This occurs even during the data generation steps, whether by economists or by laboratory scientists, since the absence of data leaves policymakers unable to function rationally. With respect to the validity of economic theories: if two economists examine the

[13] Note that efficiency decisions are not restricted to economics. For the laboratory scientist, the question of efficiency may include which chemical to investigate, as well as the experimental design. This in turn has policy implications because, in the absence of data, risk assessment is impossible.

same data set, they *should* arrive at the same conclusions. The problem is to assure that the set of information being examined is truly the same.

RISK ASSESSMENT

In toxicology, risk is often defined in terms of LD_{50}. This is, however, only the beginning of risk estimation. Beginning with the datum that the LD_{50} of a given substance is x mg/kg, the estimation of risk deriving from the failure of a landfill containing substance x must include:

1. The risk (probability) of the occurrence of a failure, such as by leaking or by breach of the barriers.
2. The risk of the transport of the toxin to either groundwater or surface water and/or to soil.
3. Contamination of an ecosystem or of humans. The contamination must be significant (i.e., above the threshold for effects). This includes the situation where the threshold is zero (i.e., there is no harmless exposure).

Thereafter, it is necessary to determine whether the consequences are immediate (including "delayed" neurotoxicity and perhaps even birth defects) or delayed (primarily the genotoxic effects of mutation and carcinogenesis).

The mere existence of a leak is not necessarily a hazard: for example, there are landfills in Alabama that are built over 500 feet of chalk; even the failure of containment measures in such a landfill is not a realistic hazard. On the other hand, "exposure" need not be direct: humans can be exposed via ecosystem contamination, as in the case of PCB contamination of fish in many lakes. The human exposure occurs via the fish, which were contaminated from the effluent of PCB plants (for example). No direct contact between humans and the source of contamination need occur. This need to evaluate the probability of a series of conditional probabilities considerably complicates the evaluation of the external costs of a given situation.

Equity Issues

Strictly economic issues are determined in terms of economic efficiency (i.e., goods go to the highest bidder). There is another question that must be asked: *Should* particular goods be allowed to go to the highest bidder? This is an *equity issue* and is not strictly the province of the economist. The answer depends on the social costs (or benefits): the total cost to society of a particular item. It must include both the private cost (what the buyer pays) and any externalities. Externalities are costs (or benefits) that occur when one agent's action *unintentionally* affects the costs (or benefits) of another agent.

In environmental situations, the major emphasis is on detrimental effects. Note that there may be a cost and a benefit (e.g., a paper company that benefits by not cleaning up its effluent causes a downstream brewery a major cost in water cleanup), no private cost (e.g., dioxins), or no private benefit (e.g., oil spills). All, nevertheless, produce externalities.

Both costs and benefits include items measured in economic units (dollars). In addition, costs and benefits can include *incommensurables,* those items that cannot be converted to an economic value, but which are nevertheless are quantifiable. One example of an incommensurable is death. Although no one pretends that money compensates for the loss of a spouse or child, it is possible to calculate what costs the survivors incur. These costs include not only lost wages, but the cost of hiring someone to do the unpaid work (housework, putting up screens) and may even include such items as "loss of companionship."

Intangibles are unquantifiable incommensurables. One example would be the lifestyle effects of oral contraceptives. Whereas one can quantify the expected cost (to family in income, to society in productivity and taxes) that results from unwanted pregnancy, one cannot put dollar values on release from *fear* of pregnancy.

Several other factors influence the decisions that industries and individuals make about costs and benefits. *Discounting* is the process of compensating for the effects of passage of time. In general, if a cost can be delayed for more than 20 years, its discounted value is zero. Similarly, if a benefit must be paid for now, but does not materialize for 20 years, it is worthless. *Marginal cost* (or marginal benefit) is the incremental change in cost (benefit) derived per unit change. For example: setting up the plates to print a book costs the same whether 2 or 2000 copies are printed; the incremental cost is only for paper and the machine operator's time. Environmentally, the cost of removing the first 95% of a pollutant may be low compared to the cost of removing each additional percent.

Applied to practical environmental issues, the concepts of discounting and marginal cost mean that it is difficult or impossible to use economic incentives to protect the environment unless there is a relatively quick "pay-off." If the cost of pollution control equipment, such as scrubbers on a chimney, is not recovered for 20 years or more, industries cannot afford to put scrubbers on chimneys unless nonmarket incentives are provided. These can take the form of benefits (tax relief) or costs (fines, taxes). Moreover, it is relatively cheap to clean up "most" of the pollution; to insist on 100% cleanup may be prohibitively expensive. Since resources are always limited, 100% cleanup of one waste stream may mean that another is not cleaned up at all, while both could have been cleaned up to the 95% level.[14]

[14] There may, of course be noneconomic motives to enforcing 100% cleanup. In the case of the Exxon Valdiz oil spill, exorbitant cleanup costs might be a deterrent to further negligence by the company. This assumes that further oil spills can be prevented by an economically competitive company.

Sociology of Public Response[15]

Unlike the recently overthrown governments of eastern Europe and the Soviet Union, our government is to a certain extent answerable for its actions. Through local government, as well as through state and federal representatives, it is possible for those who are affected by policies to make themselves heard. It is, of course, inevitable that the rich, whether they are individuals or industries, speak more loudly than the poor, the middle class, or the owners of small businesses. The latter must band together into groups before their voices are heard. As demonstrated by the various anti-landfill and anti-incinerator coalitions, however, pressure groups do influence policy.

Psychology is the study of individual human beings; sociology, of aggregate human action. Environmental psychologists and sociologists investigate the factors that cause individuals to respond positively or negatively to the siting of public projects.

Definitions

LULU.   LULU, a *l*ocally *u*nwanted *l*and *u*se, may be as mundane as a neighborhood that objects to an apartment building replacing a single-family house. More often, the term is used to denote use of land for purportedly sound environmental projects, such as incinerators, landfills, or nuclear power plants.

NIMBY.   NIMBY, *n*ot *i*n *m*y *b*ack *y*ard, is the rallying cry of people faced with a LULU. The term *NIMBY syndrome* refers to the almost universal reluctance to live near a LULU. NIMBY is often used as a derogatory term by officials trying to locate (for example) a waste disposal site or nuclear power plant. The use of NIMBY often suggests that the public response is automatic and irrational. In several cases, groups opposed to the siting of (for example) a landfill have insisted that the correct rallying cry is "Not in *anyone's* backyard," but whether they have analyzed the consequences of such an ideal is not clear.

Origins of the NIMBY response are fairly straightforward. In environmental contexts, the most frequently mentioned concerns are that there will be untoward health effects. The health fears include delayed effects, especially cancer and birth defects, as well as stress and emotional upset. Property depreciation is the second major concern. Severe pollution leads to essentially total loss of value of a property (as was shown at Love Canal, where one could hardly have given the houses away for some years even though owners continued to be required to pay mortgages).

Not only individuals, but also local governments, may cry NIMBY. Com-

---

[15] Much of the material in this section derives from a lecture given by J. Stanley Black while he was on the faculty of the Institute for Environmental Studies. Again, any errors are mine, since I have modified his lecture.

munities are most often worried about loss of revenue due to a loss of tourist attractions or of other businesses. Alternatively, towns may fear that the LULU will attract undesirable industries. For example, a waste disposal site might well attract more chemical industry to a given area. This is a mixed blessing, since an increase of jobs would be viewed as favorable, but increased air and water pollution would be undesirable. Finally, both communities and individuals cite aesthetic concerns as a reason for the NIMBY response. Especially when landfills or hazardous waste dumps are being sited, neighbors resent the occurrence of "odors" (which may be chemical fumes). Everyone agrees that open landfills are unattractive. Even model landfills increase truck traffic and the probability of spills and accidents.

If the NIMBY response coalesces to an organized opposition, less obvious concerns are voiced. These include the condition of the community after the life span of the LULU. A nuclear power plant, for instance, is projected to have a lifespan of approximately 30 years; many communities will then be faced with the existence of a blighted area or with the removal of large quantities of radioactive material. Similarly, a hazardous waste landfill remains hazardous, and retains the potential to contaminate water supplies, long after it is full and is no longer contributing jobs to the community. Lack of local control is often part of the objection to a LULU. The ability of a community to regulate its space is lost or perceived as lost. This is particularly critical if government officials choose the site, and is more intense in smaller towns, which feel threatened by outsiders' changes.

Note that a NIMBY response does not require an intrinsic hazard or offensiveness. Even when intrinsically unobjectionable projects, such as reservoirs, are planned, those nearby may object.[16] A storage reservoir in which water is pumped up to an upper reservoir during times of slack demand and released to a lower reservoir (with generation of electricity) during peak demand was recently proposed for a hill above the Catskill mountain town of Prattsville, New York. A similar storage reservoir in a nearby valley is attractive and provides a swimming pool as well as fishing. The proponents of the new reservoir argue that the project would supply jobs in a chronically depressed area, both during the construction and through increased tourism thereafter. The NIMBY response to the project was that it would appropriate good farmland, that the jobs would probably not go to local residents but to outsiders, and that each construction job tended to leave behind people who were then unemployed, adding to the town's welfare costs.

Moreover, an organized NIMBY response often consists of a composite response. In the case of the storage reservoir, environmentalists argued that there was no need for more electric power: there was already a surplus, and conservation measures could prevent future demand from rising above the

---

[16] Reservoirs take the best farmland, which is in the valleys; they also restrict the uses to which landowners may put adjoining land and land along any streams feeding the reservoir.

present capacity. Real estate agents, who earn commissions by selling land, favored the reservoir. Local landowners were divided. Those who stood to gain from selling land near the proposed reservoir were for it; those who owned land to be flooded, or who preferred fewer visitors, or who dislike the trailer parks that sprout in the wake of construction projects, opposed it.[17] And this was an essentially innocuous project: no permanent pollution, no hazardous chemicals deposited, no air pollution from factory chimneys. Even more emotions are engendered by potentially hazardous LULUs.

There is no such thing as a risk-free world. The people who cry *Not in anyone's backyard* should consider their use of plastics, automobiles, paint and solder, and batteries. Every production line engenders a waste stream, every product a waste. Therefore, it becomes essential to measure hazards and to compare the measured risks (or costs) and the benefits. Immediate questions include:

- Whose risk, and whose benefit? This is an *equity* (fairness) issue. If the people who benefit also bear the risk, equity is satisfied. If benefits accrue to one group (landfill owners) while others (landfill neighbors) bear the risk of contaminated water, adjustments must be made, preferably in the form of direct safeguards against damage, but at least of generous indemnification in the event that there is damage.
- Acceptability of risk is separate from the magnitude of risk.
- Irreversibility of damage affects the degree and type of indemnity (or protection) needed. For example, contamination of the water supply of Atlantic City, New Jersey (and many other aquifers) cannot be reversed; remedial measures to prevent the carcinogen from getting into the drinking water of residents are quite expensive, as is tying into another aquifer. Sometimes remedial action is impossible.

Two levels of analysis are needed. First, the *level of risk* must be estimated. Although uncertainties frequently attach to such a measurement, it is based on data, and the uncertainty arises in the interpretation of the facts. Second, the *acceptability* of the measured risk must be determined. This is a less factual determination and, being value based, is always arguable. Moreover, although the risk can be measured objectively for each person individually or for all "neighbors" in the aggregate, acceptability will differ between people facing exactly the same risks. For example, despite all the opposition to siting of nuclear power plants, property values near such plants actually tend to increase. There are enough people who want to take advantage of the area's low taxes (resulting from the high taxes paid by the nuclear utility) to outweigh the departure of those to whom the risk is unacceptable.

---

[17] As of this writing, the Prattsville storage reservoir has been canceled, due to unfavorable environmental impact assessment by the state (with help from environmental agencies). Since such projects have a habit of being resurrected, local residents remain alert.

One problem with discussions of environmental issues is that a given question is often discussed as a factual question by one side, while the other side views it as a value-based question (Table 13.1). Typically, neither side realizes the difference in the baselines. Moreover, it is rarely realized that risk assessments must be repeated with time. For example, the permeability of clay liners for landfills was measured using water, providing a very low estimate for the risk that a leachate will permeate 15 feet of compacted clay. When data were obtained showing that clay is up to 100-fold as permeable to organic solvents as to water, risk estimates obviously had to be recalculated. Similarly, public acceptance often changes with time (Figure 13.1). For example, the public perception of nuclear power in the 1950s was that the

Table 13.1 Comparison of Perception of Risk by "Experts" and Laypeople

Lay Rankings	Hazard	Experts' Rankings
1	Nuclear power	20
2	Motor vehicles	1
3	Handguns	4
4	Smoking	2
5	Motorcycles	6
6	Alcoholic beverages	3
7	General aviation (private planes)	12
8	Police work	17
9	Pesticides	8
10	Surgery	5
11	Firefighting	18
12	Large construction	13
13	Hunting	23
14	Spray cans	26
15	Mountain climbing	29
16	Bicycles	15
17	Commercial aviation	16
18	Electric power (nonnuclear)	9
19	Swimming	10
20	Contraceptives	11
21	Skiing	30
22	X-rays	7
23	High school/college football	27
24	Railroads	19
25	Food preservatives	14
26	Food coloring	21
27	Power mowers	28
28	Prescription antibiotics	24
29	Home appliances	22
30	Vaccinations	25

Source: Adapted from W. F. Allman, "We have nothing to fear," *Science85:*41, 1985.

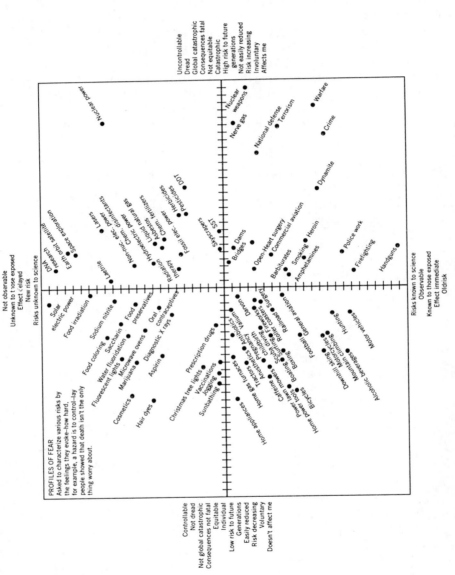

Figure 13.1 Factors determining lay reactions to risk, and the location of various risks in relation to these factors. (Adapted from W. F. Allman, *Science* 85:40–41. October 1985.)

technology was essentially risk free and would generate very cheap power. Neither perception now holds.

The role of uncertainty (concerning both facts and values) should be given far more emphasis, because it is an important component of the acceptability of a risk. The limits of the data should be defined but often are not. The lack of sensitivity measures or confidence limits on data points make data interpretation difficult.

Sources of Uncertainty in Data

There are often time lags in determining the effects of a technology, since (as any actuary would agree) the best predictions are based on past history. Thus the actual safety of nuclear power plants will be much more accurately determined when they have been running for a century. At present, estimates for the typical U.S. nuclear power plant (not the type represented by Chernobyl) range from one accident in 400 plant-years to 1 per 400,000 plant-years. Obviously, one's attitude toward nuclear power plants will differ, depending on which estimate of risk one believes.

In the case of less dramatic disasters, especially when carcinogenic chemicals are involved, there is often a problem of causal confounding, which may require large-scale epidemiological studies to evaluate. An increase in childhood leukemia near the British nuclear plant at Sellafield may represent an unexpected hazard of paternal exposure to low-level nuclear radiation, or a blip (cluster) in the data. Only additional studies will tell.

The applicability of models (whether animal or computer) to the actual situation is often disputed. This difficulty is increased when expertise involves assumptions and estimates of parameters. Again, Sellafield provides an example of the problem. Our society's emphasis on progress, which results in a rush to judgment, exacerbates the effects of uncertainty. There is tremendous pressure to adopt new technologies even with insufficient information, often based on the mere absence of information of harmful effects. (N.B.: Absence of evidence is *not* evidence of absence!) The existence of second-order (or higher) effects, which are often not even examined, also exacerbates the assumption of safety. Last—but far from least—human error and self-interest provide biases in any direction. As a result, an equitable distribution of risks and benefits is critical in situations involving local power or local option.

Given a risk level, what determines the acceptability of the risk? In summary, one's own *enlightened self-interest,* including second-order costs and benefits. Nonmonetary effects must also be considered. Nondollar effects, usually defined as *utility functions,* can get quite complex. The credibility of the institution proposing the technology is often very important in determining acceptability. When a hazardous waste incinerator was proposed in Kankakee, Illinois, opposition was vocal but relatively ineffective until it was learned that the developer was already in violation of Illinois law at another of its disposal sites. Even proponents' emphasis on the minor nature of these other violations did not help. As a local politician said, supporting the incinerator after that would have been suicide. Unfortunately, over the past 20 years, essentially *all* U.S. institutions have suffered decreased credibility. Perhaps as a result of the increased cynicism of the American public—but perhaps also because of the numerous incidents of industrial malfeasance and political acquiescence—neither hazardous waste sites nor incinerators have been sited successfully.

Acceptability of Risks

Risks are generally more acceptable if they are voluntary (smoking versus radon), if they are spread out over time (car deaths versus airplane crashes), if they are optional (alternatives exist), if they are luxuries rather than necessities (people skydive), and if they are occupational rather than nonoccupational. To a considerable extent, all of these factors measure the degree to which the risk assumer can *choose* the risk. A risk I choose myself is, statistically, more acceptable than one imposed from outside.

APPENDIX

AN ANNOTATED
BIBLIOGRAPHY

Consider the lilies of the field: they neither toil, nor do they spin; yet I assure
you that not Solomon in all his glory was arrayed like one of these.
—Matthew 6:28–29.

There are innumerable books about toxic chemicals and their effects on or
in the environment. They range from carefully documented treatises to shrill
diatribes. The following list includes only sources that I consider reliable,
although some are frankly polemic. I have separated the list into *environment,*
including related social issues (e.g., human consequences of nuclear power)
and *toxicology,* consisting of books or articles that deal with a toxicants and
their biology or chemistry. Many of these books, as well as many of the
articles that served as invaluable sources of information for me, are also
mentioned in the notes of various chapters. Unfortunately, some of the books
may be out of print. The season for topical books is short, and even libraries
now discard books on a regular basis.

I do not pretend to have included all reliable sources. New books appear
as rapidly as older books go out of print, and most are not listed here, even
if they are newer and better than the books I list, simply because I have not
come across them. The best sources for current information on books about
the environment are the weekly *Science News* and the monthly *American
Scientist,* both of which regularly comment on books meant for the general
reader. Many of the magazines listed below also review books of interest to
their readers.

I have also listed periodicals that deal with the environment. These range
from magazines meant for a general readership to highly technical, specialized journals. For more general up-to-date information on many scientific
issues, I recommend *Environment.* On a weekly basis, health and environmental toxicology are also covered in *Science News* and in the Tuesday
"Science" and the Wednesday "Health" sections of the *New York Times.*
Environmental news is often included in the "Science and Technology"
section of the *London Economist.*

339

BOOKS OF GENERAL INTEREST

Environment and Social Issues

Brown, Lester R., and members of the Worldwatch Institute. *State of the World.*
A yearly report by the Worldwatch Institute on progress toward a sustainable society.

Carson, Rachel. 1962. *Silent Spring,* Fawcett Crest Books, New York.
This is one of the classics of environmental toxicology, and worth reading and rereading. It reported many problems with the persistent and bioaccumulative insecticides before these were published in the scientific literature, leading opponents to charge inaccuracy. Although some of Carson's predictions have not [yet] come true, the facts are absolutely accurate. Carson's environmental philosophy remains highly pertinent.

Giangrande, C. 1983. *The Nuclear North.* House of Anansi Press, Toronto, Ontario, Canada.
A Canadian view of the nuclear establishment in that country, emphasizing social issues.

Illesh, A. 1987. *Chernobyl: A Russian Journalist's Eyewitness Account.* Richardson and Steirman, New York.
An eyewitness account of the immediate aftermath of the explosion at Chernobyl.

Jackson, W., W. Berry, B. Colman, eds. 1984. *Meeting the Expectations of the Land: Essays in Sustainable Agriculture and Stewardship.* North Point Press, Berkeley, Calif.
A relatively technical set of essays about sustainable agriculture.

Leopold, A. 1969. *A Sand County Almanac.* Sierra Club/Ballantine Books, New York.
Elegant essays on ecology and the human place in ecosystems. Leopold's land ethic is even more important today than when it was first published in 1949. Moreover, Leopold was one of the earliest voices to argue that there is a land ethic, an environmental ethic.

Medvedev, Z. 1992. *The Legacy of Chernobyl.* W.W. Norton, New York.
Medvedev, a biologist, was long an expatriate spokesman for the Soviet scientific community. In the post–Cold War era, he has had official cooperation from the former Soviet Union.

Rathje, W., and C. Murphy 1993. *Rubbish!* HarperPerennial, HarperCollins Publishers, New York.
A lively account of the history and findings of the Garbage Project, which has catalogued the contents of garbage cans and landfills. The book dispels some pervasive (and "politically correct") myths about

garbage and also presents perspectives on the disposal habits of societies ranging from ancient Troy to colonial New York.

Roueché, B. 1968. *What's Left: Reports on a Diminishing America*. Berkeley Medallion Books, New York.
Essays about wild places in the United States, and the people who live or work there. Not by any means as gloomy as the title.

Sigma Xi Forum Proceedings, 1991. *Global Change and the Human Prospect: Issues in Population, Science, Technology and Equity*. Sigma Xi, Research Triangle Park, N.C.
The essays are slightly revised versions of talks given at a forum of the same title and are divided into three groups: the world as it is, the world as we would like it, and how we get from here to there. Authors include economists, ecologists, and several voices for the Third World as well as the usual representatives of the United Nations, U.S. industry, and the U.S. political scene.

Stobaugh, R., and D. Yergin, 1979. *Energy Future: Report of the Energy Project at the Harvard Business School*. Random House, New York.
Although somewhat old, this book details the economic and political obstacles to energy efficiency in the United States.

World Resources Institute, 1992. *The 1993 Information Please Environmental Almanac*, Houghton Mifflin Company, New York.
A succinct collection of information about environmental matters, arranged both by country and, for the United States and Canada, by topics ranging from "grassroots activities" and "ecotourism" to the more usual air pollution and ozone depletion.

The World Resources Institute, The United Nations Environment Programme, and The United Nations Development Programme. *World Resources: A Report*.
Another yearly report on the state of the world, replete with statistics on the production, use, and status of natural resources.

Toxicology

Cairns, J. 1978. *Cancer, Science and Society*. W.H. Freeman, San Francisco.
An elegantly illustrated, straightforward account of the epidemiology of cancer and of classical and newer cancer research (although it precedes identification of oncogenes). Cairns's articles also appear in Scientific American.

Egginton, J. 1980. *The Poisoning of Michigan*. W.W. Norton, New York.
A detailed account of the farmers whose farms and lives were destroyed by PBBs.

Epstein, S. S. 1978. *The Politics of Cancer*. Sierra Club Books, San Francisco.

Epstein is one of the more strident voices in the debate over health effects of chemicals, but the data he presents are valid.

Epstein, S. S., L. O. Brown, and C. Pope. 1982. *Hazardous Waste in America*. Sierra Club Books, San Francisco.
A strident but reliable book.

Harte, J., C. Holdren, R. Schneider, and C. Shirley. 1991. *Toxics A to Z: A Guide to Everyday Pollution Hazards*. University of California Press, Berkeley.
A concise introduction to the principles of toxicology, routes of exposure, and environmental effects, followed by a commentary on four categories of exposure—to metals, petrochemicals, radiation, and pesticides—and a section on wastes and waste disposal. The final section of the book consists of an alphabetical list of chemicals present in the environment at high levels (or otherwise significant to human health) and a summary of their effects. Most of the chemicals listed have had media exposure (2,4,5-T, aldicarb, plutonium); the rest are those that toxicologists tend to worry about (aflatoxins, lead, iodine-131). Many are, of course, on both lists (radon, plutonium).

Highland, J. H., and R. H. Boyle. 1980. *Malignant Neglect*. Vintage Books, New York.
Another strident but generally reliable book.

Klaassen, C. D., M. O. Amdur, and J. Doull. 1986. *Casarett and Doull's Toxicology: The Basic Science of Poisons*, 3rd ed. Macmillan, New York.
The standard textbook in health effects toxicology. Individual chapters are written by specialists in the field. A 4th edition came out in 1991.

Knightley, P., H. Evans, E. Potter, and M. Wallace. 1979. *Suffer the Children: The Story of Thalidomide*. Viking Press, New York.
The history of the thalidomide disaster, originally published by the *Times* of London. Unusual because it documents and gives equal emphasis to the human tragedy and to the scientific and regulatory background of that tragedy.

Kruus, P., and I. M. Valeriote. 1979. *Controversial Chemicals: A Citizen's Guide*. Multiscience Publications, Montreal, Quebec, Canada.
An excellent, unbiased treatise that presents the facts about 25 chemicals, including life-style chemicals (alcohol, caffeine, saccharin, tobacco), pesticides (DBCP, DDT, fenitrothion, kepone/mirex), environmental pollutants (arsenic, cadmium, dioxin, lead, mercury, nitrates, nitrogen oxides, PCBs, phosphates, sulfur dioxide, vinyl chloride), and additives to consumer products (asbestos, benzene, fluoride, fluorocarbons, nitrites, Tris, vinyl chloride).

Lappé, M. 1991. *Chemical Deception*. Sierra Club Books, San Francisco.
A primer of health effects toxicology, from a medical doctor with a maximalist's view of environmental hazards of chemicals. Although

Lappé has a tendency to emphasize the less optimistic interpretation, and an annoying tendency to be sloppy about little details, the book is scientifically quite sound.

Roueché, B. 1982, 1986. *The Medical Detectives,* Vol. I (1982) and Vol. II (1986). Washington Square Press, New York.
A collection of curious case histories, with superb background information on topics ranging from pesticide poisoning to Dutch elm disease. Roueché's new stories appear at unpredictable intervals in the *New Yorker.*

Schardein, J. L. 1985. *Chemically Induced Birth Defects.* Marcel Dekker, New York.
A technical tome summarizing the literature on teratogenic chemicals, with a copious bibliography.

Smith, W. E., and A. Smith. 1975. *Minamata.* Holt, Rinehart and Winston, New York.
A prize-winning photographer's history of the pollution of Minamata Bay, told in words as well as in pictures.

Stellman, J. M., and S. M. Daum. 1973. *Work Is Dangerous to Your Health.* Vintage Books, Random House, New York.
A catalog of the chemicals to which one may be exposed in the workplace, and the hazards associated with those chemicals. Listings include suspected, as well as proven, hazards. For the general reader.

Ware, G. W. 1989. *The Pesticide Book.* Thomson Publications, Fresno, Calif.
A clear description of the common pesticides, including their uses and risks.

MAGAZINES: GENERAL READERSHIP

Environment
Covers all aspects of environmental issues, but emphasizes legal and economic perspectives. It is particularly valuable because it carries articles dealing with Europe, Asia, and Africa.

Garbage
A relatively new magazine dealing with all aspects of garbage—generation, transport, disposal—and with the life-styles that generate it. Moderate in tone, with advertisements for "green" products.

WorldWatch
Extremely broad coverage of environmental issues, with emphasis on social science aspects—economics, politics, sociology—rather than biology or chemistry.

Societies' publications
Many conservation organizations—including Audubon, Nature Con-

servancy, and National Wildlife Federation—publish excellent magazines for their members. Many of these are also available in libraries.

TECHNICAL JOURNALS

Ecotoxicology

Ambio
Archives of Environmental Contamination and Toxicology
Bulletin of Environmental Contamination and Toxicology
Aquatic Toxicology
Environmental Toxicology and Chemistry
Reviews in Environmental Toxicology
Reviews of Environmental Contamination and Toxicology (formerly, Residue Reviews)

Chemistry and Engineering

Chemosphere
Environmental Science and Technology
Journal of Agricultural and Food Chemistry
Journal of Pesticide Biochemistry and Physiology

Health Effects

Archives of Environmental Health
Archives of Toxicology
Environmental Research
Environmental Health Perspectives
Journal of Applied Toxicology
Journal of Toxicology and Environmental Health
Fundamental and Applied Toxicology
Molecular Toxicology
Neurotoxicology and Teratology
Regulatory Toxicology
Teratogenesis, Carcinogenesis and Mutagenesis
Teratology
Toxicology
Toxicology and Applied Pharmacology
Toxicology Letters

INDEX

Page numbers followed by letters indicate references to figures (F) and tables (T).

ENVIRONMENTAL SCIENCE
AND TECHNOLOGY

A Wiley-Interscience Series of Texts and Monographs

Edited by JERALD L. SCHNOOR, *University of Iowa*
ALEXANDER ZEHNDER, *Swiss Federal Institute for Water Resources and Water Pollution Control*